[redacted]

Algorithms for Network Programming

ALGORITHMS FOR NETWORK PROGRAMMING

Jeff L. Kennington
Richard V. Helgason

Southern Methodist University

A Wiley-Interscience Publication
JOHN WILEY & SONS
New York • Chichester • Brisbane • Toronto

***Library of Congress Cataloging in Publication Data*:**

Kennington, Jeff L
Algorithms for network programming.

Includes bibliographical references and index.
1. Programming (Mathematics) 2. Algorithms.
3. Systems analysis. I. Helgason, Richard V., joint author. II. Title.

QA402.5.K45 001.64′2 80-258
ISBN 0-471-06016-X

Printed in the United States of America

10 9 8 7 6 5 4 3 2 1

To Carolyn Kennington and Janice Helgason

Foreword

Network optimization has emerged as one of the most important *practical* branches of mathematical programming. Innovations of the past decade, and indeed of the past few years, have brought networks into focus as a fundamental planning tool for solving a wide variety of problems in industry and government. Major applications include distribution, production scheduling, transportation, resource allocation, financial planning, inventory systems, facility location, and a host of other important areas.

Ideally suited to "what if" analyses, networks are also greatly expanding the boundaries of *large-scale* optimization. Network problems involving thousands of constraints and millions of variables can now be handled routinely. Moderate-sized network problems can be solved more than 100 times faster with specialized network algorithms than with the most sophisticated of present-day linear programming solution systems.

The implications of such developments are little short of staggering. On the one hand, the need to compromise model realism by neglecting potentially important variables (because standard procedures cannot accommodate problems beyond a limited size) is largely obviated in network settings. On the other hand, problem nonlinearities and environmental uncertainties that cannot be adequately captured by a single tractable model structure can now often be treated by successive approximation and alternative scenario approaches, using a network model base, because of the remarkable efficiency with which network models can be solved.

But the effect of recent innovations in the network field does not stop here. Many practical applications that cannot be modeled as pure networks can be formulated as generalized networks or as networks with side constraints and side variables. Examples are particularly abundant in financial analysis, physical distribution, and energy fields. Extensions of network and linear programming methodology to these important problem classes have provided major computational gains and have already found practical application.

A graphic illustration of the significance of these developments is provided by the recent implementation of a specialized LP/embedded network algorithm in an integrated production, distribution, and inventory system. In its first year of operation, the system has saved more than $8 million for the company in which it has been applied. Other valuable practical applications of this extended branch of network methodology are rapidly appearing.

The question naturally arises: What reference provides the broad algorithmic details that underlie these major practical network innovations, from both a conceptual and an implementation point of view? Until Jeff Kennington and Dick Helgason wrote this book, the question had no answer. The authors have succeeded in bringing together for the first time, in a single place, numerous useful and important principles from latest advances in the network area.

This book is not an easy popularization of the network field or its methods. It is a carefully documented and often demanding algorithmic reference. This structure provides many bonuses for the serious reader. The organization contains rigorous algorithmic descriptions capable of being translated into computer programs with a minimum of elaboration. These are supplemented by numerous illustrations, diagrams, and numerical examples that help to infuse abstract characterizations with tangible meaning.

In the realm of algorithmic explication this book is unique. And in a field as rapidly growing as networks, it is a mark of distinction to achieve the notable degree of currency that Kennington and Helgason have provided.

FRED GLOVER
DARWIN KLINGMAN

Preface

This is a book about mathematical programs that may be associated with a network. These problems all involve the minimization of some function of the flows in a network subject to flow conservation constraints (flow-in equals flow-out), lower and upper bound constraints on the flow through an arc, and possibly additional constraints concerning flow on several different arcs. The basic models discussed in this text include the *minimal cost network flow model*, the *generalized network flow model*, the *multicommodity network flow model*, the *network flow with side constraints model*, and the *convex cost network flow model*. This classification also includes the following well-known special cases: the *transportation problem*, the *assignment problem*, the *maximal flow problem*, and the *shortest path problem*.

Our interest over the last 6 years in this class of mathematical programs has been motivated by applications, and our work has been in the area of algorithm design and implementation. Unfortunately, many of the details required to implement algorithms for this class of problems are not well known, nor are they easily obtained from the existing literature. Some of the techniques and data structures in current use are described only briefly in the literature and in some cases can be discerned only from examination of existing software. We have attempted to collect between two covers *clear* and *concise* descriptions of the special purpose algorithms for the various network models along with a mathematical justification for these algorithms. In addition we present the basic data structures used for implementation.

The primal simplex method, developed by George B. Dantzig in 1947, provides the framework for most of the algorithms presented here. The most important algorithm in this book is the specialization of the primal simplex algorithm for the minimal cost network flow model, which we call the *simplex method on a graph*. These basic ideas are then extended to the various other network models. In each case the results required for these specializations have been developed by the formal proposition-proof format. The result is a rigorous and organized presentation that allows the reader the maximum flexibility in what he or she chooses to read.

This book can be used both as a reference book and as a textbook for first-year graduate students in the fields of applied mathematics, computer science, engineering management, industrial engineering, management science, operations research, and other disciplines that deal with flow problems on networks. The book is designed for a one-semester course in network flows. Even though the material requires mathematical maturity, some knowledge of linear programming is the only prerequisite. The pertinent background material is summarized in the second chapter.

There has been much work on network algorithms since Ford and Fulkerson's classic book appeared in 1962, and we believe that the compendium of techniques presented here represents the current state of the art of network flow algorithms. We have a library of routines that may be used to solve many of the problems presented here, and we are willing to share these routines with others interested in network flows. We hope that this book encourages and aids even more widespread use of these models and techniques.

This manuscript has benefited from careful readings by graduate students at Southern Methodist University: Russ Baldwin, Behord Chen, Keyvan Farhangian, Kay Sneary, Tim Quilici, and John Whited. We also wish to express our appreciation to Dr. Joseph Bram of the Air Force Office of Scientific Research for research support during the years in which we learned most of what is presented in this text. Lastly, we express our appreciation to Milly Manley and Gina Snipes for their excellent typing of the manuscript.

JEFF KENNINGTON
DICK HELGASON

Dallas, Texas
April 1980

Contents

Algorithms for Network Programming

CHAPTER 1

Introduction

The most interesting and fascinating mathematical programs that we have encountered involve the minimization of a function (linear or nonlinear) of flow in (or within) a network. Since the development of the simplex algorithm by George Dantzig in 1947, network models have been extensively studied. We attribute this enormous activity to the fact that there are numerous applications for network flow models. In addition, these models all exhibit an intriguing and, we believe, elegant structure. This network structure can be exploited in the development of specialized algorithms that produce solutions in one one-hundredth the time and cost required for general algorithms. Furthermore, network geometry (or relationships) can be easily displayed in two-dimensional drawings, greatly simplifying the communication problem between the analyst and the client for whom the model is designed. The following is a short list of systems in which network models have been used:

1. Production-distribution systems.
2. Military logistics systems.
3. Urban traffic systems.
4. Railway systems.
5. Communication systems.
6. Pipe network systems.
7. Facilities location systems.
8. File merge systems.
9. Routing and scheduling systems.
10. Electrical networks.

1.1 NOTATION AND CONVENTION

The notational conventions used in this text are presented in this section. Matrices are denoted by upper case Latin letters in boldface type. Graphs and graphical sets are denoted by upper case Latin letters in script (e.g., $\mathscr{Q}$). The symbol **I** denotes the identity matrix with dimension appropriate for the given context. A matrix with all zero entries is denoted by **0**. The element in the ith row and jth column of a matrix **A** is denoted by A_{ij}, and the vector whose entries are from the jth column of **A** is denoted by $\mathbf{A}(j)$. Certain fixed integers are denoted by upper case Latin letters with a bar above, such as $\bar{K}$. Lower case Latin and Greek letters in boldface type are used to denote vectors. The symbol $\mathbf{e}^i$ denotes the vector having the ith component equal to 1 and all other components equal to 0, while the symbol **1** denotes the vector of all ones. The ith component of the vector **c** is denoted by c_i. Upper case Greek and Latin letters are used for non-graphical sets. The empty set is denoted by ϕ. Primes are used to denote matrix transposition. However, we make use of a special convention to avoid the proliferation of primes in the context of matrix multiplication involving a vector and a matrix. We adopt the viewpoint that n-component vectors are neither rows ($1 \times n$ matrices) nor columns ($n \times 1$ matrices), and they are treated in a multiplicative context with a matrix in a manner that makes the dimensions conformable for multiplication, that is, a vector is treated as a row when it appears on the left of a matrix and as a column when it appears on the right. The inner product of two vectors **a** and **b** are denoted simply by **ab**, and we shall have no occasion to use the product of a column and a row. In geometric contexts we display a vector as a row or column, whichever is more convenient. Superscripts usually denote membership within some finite class, rather than exponentiation. Sigma is used to denote the scalar signum function, defined by

$$\sigma(x) = \begin{cases} 1 & \text{if } x > 0 \\ 0 & \text{if } x = 0 \\ -1 & \text{if } x < 0. \end{cases}$$

Other mathematical notation employed is standard. We adopt the use of the symbol ■ to signify the end of a proof.

1.2 LINEAR NETWORK MODELS

We begin by presenting the concept of a *network*. A network is composed of two types of entities: *arcs* and *nodes*. The arcs may be viewed as unidirectional means of commodity transport, and the nodes may be interpreted as locations or terminals connected by the arcs and served by whatever physical means of transport are associated with the arcs. Hence arcs may represent streets and highways in an urban transportation network, pipes in a water distribution network, telephone lines in a communication network, and so on. We limit consideration to networks having only a finite number of nodes and arcs. For a network having $\bar{I}$ nodes and $\bar{J}$ arcs, we impose an ordering on the nodes and arcs, placing them into a one-to-one correspondence with $1,\ldots,\bar{I}$, and $1,\ldots,\bar{J}$, respectively. The structure of the network can be displayed by means of a labeled drawing in which nodes are represented by circles and arcs are represented by line segments connecting (incident on) two nodes. An arrowhead on the line segment indicates the arc direction. An illustration of such a network representation is given in Figure 1.1. Note that we allow for more than one arc to connect the same two nodes. For the example of Figure 1.1, both the first and second arcs are directed away from node 1 and directed toward node 2.

The structure of the network may also be described by an $\bar{I}\times\bar{J}$ matrix, defined as follows:

$$A_{ij}=\begin{cases} +1 & \text{if arc } j \text{ is directed away from node } i, \\ -1 & \text{if arc } j \text{ is directed toward node } i, \\ 0 & \text{otherwise.} \end{cases}$$

The matrix **A** defined above is called a *node-arc incidence matrix*. The node-arc incidence matrix corresponding to the network of Figure 1.1 is

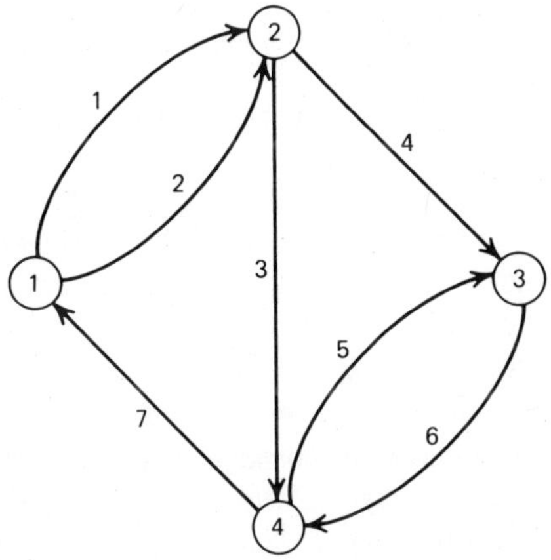

FIGURE 1.1 Labeled drawing representing a network.

given as follows:

$$
\begin{array}{c}
\text{nodes} \\
1 \\ 2 \\ 3 \\ 4
\end{array}
\overset{\displaystyle \overbrace{\begin{array}{ccccccc} 1 & \;\;2 & \;\;3 & \;\;4 & \;\;5 & \;\;6 & \;\;7 \end{array}}^{\text{arcs}}}{
\begin{bmatrix}
1 & 1 & & & & & -1 \\
-1 & -1 & 1 & 1 & & & \\
 & & & -1 & -1 & 1 & \\
 & & -1 & & 1 & -1 & 1
\end{bmatrix}}
$$

A characteristic of this matrix is that each column has exactly two nonzero entries, one being a $+1$ and the other a -1. Any matrix (regardless of origin) having this characteristic will be called a node-arc incidence matrix.

We now define the other quantities associated with the network that enable us to describe a linear network program. The decision variable x_j denotes the amount of flow through arc j, and the $\bar{J}$-component vector of all flows is denoted by $\mathbf{x}$. The unit cost for flow through arc j is denoted by c_j, and the corresponding $\bar{J}$-component vector of costs by $\mathbf{c}$. The arc capacity for flow through arc j is given by u_j with corresponding $\bar{J}$-component vector $\mathbf{u}$. The requirement at node i is denoted by r_i with corresponding $\bar{I}$-component vector $\mathbf{r}$. If $r_i > 0$, then node i is said to be a *supply point* with supply equal to r_i. If $r_i < 0$, then node i is said to be a *demand point* with demand equal to $|r_i|$. Nodes having $r_i = 0$ are called *transshipment points*. Mathematically, the *minimal cost network flow problem* may be stated as follows:

$$
\left.\begin{array}{ll}
\min & \mathbf{cx} \\
\text{s.t.} & \mathbf{Ax} = \mathbf{r} \\
 & \mathbf{0} \leqslant \mathbf{x} \leqslant \mathbf{u}
\end{array}\right\} \quad \text{NP},
$$

where $\mathbf{A}$ is a node-arc incidence matrix, and s.t. is the abbreviation for "subject to."

Many special cases of NP have been studied. We give the name of each of these and indicate the specialization.

1. *Uncapacitated Transshipment Problem.* This problem is a specialization of NP in which $u_j = \infty$ for all j, that is, the arcs have infinite capacity.
2. *Capacitated Transportation Problem.* This is a special case of NP in which every node is either a supply point or a demand point. Furthermore, all arcs are directed away from supply nodes and are directed toward demand nodes. Figure 1.2 illustrates a typical structure for a capacitated transportation problem.

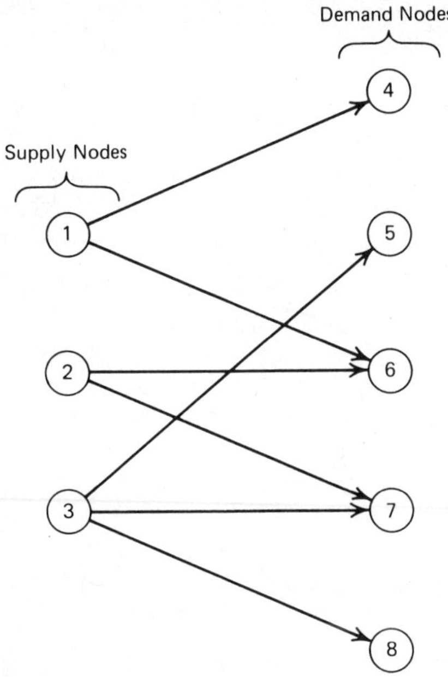

FIGURE 1.2 Transportation network structure.

3 *Transportation Problem.* This problem is a special case of the capacitated transportation problem in which $u_j = \infty$ for all j.

4 *Assignment Problem.* This is a special case of the transportation problem in which the requirements all equal ± 1 (i.e., $|r_i| = 1$ for $i = 1,\ldots,\bar{I}$).

5 *Maximal Flow Problem.* Given a network with arc capacities u_j for $j = 1,\ldots,\bar{J}$, the maximal flow problem is to find the maximal continuous flow from node s to node t that does not violate the capacity constraints $\mathbf{0} \leqslant \mathbf{x} \leqslant \mathbf{u}$, where s and t may be any given node pair. We view this problem as a special case of NP in which $\mathbf{r} = \mathbf{0}$ and exactly one arc has nonzero cost. To illustrate this view, suppose we wish to determine the maximal continuous flow from node 1 to node 4 in the network of Figure 1.1. We simply append a return arc, say 8, from node 4 to node 1 having $u_8 = \infty$ and $c_8 = -1$. Setting $r_i = 0$ for $i = 1,\ldots,4$, and $c_j = 0$ for $j = 1,\ldots,7$ results in the prescribed form. The revised network is illustrated in Figure 1.3.

6 *Shortest Path Problem.* Given a network whose arcs each have lengths $c_j \geqslant 0$ for $j = 1,\ldots,\bar{J}$, the shortest path problem is to find the shortest path in the network from node s to node t, where s

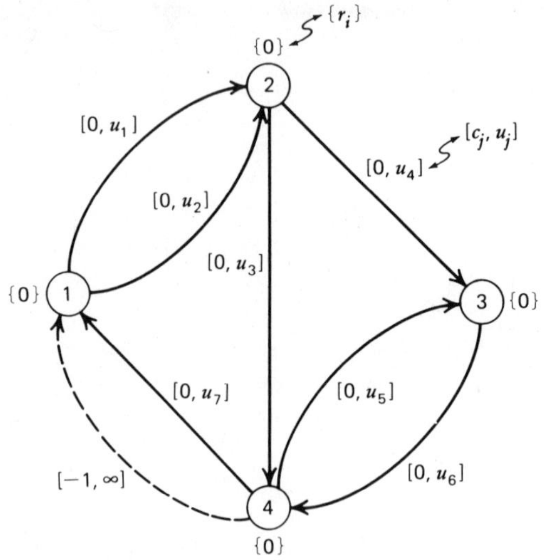

FIGURE 1.3 Maximal flow problems.

and t may be any given node pair. This problem may be viewed as a special case of the uncapacitated transshipment problem in which $r_s = 1$, $r_t = -1$, and all other requirements are zero. Suppose we wish to find the shortest path from 1 to 3 in the network of Figure 1.1. The NP corresponding to this problem is given in Figure 1.4. At optimality a unit of flow will take the shortest path

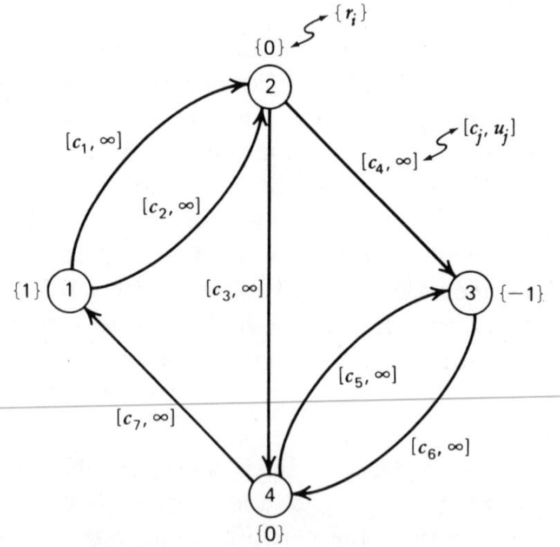

FIGURE 1.4 Shortest path problems.

from s to t. If the unit of flow splits into fractional amounts on some arcs, there exist multiple shortest paths.

1.3 HISTORY OF NETWORK PROGRAMMING

In 1975 the Royal Swedish Academy of Science awarded the Nobel Prize in Economic Science in equal shares to Professor Leonid Kantorovich of the USSR and Professor Tjalling C. Koopmans of the United States for their contributions to the theory of optimum allocation of resources. Although these distinguished professors investigated a wide variety of optimization problems, it is interesting to note that both are associated with some of the first papers describing network flow problems as we know them today. In 1939, in an expository paper, Kantorovich discussed a class of optimization models along with specific examples. The underlying idea of each example was the achievement of the highest possible production on the basis of the optimum utilization of existing resources. One of these examples involved the distribution of freight flows (given in units of carloads) among the different routes of a railroad network in such a way as to satisfy the requirements and capacity restrictions on the routes while minimizing the expenditure of fuel.

Further discussion of the flow of freight in the USSR was given by Kantorovich and Gavurin. This work did not become known in the West until the late 1950s. Meanwhile, in the United States, Frank L. Hitchcock presented what is now the standard formulation of the transportation problem. Working independently, Professor Koopmans formulated the same problem in connection with his work with the Combined Shipping Adjustment Board. Because of this work, the transportation problem is often referred to as the *Hitchcock transportation problem* or the *Hitchcock-Koopmans transportation problem*.

In 1956 Alex Orden proposed a generalization of the transportation model in which transshipment points were allowed. This formulation is known today as the uncapacitated transshipment problem. At approximately the same time, both the maximal flow problem and the minimal cost network flow problem were formulated and investigated by the famous team of Lester Ford and Delbert Fulkerson.

From 1950 through 1965 much activity was directed toward developing algorithms for linear network flow models. The algorithms developed may be classified into two types as follows: (1) specializations of the primal simplex method and (2) primal-dual methods. The primal simplex specializations began with the work of Dantzig and culminated with the paper of Ellis Johnson. The basis for Johnson's paper can also be found in two basic books: one by Dantzig and the other by Charnes and Cooper. The primal-dual methods originated with Harold Kuhn's *Hungarian algorithm* for the assignment problem and culminated with Delbert Fulkerson's *out-of-kilter algorithm*. Most of the activity since those classical papers has involved the efficient implementation of these basic techniques and the extension of this technology to linear network problems that involve gains, to linear multicommodity problems, to linear network problems with additional general constraints, and to network problems having nonlinear convex cost functions.

1.4 THE ART OF NETWORK PROGRAMMING

The subject of this book may be dichotomized into the *art* and the *science* of network programming. We view the process of developing algorithms for solving network programs as a science. However, the process of building a network model to help with the analysis of a real problem must be viewed as an art. This book is primarily directed toward developing the science of network programming. However, throughout the text we present applications that help to illustrate the scope and usefulness of network formulations.

1.4.1 Production-Distribution at Detroit Motors

The vice-president for production at Detroit Motors, a large automobile producer, faces the problem each month of determining the production level (in vehicles) for each of four assembly plants. The production-distribution plan is usually made 2 months in advance, and the vice-president is

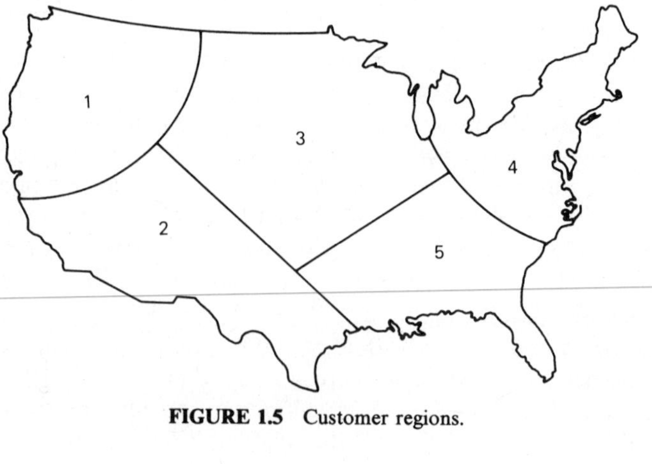

FIGURE 1.5 Customer regions.

Table 1.1 April Demand

Customer Region	Vehicles Requested for April
1	2840
2	2800
3	2600
4	2820
5	2750

currently working on the plan to meet the April demand. There are hundreds of customers (i.e., dealers), who have been grouped by location into five regions as shown in Figure 1.5. The April demand for each of these five regions is given in Table 1.1. Detroit Motors has four assembly plants, which are located in Michigan, California, Texas, and Georgia (see Figure 1.6). Because of the difference in assembly equipment utilized in the

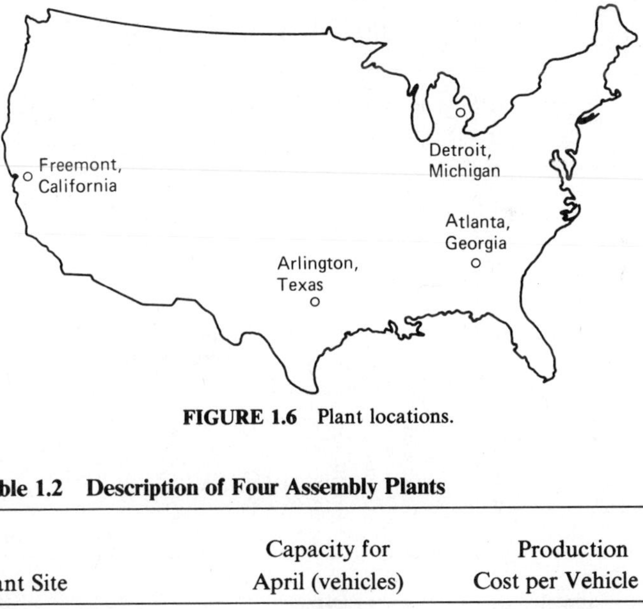

FIGURE 1.6 Plant locations.

Table 1.2 Description of Four Assembly Plants

Plant Site	Capacity for April (vehicles)	Production Cost per Vehicle ($)
Detroit, Michigan	4000	2100
Freemont, California	4500	2000
Arlington, Texas	2700	1600
Atlanta, Georgia	3000	1700

Table 1.3 Transportation Cost and Capacity for April

	Plant Site							
	Detroit		Freemont		Arlington		Atlanta	
Sales Region	Cost ($/vehicle)	Capacity (vehicles)	Cost ($/vehicle)	Capacity (vehicles)	Cost ($/vehicle)	Capacity (vehicles)	Cost ($/vehicle)	Capacity (vehicles)
1	800	0	200	3000	500	1000	0	0
2	600	2000	400	2000	200	2000	400	1000
3	300	2000	500	1000	200	1000	400	1000
4	200	2000	900	200	200	1000	300	1000
5	400	500	0	0	400	300	100	2000

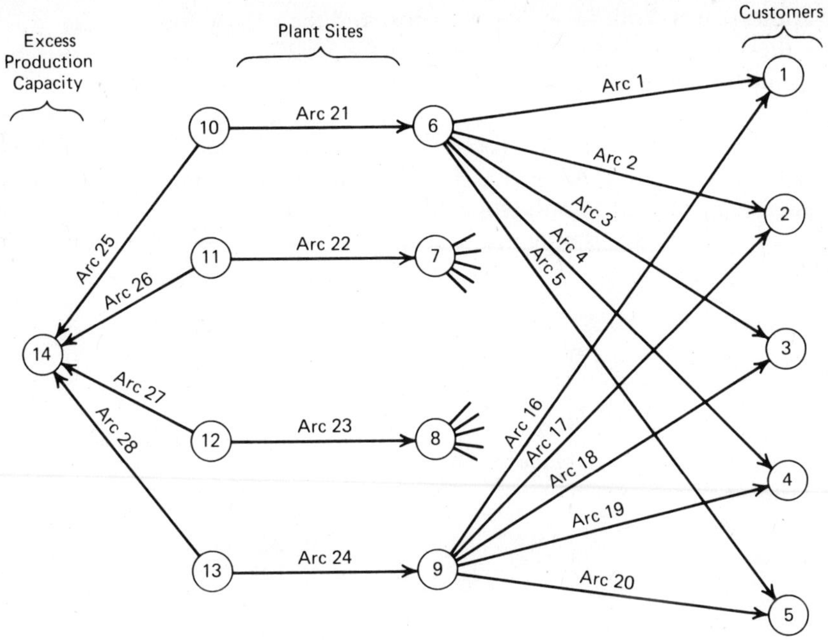

FIGURE 1.7 Network structure for Detroit Motors.

Table 1.4 Node Requirements (Figure 1.7)

Node Number	Requirement
1	−2840
2	−2800
3	−2600
4	−2820
5	−2750
6	0
7	0
8	0
9	0
10	4000
11	4500
12	2700
13	3000
14	−390

various plants, both the production cost and the production capacity differ at the four sites (see Table 1.2). The $4 \times 5 = 20$ transportation costs for shipments between the four plants and the five customer regions are as shown in Table 1.3. This table also gives the capacity of the distribution system. Therefore the total cost for producing a vehicle in Atlanta to be sold in region 3 is $1700 + $400 = $2100. The vice-president's problem is to assign the April sales to the four plants so that the sum of production and transportation costs is minimal.

Table 1.5 Arc Specifications (Figure 1.7)

Arc Number	"From" Node	"To" Node	Unit Cost	Arc Capacity
1	6	1	800	0
2	6	2	600	2000
3	6	3	300	2000
4	6	4	200	2000
5	6	5	400	500
6	7	1	200	3000
7	7	2	400	2000
8	7	3	500	1000
9	7	4	900	200
10	7	5	0	0
11	8	1	500	1000
12	8	2	200	2000
13	8	3	200	1000
14	8	4	200	1000
15	8	5	400	300
16	9	1	0	0
17	9	2	400	1000
18	9	3	400	1000
19	9	4	300	1000
20	9	5	100	2000
21	10	6	2100	4000
22	11	7	2000	4500
23	12	8	1600	2700
24	13	9	1700	3000
25	10	14	0	∞
26	11	14	0	∞
27	12	14	0	∞
28	13	14	0	∞

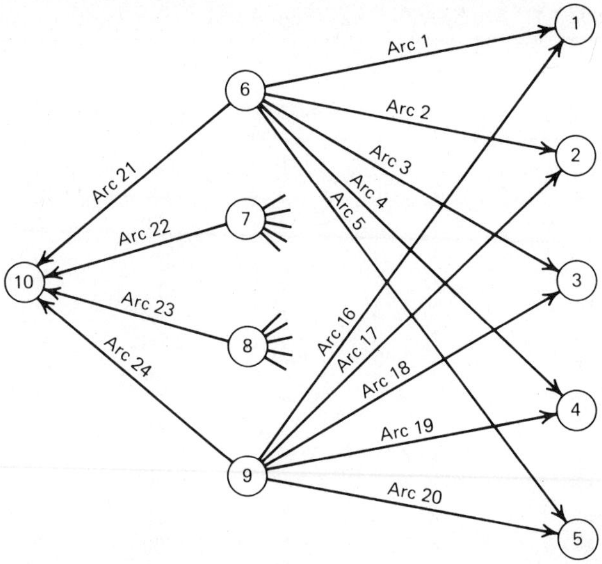

FIGURE 1.8 Alternative network structure for Detroit Motors.

The underlying network structure for this production-distribution system is shown in Figure 1.7. Note that the excess production capacity is absorbed by a node that is connected to the four plants by arcs with zero unit cost. The requirements associated with the 14 nodes are given in Table 1.4, and the arcs are listed in Table 1.5. An alternative formulation as a transportation problem is presented in Figure 1.8 and Tables 1.6 and 1.7.

Table 1.6 Node Requirements (Figure 1.8)

Node Number	Requirement
1	−2840
2	−2800
3	−2600
4	−2820
5	−2750
6	4000
7	4500
8	2700
9	3000
10	−390

Table 1.7 Arc Specifications (Figure 1.8)

Arc Number	"From" Node	"To" Node	Unit Cost	Arc Capacity
1	6	1	2900	0
2	6	2	2700	2000
3	6	3	2400	2000
4	6	4	2300	2000
5	6	5	2500	500
6	7	1	2200	3000
7	7	2	2400	2000
8	7	3	2500	1000
9	7	4	2900	200
10	7	5	0	0
11	8	1	2100	1000
12	8	2	1800	2000
13	8	3	1800	1000
14	8	4	1800	1000
15	8	5	2000	300
16	9	1	0	0
17	9	2	2100	1000
18	9	3	2100	1000
19	9	4	2000	1000
20	9	5	1800	2000
21	6	10	0	∞
22	7	10	0	∞
23	8	10	0	∞
24	9	10	0	∞

1.4.2 Scheduling at Texas Construction

Texas Construction Company has been given a $10 million contract to build a new indoor swimming pool for Southern Methodist University. The construction manager for this job has described the project in terms of a dozen major activities. The construction time and immediate predecessors for each of the 12 activities are given in Table 1.8. For example, activity *D* cannot be started until both activities *A* and *C* have been completed. Furthermore, it will take 5 months to complete activity *D*.

A precedence diagram illustrating the relationships among the 12 activities is shown in Figure 1.9 (the dashed arcs are dummy activities that are used to enforce precedence relationships), that is, activity *D* cannot be

Table 1.8 Activity Specifications

Activity	Predecessor Activities	Activity Time (months)
A	None	1
B	*C*	2
C	None	4
D	*A*, *C*	5
E	*A*	2
F	*C*	1
G	*B*, *D*	2
H	*D*	3
I	*B*	1
J	*C*, *E*	3
K	*B*, *F*	1
L	*I*	2

begun until both *M* and *N* have been completed. Node 1 is called the starting node, and node 11 the ending node, for the construction project. Hence the project is complete when all activities leading into node 11 have been completed.

Suppose the president of Southern Methodist University wishes to know the earliest time that the pool could be completed if construction

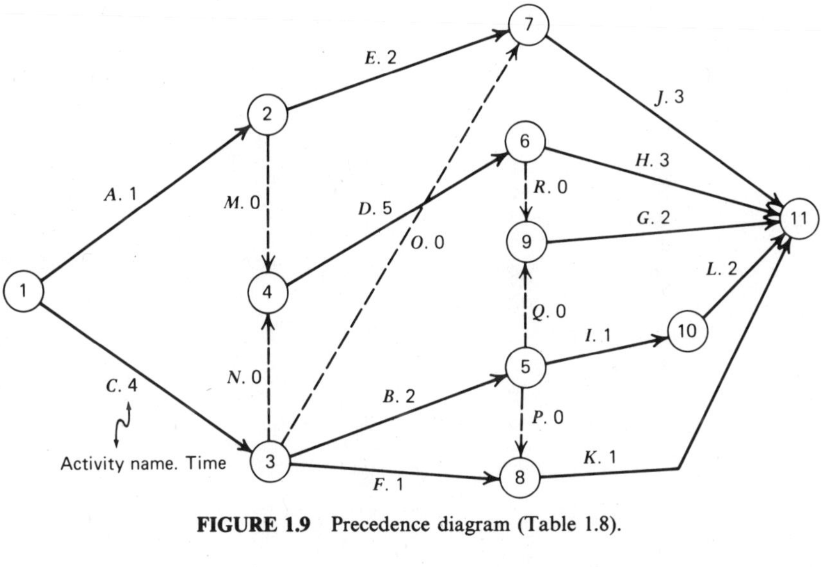

FIGURE 1.9 Precedence diagram (Table 1.8).

Table 1.9 Node Requirements (Figure 1.9)

Node	Requirement
1	1
2	0
3	0
4	0
5	0
6	0
7	0
8	0
9	0
10	0
11	−1

Table 1.10 Arc Specifications (Figure 1.9)

Arc Number	"From" Node	"To" Node	Unit Cost	Arc Name
1	1	2	−1	*A*
2	3	5	−2	*B*
3	1	3	−4	*C*
4	4	6	−5	*D*
5	2	7	−2	*E*
6	3	8	−1	*F*
7	9	11	−2	*G*
8	6	11	−3	*H*
9	5	10	−1	*I*
10	7	11	−3	*J*
11	8	11	−1	*K*
12	10	11	−2	*L*
13	2	4	0	*M*
14	3	4	0	*N*
15	3	7	0	*O*
16	5	8	0	*P*
17	5	9	0	*Q*
18	6	9	0	*R*

were to begin on January 1. The earliest completion date for this construction project is determined by the longest sequence of activities connecting node 1 to node 11 in the precedence diagram (Figure 1.9). Consider the sequence of activities (C,B,I,L) and (A,M,D,R,G) with times of $4+2+1+2=9$ months and $1+0+5+0+2=8$ months, respectively. Clearly, the pool cannot be completed in fewer than 9 months and cannot be ready before October 1. The problem of finding the shortest completion time for a construction project can be formulated as a minimal cost network flow problem as illustrated in Tables 1.9 and 1.10, where each of the arc capacities is any number that is no less than 1. The unit of supply will seek the least cost (longest time) path from node 1 to node 11.

1.4.3 Flight Scheduling on Gamma Airlines

A football fan who lives in San Francisco wishes to see a football game in Dallas, Texas. This fan wants to leave the day of the game and arrive in Dallas no later than 7:00 PM. Unfortunately, the fan is on a tight budget and must take the least costly flights available on Gamma Airlines, even if circuitous routing and long layovers are involved. The flights available are shown in Table 1.11. To change flights (i.e., make connecting flights) Gamma requires at least a 1 hour connection time. Hence the fan cannot take Flight 2 to Atlanta and transfer there to Flight 4. However, transfers can be made from Flight 1 at Atlanta to any of the other flights.

The network corresponding to this schedule is displayed in Figure 1.10. Note that Flight 1 makes connections in Atlanta with the other three

Table 1.11 Flight Schedule for Gamma Airlines

Flight Number	Origin	Destination	Departure Time	Arrival Time	Cost ($)
1	San Francisco	Chicago	8:00 AM	1:00 PM	100
	Chicago	Atlanta	2:00 PM	3:00 PM	100
	Atlanta	Dallas	3:40 PM	6:00 PM	250
2	San Francisco	Atlanta	11:00 AM	4:00 PM	250
	Atlanta	Chicago	4:00 PM	5:00 PM	150
	Chicago	Dallas	5:00 PM	7:00 PM	100
3	Atlanta	Miami	4:00 PM	5:00 PM	100
	Miami	Dallas	5:00 PM	7:00 PM	100
4	San Francisco	New York	8:00 AM	2:00 PM	240
	New York	Atlanta	2:00 PM	4:00 PM	50
	Atlanta	Dallas	4:00 PM	6:00 PM	210

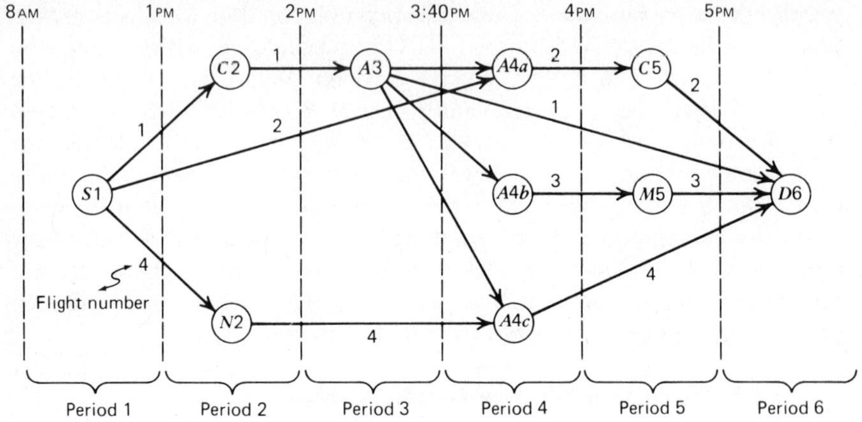

FIGURE 1.10 Network structure of the route problem (periods refer to time periods).

flights but Flights 2, 3, and 4 do not make connections. The node requirements and arc specifications are given in Tables 1.12 and 1.13. The capacity for each arc is any number that is no less than 1.

Suppose that on game day the president of Gamma Airlines offers free transportation to all passengers who board in San Francisco with Dallas as their final destination. Suppose the aircraft used on Flight 1 has seating for 500 passengers whereas each of the aircraft for the other three flights can accomodate only 250. Also suppose confirmed reservations have already been established on various legs of the flights, as shown in Table 1.14. The problem of determining how many passengers can obtain

Table 1.12 Node Requirements (Figure 1.10)

Node	Requirement
$S1$	1
$C2$	0
$N2$	0
$A3$	0
$A4a$	0
$A4b$	0
$A4c$	0
$C5$	0
$M5$	0
$D6$	−1

Table 1.13 Arc Specifications (Figure 1.10)

Arc Number	"From" Node	"To" Node	Unit Cost
1	$S1$	$C2$	100
2	$S1$	$A4a$	250
3	$S1$	$N2$	240
4	$C2$	$A3$	100
5	$N2$	$A4c$	50
6	$A3$	$D6$	250
7	$A3$	$A4a$	0
8	$A3$	$A4b$	0
9	$A3$	$A4c$	0
10	$A4a$	$C5$	150
11	$A4b$	$M5$	100
12	$A4c$	$D6$	210
13	$C5$	$D6$	100
14	$M5$	$D6$	100

Table 1.14 Confirmed Passenger Details

Flight Number	Origin	Destination	Number of Confirmed Passengers
1	San Francisco	Chicago	100
	Chicago	Atlanta	50
	Atlanta	Dallas	300
2	San Francisco	Atlanta	50
	Atlanta	Chicago	100
	Chicago	Dallas	200
3	Atlanta	Miami	100
	Miami	Dallas	150
4	San Francisco	New York	100
	New York	Atlanta	200
	Atlanta	Dallas	0

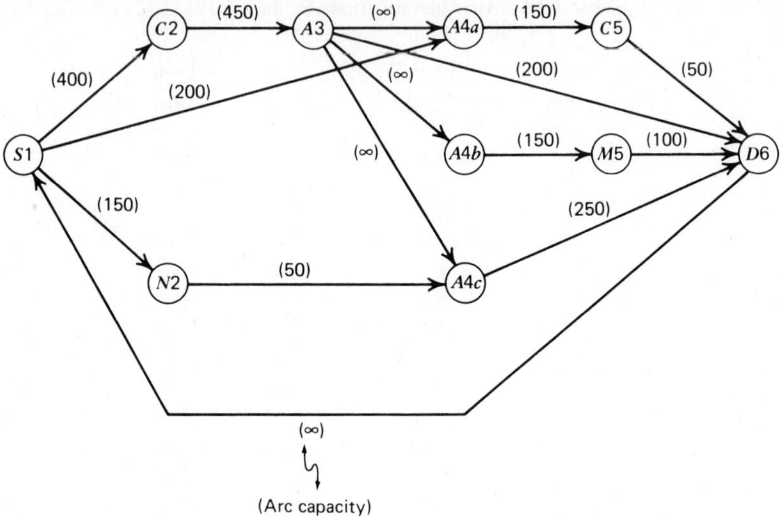

FIGURE 1.11 Network structure for maximum trip problem.

Table 1.15 Arc Specifications (Figure 1.11)

Arc Number	"From" Node	"To" Node	Unit Cost	Arc Capacity
1	$S1$	$C2$	0	400
2	$S1$	$A4a$	0	200
3	$S1$	$N2$	0	150
4	$C2$	$A3$	0	450
5	$N2$	$A4c$	0	50
6	$A3$	$D6$	0	200
7	$A3$	$A4a$	0	∞
8	$A3$	$A4b$	0	∞
9	$A3$	$A4c$	0	∞
10	$A4a$	$C5$	0	150
11	$A4b$	$M5$	0	150
12	$A4c$	$D6$	0	250
13	$C5$	$D6$	0	50
14	$M5$	$D6$	0	100
15	$D6$	$S1$	−1	∞

free transportation on game day can also be modeled as a network program, as shown in Figure 1.11 and Table 1.15. The node requirements are all zero.

EXERCISES

1.1 Place the following constraints in the NP form:

$$\begin{bmatrix} 1 & 1 & & & & \\ -1 & & -1 & 1 & 1 & \\ & -1 & 1 & -1 & & 1 \\ & & & & -1 & -1 \end{bmatrix} \begin{bmatrix} x_1 \\ x_2 \\ x_3 \\ x_4 \\ x_5 \\ x_6 \end{bmatrix} \begin{matrix} \leqslant \\ = \\ \leqslant \\ \leqslant \end{matrix} \begin{bmatrix} 10 \\ 0 \\ -3 \\ -5 \end{bmatrix},$$

$$0 \leqslant x_i \leqslant 3, \quad i=1,\ldots,6.$$

(*Hint*: Add slack variables and then append a new constraint that is the negative of the sum of the other constraints.)

1.2 Consider the five-node network given by Table E1.2.

Table E1.2

Node Number	Supply	Demand	Arc Number	Directed Away from	Directed Toward	Unit Cost	Arc Capacity
1	10	0	1	1	2	0	5
2	12	0	2	1	3	2	9
3	4	0	3	2	5	1	7
4	0	11	4	2	4	7	6
5	0	8	5	3	4	4	12
			6	3	5	3	15
			7	5	4	1	20

(a) Illustrate this network by a drawing, and indicate a feasible solution along with the cost of your solution.
(b) Give the node-arc incidence matrix associated with this network.
(c) Reformulate this problem by appending a dummy demand point and appropriate arcs such that total supply equals total demand. Display your answer on a drawing.
(d) Reformulate this problem by appending appropriate nodes and arcs so that there is a single supply point and a single demand point.

1.3 Consider the six-node network given by Table E1.3.

Table E1.3

Arc Number	Directed Away from	Directed Toward	Arc Length
1	1	2	1
2	1	4	8
3	2	3	6
4	2	5	2
5	2	6	9
6	3	4	1
7	4	6	2
8	5	3	3
9	5	6	7

(a) Give the node-arc incidence matrix for this network.
(b) Determine the shortest path from node 1 to node 6 by inspection.
(c) Formulate the problem of determining the shortest path from node 1 to node 6 as a minimal cost network flow problem by giving the vectors **c**, **r**, and **u**. Display your formulation on a drawing.

1.4 Consider the seven-node network given by Table E1.4.

Table E1.4

Arc Number	Directed Away from	Directed Toward	Arc Capacity
1	1	2	20
2	1	3	5
3	2	5	15
4	2	6	17
5	3	4	19
6	3	6	4
7	4	6	3
8	4	7	10
9	5	6	19
10	5	7	1
11	6	7	2

(a) Give the node-arc incidence matrix for this network.
(b) Determine the maximal continuous flow from node 1 to node 7.
(c) Formulate the problem of determining the maximal continuous flow from node 1 to node 7 as a minimal cost network flow problem by specifying **c**, **r**, and **u**. Display your formulation on a drawing. (*Hint*: One new arc must be appended to the network.)

1.5 Prove the following: If $\Sigma_i r_i \neq 0$, then NP has no solution.

1.6 Place the following problem in the NP form:

$$\begin{array}{ll} \min & \mathbf{cx} \\ \text{s.t.} & \mathbf{Ax}=\mathbf{r} \\ & \mathbf{v} \leqslant \mathbf{x} \leqslant \mathbf{u}, \end{array}$$

where $\mathbf{0} < \mathbf{v} \leqslant \mathbf{u}$ and **A** is a node-arc incidence matrix.

1.7 Let **A** be any node-arc incidence matrix, and let $\mathbf{C}=\mathbf{AA}'$. Give an interpretation for C_{ij}, where $i \neq j$.

1.8 The U.S. Air Force contracts with civilian carriers to fly specified daily routes among its bases to meet the demands for supplies and equipment that are stored only at certain bases. Wright-Patterson (FFO), Tinker (TIK), and Robins (WRB) serve as warehouses, and each has a

Table E1.8

Route Number	Origin	Destination	Distance (miles)
1	TIK	SZL	312
	SZL	BLV	200
	BLV	FFO	324
	FFO	BYH	409
	BYH	CBM	189
2	TIK	LRF	326
	LRF	BYH	103
	BYH	CBM	182
	CBM	WRB	404
	WRB	CBM	404
3	WRB	BYH	465
	BYH	BLV	113
	BLV	TIK	473

Table E1.9

	Faculty Member						
Course	Dr. Bhat	Dr. L. Cooper	Dr. M. Cooper	Dr. Kennington	Dr. LeBlanc	Dr. Schmeiser	Number of Times Course Will Be Offered per Year
Linear Programming	7	1	2	2	2	7	2
Nonlinear Programming	8	4	4	4	1	8	2
Integer Programming	9	5	1	3	4	9	2
Network Programming	10	8	6	1	6	10	2
Dynamic Programming	11	2	3	7	5	11	1
Optimization Theory	12	3	5	5	3	12	1
Statistics	3	6	7	6	7	1	3
Simulation	6	9	12	12	11	2	2
Queueing	2	10	10	10	10	4	1
Stochastic Processes	1	11	11	8	9	3	2
Inventory	5	7	8	9	8	5	1
Reliability	4	12	9	11	12	6	1
Number of courses available to teach per year	2	2	4	4	4	4	20

capacity of 25 tons per day. These three warehouses serve the following bases with the given daily demands:

Base	Daily Demand (lb)
Whiteman (SZL)	15,000
Scott (BLV)	10,000
Little Rock (LRF)	18,000
Blytheville (BYH)	18,000
Columbus (CBM)	25,000

The three routes shown in Table E1.8 are flown daily. The capacity of each aircraft is 25,000 pounds. We assume that units may be transferred from one aircraft to another without incurring extra cost. Present a formulation of the problem of meeting the demands at the five bases that minimizes the pound-miles for the system.

1.9 In the Operations Research Department at Southern Methodist University a faculty of 6 will offer 20 courses during an academic year. The faculty preferences for teaching these courses are listed in Table E1.9, along with the number of courses each faculty member will teach and the number of offerings of each course. The lower the preference number, the more desirable the faculty member considers that teaching assignment. Present a network model that will minimize the sum of all preferences for all assignments under the restriction that no faculty member may teach the same course twice in 1 year.

1.10 Suppose that for the Detroit Motors model described in Section 1.4.1 you incur the penalties shown in Table E1.10 for unused production capacity. Reformulate the model to account for these penalties.

Table E1.10

	Penalty for Unused Production Capacity	
Plant Site	Range of 0–100 ($/vehicle)	Range of 100–∞ ($/vehicle)
Detroit	1000	2000
Freemont	1200	1500
Arlington	700	1000
Atlanta	900	2000

1.11 Formulate the problem of finding the shortest completion time for the construction project listed in Table E1.11.

Table E1.11

Activity	Predecessor Activities	Activity Time (months)
M	—	5
N	M	3
O	M	12
P	N	15
Q	O, P	11
R	M, P	20
S	P	8
T	P	10
U	O, P	9
V	S, T	6
W	S	5
X	T	4
Y	W, T	1
Z	X	7

(*Hint*: See Section 1.4.2.)

1.12 Reformulate the two problems presented in Section 1.4.3, using the flight schedule shown in Table E1.12.

Table E1.12

Flight Number	Origin	Destination	Departure Time	Arrival Time	Cost ($)
1	San Francisco	Chicago	8:00 AM	1:00 PM	200
	Chicago	Atlanta	2:30 PM	3:30 PM	100
	Atlanta	Dallas	4:00 PM	6:00 PM	50
2	San Francisco	Atlanta	10:00 AM	2:45 PM	300
	Atlanta	Chicago	3:45 PM	4:45 PM	75
	Chicago	Dallas	5:00 PM	7:00 PM	65
3	Atlanta	Miami	4:00 PM	5:00 PM	43
	Miami	Dallas	5:00 PM	7:00 PM	57
4	San Francisco	New York	8:00 AM	2:00 PM	220
	New York	Atlanta	2:00 PM	4:00 PM	98
	Atlanta	Dallas	4:00 PM	6:00 PM	66

NOTES AND REFERENCES

Section 1.1

A host of problems that have been solved using network programming technology by the team of Professors Glover and Klingman can be found in Refs. [82] and [78].

Section 1.3

The original papers used in our research for this section are as follows: Charnes and Cooper [30], Dantzig [44,47], Ford and Fulkerson [65,66], Fulkerson [70], Hitchcock [103], Johnson [112], Kantorovich [115], Koopmans and Reiter [126], Kuhn [129], and Orden [149].

CHAPTER 2

Linear Programming

Linear network flow models are special cases of the linear program. Therefore a study of network flows must naturally begin with a review of the theory of linear programming. Since there are numerous textbooks devoted to the study of linear programming, this chapter presents only the portions of this rich theory that pertain to network flows.

2.1 THE LINEAR PROGRAM

Let **A** be an $m \times n$ matrix, let **c** and **u** be n-component vectors, and let **b** be an m-component vector. Without loss of generality, the *linear program* may be stated mathematically as follows:

$$\left.\begin{array}{ll}\min & \mathbf{cx} \\ \text{s.t.} & \mathbf{Ax}=\mathbf{b} \\ & \mathbf{0} \leqslant \mathbf{x} \leqslant \mathbf{u}\end{array}\right\} LP,$$

where it is assumed that $m<n$, the rank of **A** is m, and $\mathbf{u} \geqslant \mathbf{0}$. Components of **c** and **u** may be allowed to assume arbitrarily large finite values. Problems having inequality constraints may be placed in the form prescribed above via the addition of what are traditionally called slack or surplus variables. The rules for accomplishing this transformation can be found in any linear programming textbook.

The case where $\mathbf{c}=\mathbf{0}$ is not of interest to us, since any solution to the system of constraints will be an optimum for LP. Thus we shall assume $\mathbf{c}\neq\mathbf{0}$ for our development.

2.2 PROPERTIES OF A EUCLIDEAN VECTOR SPACE

In this section we review fundamental topological and geometric notions of interest in mathematical programming. Let R^n denote the n-dimensional Euclidean vector space with the Euclidean norm of $\mathbf{x}\varepsilon R^n$ denoted by $\|\mathbf{x}\|$. We define a *direction* in R^n to be any point $\mathbf{x}\varepsilon R^n$ with $\mathbf{x}\neq\mathbf{0}$. We say that $\mathbf{x}$ is a normalized direction if $\|\mathbf{x}\|=1$. Given two distinct points $\mathbf{x}^1$ and $\mathbf{x}^2$ of R^n and a set of scalars $\Lambda\subset R^1$, we define the set of linear combinations of the points $\mathbf{x}^1$ and $\mathbf{x}^2$ induced by Λ to be $\Gamma^{\Lambda}=\{\mathbf{x}:\mathbf{x}=(1-\lambda)\mathbf{x}^1+\lambda\mathbf{x}^2, \lambda\varepsilon\Lambda\}$. If $\Lambda=R^1$, then Γ^{Λ} is called the *line through* $\mathbf{x}^1$ *and* $\mathbf{x}^2$. If $\Lambda=\{\lambda:\lambda\geqslant 0\}$, then Γ^{Λ} is called the *closed half-ray at* $\mathbf{x}^1$ *in the direction* $\mathbf{x}^2-\mathbf{x}^1$. If $\Lambda=\{\lambda:0\leqslant\lambda\leqslant 1\}$, then Γ^{Λ} is called the *closed segment with* endpoints $\mathbf{x}^1$ and $\mathbf{x}^2$. *Open half-ray* and *open segment* are defined analogously by replacing the relations $\leqslant$ with $<$ in the above definitions. A set $X\subset R^n$ is said to be *convex* if the closed segment with endpoints $\mathbf{x}^1$ and $\mathbf{x}^2$ is a subset of X for any two distinct points $\mathbf{x}^1$ and $\mathbf{x}^2$ in X. A point $\mathbf{x}\varepsilon X$ is said to be an *extreme point* of the convex set X if no distinct points $\mathbf{x}^1$ and $\mathbf{x}^2$ exist in X such that $\mathbf{x}$ is in the open segment with endpoints $\mathbf{x}^1$ and $\mathbf{x}^2$. More generally, if $X\subset R^n$ is convex, any subset $Y\subset X$ is said to be an *extreme subset* of X if the set difference, $X\sim Y\equiv\{\mathbf{x}:\mathbf{x}\varepsilon X, \mathbf{x}\not\varepsilon Y\}$, is convex.

Given a point $\mathbf{x}^0\varepsilon R^n$ and a scalar $\epsilon>0$, the set $B_\epsilon(\mathbf{x}^0)=\{\mathbf{x}:\mathbf{x}\varepsilon R^n, \|\mathbf{x}-\mathbf{x}^0\|<\epsilon\}$ is called the *open ball* with center $\mathbf{x}^0$ and radius ϵ. A point $\mathbf{x}$ is said to be *interior to* the set $X\subset R^n$ if there exists an $\epsilon>0$ such that $B_\epsilon(\mathbf{x})\subset X$. A point $\mathbf{x}$ is said to be *exterior to* the set $X\subset R^n$ if there exists an $\epsilon>0$ such that $B_\epsilon(\mathbf{x})\cap X=\phi$. Any point $\mathbf{x}$ that is neither interior nor exterior to the set $X\subset R^n$ is said to be a *boundary point* of X. Any point of X that is not an interior point is obviously a boundary point of X. Any point that is not a point of X and is not exterior to X is obviously a boundary point of X also. Clearly, an extreme point of a convex set is a boundary point, but there may be boundary points for a convex set that are not extreme points. A set $X\subset R^n$ is said to be *open* if all its points are interior points. A set $X\subset R^n$ is said to be *closed* if it contains all of its boundary points. A set $X\subset R^n$ is said to be *bounded* if there exists some $\epsilon>0$ such that $X\subset B_\epsilon(\mathbf{0})$.

A real-valued function f is *continuous* on a set $X\subset R^n$ if $f(\mathbf{x})$ is defined for each $\mathbf{x}\varepsilon X$, and if, corresponding to any $\epsilon>0$ and $\mathbf{x}^0\varepsilon X$, there is some $\delta>0$ such that $|f(\mathbf{x})-f(\mathbf{x}^0)|<\epsilon$ whenever $\mathbf{x}\varepsilon[X\cap B_\delta(\mathbf{x}^0)]\subset R^n$. A real-valued function of f is *convex* on a convex set $X \subset R^n$ if, for any two points $\mathbf{x}^1$ and $\mathbf{x}^2$ of X and scalar λ such that $0 \leqslant \lambda \leqslant 1$, $f[(1-\lambda)\mathbf{x}^1 + \lambda\mathbf{x}^2] \leqslant (1-\lambda)f(\mathbf{x}^1) + \lambda f(\mathbf{x}^2)$. A real-valued function f is *concave* on a convex set $X \subset R^n$ if, for any two points $\mathbf{x}^1$ and $\mathbf{x}^2$ of X and scalar λ such that $0 \leqslant \lambda \leqslant 1$, $f[(1-\lambda)\mathbf{x}^1 + \lambda\mathbf{x}^2] \geqslant (1-\lambda)f(\mathbf{x}^1) + \lambda f(\mathbf{x}^2)$.

We now state without proof some properties of continuous functions that are of interest in mathematical programming. Proofs may be found in any standard analysis textbook.

Proposition 2.1

Any function that is continuous on a closed and bounded set $X \subset R^n$ is bounded on X.

Proposition 2.2

Any function that is continuous on a closed and bounded set $X \subset R^n$ takes a maximum and a minimum value on X.

The linear combination of a finite set of points $\{\mathbf{x}^1, \ldots, \mathbf{x}^m\}$ of R^n, given by $\sum_{i=1}^m c_i \mathbf{x}^i$, is said to be a *convex combination* if $\sum_{i=1}^m c_i = 1$ and $\mathbf{c} \geqslant \mathbf{0}$. Given a set $X \subset R^n$, the *convex hull* of X is defined to be the union of all convex combinations of finite subsets of X.

Sets of the form $H(\mathbf{d}, w) = \{\mathbf{x} : \mathbf{x} \varepsilon R^n, \mathbf{dx} = w\}$ for a given vector $\mathbf{d} \varepsilon R^n$, $\mathbf{d} \neq \mathbf{0}$, and given scalar w are called *hyperplanes*. We also associate with $H(\mathbf{d}, w)$ the two *closed half-spaces* defined by $H^L(\mathbf{d}, w) = \{\mathbf{x} : \mathbf{x} \varepsilon R^n, \mathbf{dx} \leqslant w\}$ and $H^U(\mathbf{d}, w) = \{\mathbf{x} : \mathbf{x} \varepsilon R^n, \mathbf{dx} \geqslant w\}$. A *polytope* is defined to be the intersection of a finite number of closed half-spaces. A bounded polytope is called a *polyhedron*. An extreme point of a polytope is called a *vertex*. A maximal extreme closed segment of a polytope is called a *finite edge*. An extreme line or maximal extreme closed half-ray of a polytope is called an *infinite edge*.

We now state without proof a number of properties of polytopes and polyhedra that are of interest in linear programming.

Proposition 2.3

The number of vertices and edges of a polytope is finite.

Proposition 2.4

The number of vertices of a nonempty polyhedron is nonzero.

Proposition 2.5

A nonempty polyhedron is the convex hull of its vertices.

Proposition 2.6

A polytope is a closed convex set.

2.3 CHARACTERIZATION OF SOLUTIONS

In this section we develop a characterization of the optimal solutions to *LP*. This characterization is then exploited in the development of the primal simplex algorithm.

Let the set of feasible solutions for *LP* be denoted by $Q=\{\mathbf{x}:\mathbf{Ax}=\mathbf{b},\mathbf{0}\leqslant\mathbf{x}\leqslant\mathbf{u}\}$. For this development we assume $Q\neq\phi$. This assumption is not restrictive in any sense, since for both the linear program and for network problems we may append artificial variables to the problem to guarantee a nonempty Q in a space of higher dimension.

Under the assumption that $Q\neq\phi$ we now show that *LP* has an extreme point optimum.

Proposition 2.7

There exists an extreme point optimum for *LP*.

Proof. First we show that Q is a polyhedron. $\mathbf{Ax}=\mathbf{b}$ is equivalent to the system

$$\left\{\begin{matrix}\mathbf{Ax}\leqslant\mathbf{b}\\ \mathbf{Ax}\geqslant\mathbf{b}\end{matrix}\right\}. \tag{2.1}$$

Likewise $\mathbf{0}\leqslant\mathbf{x}\leqslant\mathbf{u}$ is equivalent to the system

$$\left\{\begin{matrix}\mathbf{x}\geqslant\mathbf{0}\\ \mathbf{x}\leqslant\mathbf{u}\end{matrix}\right\}. \tag{2.2}$$

Since (2.1) and (2.2) define half-spaces and Q is their intersection, Q is a polytope. Because of (2.2), Q is bounded and is thus a polyhedron.

The linear function $\mathbf{cx}$ is continuous on Q. By Propositions 2.2 and 2.6, $\mathbf{cx}$ takes on a minimum value, say z^*, at some point $\mathbf{x}^*\varepsilon Q$. By our assumption that $Q\neq\phi$ and Propositions 2.3–2.5, Q has a nonzero finite number of vertices, say $\{\mathbf{x}^1,\ldots,\mathbf{x}^m\}$, and Q is the convex hull of these vertices. Thus there is a vector $\mathbf{d}\varepsilon R^m$ such that $\mathbf{x}^* = \sum_{i=1}^m d_i\mathbf{x}^i$, where $\sum_{i=1}^m d_i = 1$ and $\mathbf{d} \geqslant \mathbf{0}$. Let $J = \{i : d_i > 0\}$. Also, $J \neq \phi$ since $\mathbf{d} \geqslant \mathbf{0}$ and $\sum_{i=1}^m d_i = 1$. Then $\mathbf{x}^* = \sum_{i\varepsilon J} d_i x^i$ and $\sum_{i\varepsilon J} d_i = 1$.

We now show that, for any $\mathbf{x}^i$ such that $i\varepsilon J, \mathbf{c}\mathbf{x}^i = z^*$. Since z^* is the minimum of $\mathbf{cx}$ on Q, $\mathbf{cx}^i \geqslant z^*$ for all $i\varepsilon J$. Suppose that, for some $k\varepsilon J, \mathbf{cx}^k > z^*$. Then

$$\begin{aligned} z^* &= \mathbf{cx}^* = \mathbf{c}\sum_{i\varepsilon J} d_i\mathbf{x}^i = \sum_{i\varepsilon J} d_i(\mathbf{cx}^i) \\ &= d_k(\mathbf{cx}^k) + \sum_{i\varepsilon J\sim\{k\}} d_i(\mathbf{cx}^i) \geqslant d_k(\mathbf{cx}^k) + \sum_{i\varepsilon J\sim\{k\}} d_i z^* \\ &> d_k z^* + \sum_{i\varepsilon J\sim\{k\}} d_i z^* = z^* \sum_{i\varepsilon J} d_i = z^*, \qquad \text{a contradiction.} \end{aligned}$$

Thus, for each $i\varepsilon J, \mathbf{cx}^i = z^*$.

Since a vertex of a polyhedron is an extreme point, the existence of an extreme point optimum is demonstrated. ■

Recall that $\mathbf{A}$ is an $m\times n$ matrix having $m<n$ and rank $(\mathbf{A}) = m$. Then, via column interchanges, $\mathbf{A}$ can be partitioned as follows: $\mathbf{A} = [\mathbf{B} \vdots \mathbf{N}]$, where $\mathbf{B}$ is square and Det($\mathbf{B}$)$\neq 0$. Likewise, $\mathbf{x}$ and $\mathbf{u}$ may be partitioned similarly, so that

$$\mathbf{Ax} = \mathbf{b} \Leftrightarrow \mathbf{Bx}^B + \mathbf{Nx}^N = \mathbf{b}$$

$$\mathbf{0} \leqslant \mathbf{x} \leqslant \mathbf{u} \Leftrightarrow \left\{ \begin{array}{c} \mathbf{0} \leqslant \mathbf{x}^B \leqslant \mathbf{u}^B \\ \mathbf{0} \leqslant \mathbf{x}^N \leqslant \mathbf{u}^N \end{array} \right\}.$$

The variables corresponding to $\mathbf{B}$, that is, $\mathbf{x}^B$, are called the *basic variables*, while the remaining variables, $\mathbf{x}^N$, are referred to as *nonbasic variables*. A *basic feasible solution* for Q is one for which

$$\left\{ \begin{array}{c} [\mathbf{x}^B \vdots \mathbf{x}^N]\varepsilon Q \\ \text{and} \\ x_i^N \varepsilon \{0, u_i^N\} \qquad \text{for all } i \end{array} \right\}.$$

In other words, each nonbasic variable must assume either the value of zero or its upper bound.

We now show that extreme points of Q correspond to basic feasible solutions.

Proposition 2.8

Each extreme point of Q forms a basic feasible solution.

Proof. Let $\mathbf{x}^*$ be any extreme point of Q, and let the variables be ordered so that $0<x_i^*<u_i$ for $i=1,\ldots,k$ and $x_i^* \varepsilon \{0,u_i\}$ for $i=k+1,\ldots,n$. Reorder the columns of $\mathbf{A}$ to correspond to the order of $\mathbf{x}^*$. Suppose the first k columns of $\mathbf{A}$ are linearly dependent. Then there exist scalars λ_i, not all zero, such that

$$\sum_{i=1}^{i=k} \lambda_i \mathbf{A}(i)=\mathbf{0}.$$

Let

$$\alpha_1 = \min_{1 \leqslant i \leqslant k} \left\{ \frac{x_i^*}{|\lambda_i|} : \lambda_i \neq 0 \right\},$$

and let

$$\alpha_2 = \min_{1 \leqslant i \leqslant k} \left\{ \frac{u_i - x_i^*}{|\lambda_i|} : \lambda_i \neq 0 \right\}.$$

Select ϵ such that $0<\epsilon<\min(\alpha_1,\alpha_2)$. Then

$$0<x_i^* \pm \epsilon\lambda_i<u_i \quad \text{for } 1 \leqslant i \leqslant k. \tag{2.3}$$

Let the n-component vector $\boldsymbol{\mu}$ be given by $[\lambda_1,\ldots,\lambda_k,0,\ldots,0]$. Let

$$\mathbf{x}^1=\mathbf{x}^*+\epsilon\boldsymbol{\mu}$$

and

$$\mathbf{x}^2=\mathbf{x}^*-\epsilon\boldsymbol{\mu}.$$

Because of (2.3), $\mathbf{0}<\mathbf{x}^1<\mathbf{u}$ and $\mathbf{0}<\mathbf{x}^2<\mathbf{u}$. Furthermore, from our linear dependency assumption and the fact that $\mu_i=0$ for $i \geqslant k+1, \mathbf{A}\boldsymbol{\mu}=\Sigma_{i=1}^{i=n} \mu_i \mathbf{A}(i)=\mathbf{0}$. Hence $\mathbf{A}\mathbf{x}^1=\mathbf{A}\mathbf{x}^*=\mathbf{b}$ and $\mathbf{A}\mathbf{x}^2=\mathbf{A}\mathbf{x}^*=\mathbf{b}$.

Since $\boldsymbol{\mu}\neq\mathbf{0}, \mathbf{x}^1$ and $\mathbf{x}^2$ are feasible solutions different from each other and from $\mathbf{x}^*$. Furthermore, $\mathbf{x}^*=\frac{1}{2}\mathbf{x}^1+\frac{1}{2}\mathbf{x}^2$. This contradicts the hypothesis that $\mathbf{x}^*$ is an extreme point, and it follows that the first k columns of $\mathbf{A}$ must be linearly independent. Since rank $(\mathbf{A})$ is $m, k \leqslant m$. If $k=m$, then $\mathbf{B}$ is precisely the first m columns of $\mathbf{A}$. If $k<m$, then $\mathbf{B}$ is formed from the first

k columns of $\mathbf{A}$ together with $m-k$ columns from $\{\mathbf{A}(k+1),\ldots,\mathbf{A}(n)\}$, which form a set of m linearly independent columns. Since rank $(\mathbf{A})=m$, such a selection is always possible. ■

This proposition allows one to limit the search for an optimal solution to *LP* to basic feasible solutions for *LP* since, by Proposition 2.7, one of these must be an optimum.

2.4 PRIMAL SIMPLEX METHOD

We take the view that the primal simplex method is a member of the class of algorithms known as feasible direction methods. A direction $\mathbf{d}$ is said to be a *feasible direction* at $\mathbf{y}$ if $\mathbf{y}+\alpha\mathbf{d}$ is feasible for some $\alpha>0$. A direction $\mathbf{d}$ is said to be an *improving feasible direction* if $\mathbf{d}$ is a feasible direction and the directional derivative of $\mathbf{cx}$ in direction $\mathbf{d}$ at point $\mathbf{y}$ is negative. Given the above definitions, the general feasible direction method may be described as follows.

ALG 2.1 GENERAL FEASIBLE DIRECTION METHOD

0 *Initialization.* Let $\mathbf{y}^0$ be any feasible point, and set $k \leftarrow 0$.

1 *Find an Improving Feasible Direction.* Let $\mathbf{d}$ be an improving feasible direction at $\mathbf{y}^k$. If no such $\mathbf{d}$ exists, then terminate with the optimum, $\mathbf{y}^k$.

2 *Obtain a New Solution.* Find α^* such that

$$\mathbf{c}(\mathbf{y}^k+\alpha^*\mathbf{d})=\min_{\alpha>0}\{\mathbf{c}(\mathbf{y}^k+\alpha\mathbf{d}):\mathbf{y}^k+\alpha\mathbf{d}\varepsilon Q\}.$$

Set $\mathbf{y}^{k+1}\leftarrow\mathbf{y}^k+\alpha^*\mathbf{d}, k\leftarrow k+1$, and return to step 1.

The primal simplex method is a specialization of ALG 2.1 in which $\mathbf{y}^k$ for each k is an extreme point. Hence the simplex method is often called an *extreme point method.* This requires that direction $\mathbf{d}$ be selected in such a way that the solution in Step 2 will result in a new extreme point.

Given a basic feasible solution for the linear program, we may partition $\mathbf{A}$, $\mathbf{c}$, $\mathbf{x}$, and $\mathbf{u}$ into basic and nonbasic components, that is, $\mathbf{A}=[\mathbf{B}\,|\,\mathbf{N}]$, $\mathbf{c}=[\mathbf{c}^B\,|\,\mathbf{c}^N]$, $\mathbf{x}=[\mathbf{x}^B\,|\,\mathbf{x}^N]$, and $\mathbf{u}=[\mathbf{u}^B\,|\,\mathbf{u}^N]$. Hence *LP* may be stated as follows:

$$\min \quad \mathbf{c}^B\mathbf{x}^B+\mathbf{c}^N\mathbf{x}^N \tag{2.4}$$

$$\text{s.t.} \quad \mathbf{B}\mathbf{x}^B+\mathbf{N}\mathbf{x}^N=\mathbf{b} \tag{2.5}$$

$$\mathbf{0}\leqslant\mathbf{x}^B\leqslant\mathbf{u}^B \tag{2.6}$$

$$\mathbf{0}\leqslant\mathbf{x}^N\leqslant\mathbf{u}^N. \tag{2.7}$$

From (2.5) we obtain $\mathbf{x}^B = \mathbf{B}^{-1}\mathbf{b} - \mathbf{B}^{-1}\mathbf{N}\mathbf{x}^N$. Substituting for $\mathbf{x}^B$, we obtain the problem;

$$\min \quad \mathbf{c}^B\mathbf{B}^{-1}\mathbf{b} + (\mathbf{c}^N - \mathbf{c}^B\mathbf{B}^{-1}\mathbf{N})\mathbf{x}^N \tag{2.8}$$

$$\text{s.t.} \quad \mathbf{0} \leqslant \mathbf{B}^{-1}\mathbf{b} - \mathbf{B}^{-1}\mathbf{N}\mathbf{x}^N \leqslant \mathbf{u}^B \tag{2.9}$$

$$\mathbf{0} \leqslant \mathbf{x}^N \leqslant \mathbf{u}^N. \tag{2.10}$$

Assumption

We assume for this development that $\mathbf{0} < \mathbf{B}^{-1}\mathbf{b} - \mathbf{B}^{-1}\mathbf{N}\mathbf{x}^N < \mathbf{u}^B$. (This corresponds to what is often called the nondegeneracy assumption.)

Consider the set of directions in nonbasic space given by $D = \{\mathbf{e}^1, \mathbf{e}^2, \ldots, \mathbf{e}^{n-m}, -\mathbf{e}^1, -\mathbf{e}^2, \ldots, -\mathbf{e}^{n-m}\}$. In the development of the primal simplex algorithm we restrict attention to directions from D. Because of the assumption, we need consider (2.7) only when selecting a feasible direction. Hence $\mathbf{e}^i$ is a feasible direction if $x_i^N < u_i^N$, and $-\mathbf{e}^i$ is a feasible direction if $x_i^N > 0$.

By consideration of the directional derivative in nonbasic space, we indicate how one may select an improving feasible direction. Recall that the directional derivative of $g(\mathbf{x})$ at $\mathbf{y}$ in the direction $\mathbf{d}$ is given by

$$D_{\mathbf{d}}g(\mathbf{y}) = \lim_{t \to 0^+} \frac{g\left(\mathbf{y} + t\dfrac{\mathbf{d}}{\|\mathbf{d}\|}\right) - g(\mathbf{y})}{t},$$

provided that the limit exists. A more computationally useful form of the directional derivative is given by

$$D_{\mathbf{d}}g(\mathbf{y}) = \frac{\nabla g(\mathbf{y})\mathbf{d}}{\|\mathbf{d}\|},$$

where $\nabla g(\mathbf{y})$ is the gradient of $g(\cdot)$ evaluated at the point $\mathbf{y}$. Hence, for the program given by (2.8)–(2.10), $g(\mathbf{x}^N) = \mathbf{c}^B\mathbf{B}^{-1}\mathbf{b} + (\mathbf{c}^N - \mathbf{c}^B\mathbf{B}^{-1}\mathbf{N})\mathbf{x}^N$ and $\nabla g(\mathbf{y}) = \mathbf{c}^N - \mathbf{c}^B\mathbf{B}^{-1}\mathbf{N}$. Therefore the set of improving feasible directions may be obtained as follows:

$$\left\{\begin{matrix} \mathbf{e}^i : x_i^N < u_i^N & \text{and} & (\mathbf{c}^N - \mathbf{c}^B\mathbf{B}^{-1}\mathbf{N})_i < 0 \\ -\mathbf{e}^i : x_i^N > 0 & \text{and} & (\mathbf{c}^N - \mathbf{c}^B\mathbf{B}^{-1}\mathbf{N})_i > 0 \end{matrix}\right\}.$$

If we restrict attention to basic solutions, $x_i^N \varepsilon \{0, u_i^N\}$, and the set of improving feasible directions may be written as follows:

$$\left\{\begin{matrix} \mathbf{e}^i : x_i^N = 0 & \text{and} & (\mathbf{c}^N - \mathbf{c}^B\mathbf{B}^{-1}\mathbf{N})_i < 0 \\ -\mathbf{e}^i : x_i^N = u_i^N & \text{and} & (\mathbf{c}^N - \mathbf{c}^B\mathbf{B}^{-1}\mathbf{N})_i > 0 \end{matrix}\right\}.$$

We now show that solving the Step 2 program,

$$\min_{\alpha>0} \quad \mathbf{c}^B\mathbf{B}^{-1}\mathbf{b}+(\mathbf{c}^N-\mathbf{c}^B\mathbf{B}^{-1}\mathbf{N})(\mathbf{x}^N+\alpha\mathbf{d})$$

$$\text{s.t.} \quad \mathbf{0}\leqslant \mathbf{x}^N+\alpha\mathbf{d}\leqslant \mathbf{u}^N \tag{2.11}$$

$$\mathbf{x}^B=\mathbf{B}^{-1}\mathbf{b}-\mathbf{B}^{-1}\mathbf{N}(\mathbf{x}^N+\alpha\mathbf{d})$$

$$\mathbf{0}\leqslant \mathbf{x}^B\leqslant \mathbf{u}^B, \tag{2.12}$$

leads to a new basic feasible solution. Note that, because of the linearity of $g(\mathbf{x}^N)$, any increase in α results in a decrease in $g(\mathbf{x}^N)$. Hence we wish to make α as large as possible subject to (2.11) and (2.12). Let $\mathbf{d}\varepsilon\Gamma$. Then movement in direction $\mathbf{d}$ may eventually force some variable x_j to assume a value outside the range $0\leqslant x_j\leqslant u_j$. The first such variable for which this occurs will be called the *blocking variable*. Let

$$\alpha_1=\max_{\alpha>0}\{\alpha:\mathbf{0}\leqslant \mathbf{x}^N+\alpha\mathbf{d}\leqslant \mathbf{u}^N\},$$

and let

$$\alpha_2=\max_{\alpha>0}\{\alpha:\mathbf{0}\leqslant \mathbf{B}^{-1}\mathbf{b}-\mathbf{B}^{-1}\mathbf{N}(\mathbf{x}^N+\alpha\mathbf{d})\leqslant \mathbf{u}^B\}.$$

Suppose $\alpha_1\leqslant\alpha_2$. Then the blocking variable is the nonbasic corresponding to $\mathbf{d}$, and $\mathbf{B}$ remains unchanged. Hence Step 2 results in a new basic feasible solution. Suppose $\alpha_2<\alpha_1$. Then changing $\mathbf{x}^N$ by $\alpha_2\mathbf{d}$ forces some basic variable to either zero or its upper bound. Hence we simply revise the partitioning so that this basic variable is nonbasic and the nonbasic variable corresponding to $\mathbf{d}$ becomes a member of the basic set. Hence a new basic feasible solution is obtained. Using the above formulae for a feasible direction, we may state the primal simplex algorithm as follows.

ALG 2.2 PRIMAL SIMPLEX ALGORITHM

0 *Initialization.* Let $[\mathbf{x}^B \,\vdots\, \mathbf{x}^N]$ be a basic feasible solution with $\mathbf{A}=[\mathbf{B}\,\vdots\,\mathbf{N}]$.

1 *Pricing.* Let

$$\psi_1=\{i:x_i^N=0 \quad\text{and}\quad (\mathbf{c}^N-\mathbf{c}^B\mathbf{B}^{-1}\mathbf{N})_i<0\},$$

$$\psi_2=\{i:x_i^N=u_i^N \quad\text{and}\quad (\mathbf{c}^N-\mathbf{c}^B\mathbf{B}^{-1}\mathbf{N})_i>0\}.$$

If $\psi_1\cup\psi_2=\phi$, terminate with $[\mathbf{x}^B \,\vdots\, \mathbf{x}^N]$ optimal; otherwise, select

$k \in \psi_1 \cup \psi_2$ and set

$$\delta \leftarrow \begin{cases} 1, & \text{if } k \in \psi_1 \\ -1, & \text{if } k \in \psi_2 \end{cases}.$$

2 *Ratio Test.* Set $\mathbf{y} \leftarrow \mathbf{B}^{-1}\mathbf{N}(k)$. Set

$$\Delta_1 \leftarrow \min_{\substack{1 \leq j \leq m \\ \sigma(y_j) = \delta}} \left\{ \frac{x_j^B}{|y_j|}, \infty \right\},$$

$$\Delta_2 \leftarrow \min_{\substack{1 \leq j \leq m \\ -\sigma(y_j) = \delta}} \left\{ \frac{u_j^B - x_j^B}{|y_j|}, \infty \right\}.$$

Set $\Delta \leftarrow \min\{\Delta_1, \Delta_2, u_k^N\}$.

3 *Update.* Set $x_k^N \leftarrow x_k^N + \Delta\delta$ and $\mathbf{x}^B \leftarrow \mathbf{x}^B - \Delta\delta\mathbf{y}$. If $\Delta = u_k^N$, return to Step 1.

4 *Pivot.* Let

$$\psi_3 = \{ j : \sigma(y_j) = \delta \quad \text{and} \quad x_j^B = 0 \},$$

$$\psi_4 = \{ j : -\sigma(y_j) = \delta \quad \text{and} \quad x_j^B = u_j^B \}.$$

Select any $l \in \psi_3 \cup \psi_4$. In the basis, replace $\mathbf{B}(l)$ with $\mathbf{N}(k)$ and return to Step 1.

Under the nondegeneracy assumption, $\Delta > 0$ at each iteration and convergence is guaranteed. The amazing feature of this algorithm is that, even when the nondegeneracy assumption is false, the algorithm usually converges to the optimal solution anyway. Relaxation of this assumption greatly complicates the proof of convergence but provides no computational difficulties; if Δ is zero, we simply revise the partitioning to obtain a new **B** and continue. The new basis will correspond to the same extreme point so that no progress toward optimality has been made. The only difficulty arises if the same basis reappears after some number of iterations each having $\Delta = 0$. Then the algorithm is caught in a loop, and we say that the method is *cycling*. Fortunately, real linear programs seldom cycle, and for the few that do, small changes in the tuning parameters of the code involved will usually prevent cycling. Simple rules have been devised that prevent cycling, but these are seldom used in practice.

Note that the information required to implement the primal simplex algorithm is the original data (i.e., **c**, **u**, and **A**), the current solution

$[\mathbf{x}^B \,\vdots\, \mathbf{x}^N]$, and index sets giving the column order of **B** and **N**. Furthermore, since $\mathbf{B}^{-1}$ is used in both the pricing operation and the ratio test, $\mathbf{B}^{-1}$ is also maintained. Hence at Step 4 we actually update $\mathbf{B}^{-1}$ after a column interchange with **B**.

Consider the following example.

Example 2.1

$$\begin{array}{llllllll}
\min & x_1 + x_2 + 3x_3 + 10x_4 & & & & & & \\
\text{s.t.} & x_1 & & + x_3 + & x_4 & & = & 5 \\
& -x_1 + x_2 & & & & +x_5 = & & 0 \\
& & -x_2 - & x_3 - & x_4 & & = & -5
\end{array}$$

$$0 \leqslant x_1 \leqslant 4, 0 \leqslant x_2 \leqslant 2, 0 \leqslant x_3 \leqslant 4, 0 \leqslant x_4 \leqslant 10, 0 \leqslant x_5 \leqslant 1.$$

0 (Initialization). Let

$$\mathbf{x}^B = [x_1 \quad x_4 \quad x_5] = [0 \quad 1 \quad 0] \quad \text{and} \quad \mathbf{x}^N = [x_2 \quad x_3] = [0 \quad 4].$$

Then

$$\mathbf{B} = \begin{bmatrix} 1 & 1 & 0 \\ -1 & 0 & 1 \\ 0 & -1 & 0 \end{bmatrix}, \quad \mathbf{B}^{-1} = \begin{bmatrix} 1 & 0 & 1 \\ 0 & 0 & -1 \\ 1 & 1 & 1 \end{bmatrix}, \quad \text{and}$$

$$\mathbf{N} = \begin{bmatrix} 0 & 1 \\ 1 & 0 \\ -1 & -1 \end{bmatrix}.$$

1 (Pricing at Iteration 1 with $[x_1 \quad x_2 \quad x_3 \quad x_4 \quad x_5] = [0 \quad 0 \quad 4 \quad 1 \quad 0]$)

$$\mathbf{c}^N - \mathbf{c}^B\mathbf{B}^{-1}\mathbf{N} = [1 \quad 3] - [1 \quad 10 \quad 0] \begin{bmatrix} 1 & 0 & 1 \\ 0 & 0 & -1 \\ 1 & 1 & 1 \end{bmatrix} \begin{bmatrix} 0 & 1 \\ 1 & 0 \\ -1 & -1 \end{bmatrix}$$
$$= [-8 \quad -7].$$

Then, $\psi_1 = \{1\}, \psi_2 = \phi, k = 1$, and $\delta = 1$.

2 (Ratio Test)

$$\mathbf{y} = \mathbf{B}^{-1}\mathbf{N}(1) = \begin{bmatrix} 1 & 0 & 1 \\ 0 & 0 & -1 \\ 1 & 1 & 1 \end{bmatrix} \begin{bmatrix} 0 \\ 1 \\ -1 \end{bmatrix} = \begin{bmatrix} -1 \\ 1 \\ 0 \end{bmatrix}.$$

Then, $\Delta_1 \leftarrow 1, \Delta_2 \leftarrow (4-0)/1 = 4, \Delta \leftarrow \min\{1, 4, 2\} = 1.$

3 (Update)

$$x_1^N = x_2 = 0 + 1(1) = 1,$$

$$\mathbf{x}^B = [x_1 \quad x_4 \quad x_5] = [0 \quad 1 \quad 0] - 1(1)[-1 \quad 1 \quad 0] = [1 \quad 0 \quad 0].$$

4 (Pivot). Now $\psi_3 = \{2\}, \psi_4 = \phi$; then $\mathbf{N}(1)$ replaces $\mathbf{B}(2)$. Hence

$$\mathbf{B} = \begin{bmatrix} 1 & 0 & 0 \\ -1 & 1 & 1 \\ 0 & -1 & 0 \end{bmatrix}, \quad \text{and} \quad \mathbf{B}^{-1} = \begin{bmatrix} 1 & 0 & 0 \\ 0 & 0 & -1 \\ 1 & 1 & 1 \end{bmatrix}.$$

$$\mathbf{x}^B = [x_1 \quad x_2 \quad x_5] \quad \text{and} \quad \mathbf{x}^N = [x_3 \quad x_4].$$

1 (Pricing at Iteration 2 with $[x_1 \quad x_2 \quad x_3 \quad x_4 \quad x_5] = [1 \quad 1 \quad 4 \quad 0 \quad 0]$)

$$\mathbf{c}^N - \mathbf{c}^B\mathbf{B}^{-1}\mathbf{N} = [3 \quad 10] - [1 \quad 10]\begin{bmatrix} 1 & 0 & 0 \\ 0 & 0 & -1 \\ 1 & 1 & 1 \end{bmatrix}\begin{bmatrix} 1 & 1 \\ 0 & 0 \\ -1 & -1 \end{bmatrix} = [1 \quad 8].$$

Then, $\psi_1 = \phi, \psi_2 = \{1\}$, $k = 1$, and $\delta = -1$.

2 (Ratio Test)

$$\mathbf{y} = \mathbf{B}^{-1}\mathbf{N}(1) = \begin{bmatrix} 1 & 0 & 0 \\ 0 & 0 & -1 \\ 1 & 1 & 1 \end{bmatrix}\begin{bmatrix} 1 \\ 0 \\ -1 \end{bmatrix} = \begin{bmatrix} 1 \\ 1 \\ 0 \end{bmatrix}.$$

Then, $\Delta_1 = \infty, \Delta_2 = \min\{3, 1\} = 1, \Delta = \min\{\infty, 1, 4\} = 1$.

3 (Update)

$$x_1^N = x_3 = 4 + 1(-1) = 3,$$

$$\mathbf{x}^B = [x_1 \quad x_2 \quad x_3] = [1 \quad 1 \quad 0] - 1(-1)[1 \quad 1 \quad 0] = [2 \quad 2 \quad 0].$$

4 (Pivot). Now $\psi_3 = \phi, \psi_4 = \{2\}$. Then $\mathbf{N}(1)$ replaces $\mathbf{B}(2)$. Hence

$$\mathbf{B} = \begin{bmatrix} 1 & 1 & 0 \\ -1 & 0 & 1 \\ 0 & -1 & 0 \end{bmatrix} \quad \text{and} \quad \mathbf{B}^{-1} = \begin{bmatrix} 1 & 0 & 1 \\ 0 & 0 & -1 \\ 1 & 1 & 1 \end{bmatrix}.$$

$$\mathbf{x}^B = [x_1 \quad x_3 \quad x_5] \quad \text{and} \quad \mathbf{x}^N = [x_2 \quad x_4].$$

1 (Pricing at Iteration 3 with $[x_1 \quad x_2 \quad x_3 \quad x_4 \quad x_5]=[2 \quad 2 \quad 3 \quad 0 \quad 0]$)

$$\mathbf{c}^N-\mathbf{c}^B\mathbf{B}^{-1}\mathbf{N}=[1 \quad 10]-[1 \quad 3 \quad 0]\begin{bmatrix}1 & 0 & 1\\ 0 & 0 & -1\\ 1 & 1 & 1\end{bmatrix}\begin{bmatrix}0 & 1\\ 1 & 0\\ -1 & -1\end{bmatrix}$$

$$=[-1 \quad 7].$$

Then, $\psi_1=\psi_2=\phi$—optimum found!

2.5 DUALITY THEORY

Associated with every linear program is a closely related problem called the dual program. The dual program has several important properties that will become useful in our study of network programming. The linear program (also called the primal problem) and its dual are given as follows:

$$\left.\begin{array}{ll}\min & \mathbf{cx}\\ \text{s.t.} & \mathbf{Ax}=\mathbf{b}\\ & \mathbf{0}\leqslant\mathbf{x}\leqslant\mathbf{u}\end{array}\right\}LP \qquad \left.\begin{array}{ll}\max & \mathbf{b}\boldsymbol{\pi}-\mathbf{u}\boldsymbol{\mu}\\ \text{s.t.} & \boldsymbol{\pi}\mathbf{A}-\boldsymbol{\mu}\leqslant\mathbf{c}\\ & \boldsymbol{\mu}\geqslant\mathbf{0}\end{array}\right\}DLP.$$

Under the assumption that the feasible region for *LP* is not empty, these problems are equivalent in the sense that, if we have an optimal basic solution to one, we can easily generate an optimal solution to the other. We now give two well-known results concerning these two problems.

Proposition 2.9

Let $\mathbf{x}^*$ be an optimal solution to *LP*, and let $(\boldsymbol{\pi}^*,\boldsymbol{\mu}^*)$ be an optimal solution to *DLP*. Then $\mathbf{cx}^*=\mathbf{b}\boldsymbol{\pi}^*-\mathbf{u}\boldsymbol{\mu}^*$.

Proposition 2.10

Let $\mathbf{x}^*$ be an optimal basic solution to *LP* with $\mathbf{B}$ the corresponding basis. Let $\mathbf{c}^B$ be a vector whose components are the components of $\mathbf{c}$ corresponding to columns of $\mathbf{B}$. Let $\boldsymbol{\pi}^*=\mathbf{c}^B\mathbf{B}^{-1}$ and let

$$\mu_j^*=\begin{cases}\boldsymbol{\pi}^*\mathbf{A}(j)-c_j & \text{if } x_j^*=u_j,\\ 0 & \text{otherwise.}\end{cases}$$

Then $(\boldsymbol{\pi}^*,\boldsymbol{\mu}^*)$ solves *DLP*.

Proofs of the above propositions may be found in any standard textbook on linear programming.

2.6 KUHN-TUCKER CONDITIONS FOR THE LINEAR PROGRAM

In this section we give a set of conditions (known as the Kuhn-Tucker conditions) that are both necessary and sufficient for the optimality of a feasible solution to *LP*. These conditions are stated without proof in the following proposition.

Proposition 2.11

Let $\mathbf{x}^*$ be such that $\mathbf{A}\mathbf{x}^*=\mathbf{b}$ and $\mathbf{0}\leqslant\mathbf{x}^*\leqslant\mathbf{u}$. Then $\mathbf{x}^*$ is an optimum for *LP* if and only if there exist vectors $\boldsymbol{\pi}$, $\boldsymbol{\mu}$, and $\boldsymbol{\lambda}$ such that $\boldsymbol{\pi}\mathbf{A}-\mathbf{c}=\boldsymbol{\mu}-\boldsymbol{\lambda}$, $(\mathbf{x}^*-\mathbf{u})\boldsymbol{\mu}=0$, $\mathbf{x}^*\boldsymbol{\lambda}=0$, $\boldsymbol{\mu}\geqslant\mathbf{0}$, and $\boldsymbol{\lambda}\geqslant\mathbf{0}$.

Therefore solving the system

$$\left\{\begin{array}{ll}\mathbf{A}\mathbf{x}=\mathbf{b} & (\mathbf{x}-\mathbf{u})\boldsymbol{\mu}=0\\ \mathbf{0}\leqslant\mathbf{x}\leqslant\mathbf{u} & \mathbf{x}\boldsymbol{\lambda}=0\\ \boldsymbol{\pi}\mathbf{A}-\mathbf{c}=\boldsymbol{\mu}-\boldsymbol{\lambda} & \boldsymbol{\mu}\geqslant 0,\quad \boldsymbol{\lambda}\geqslant\mathbf{0}\end{array}\right\} \tag{2.13}$$

in the unknowns $\mathbf{x}$, $\boldsymbol{\pi}$, $\boldsymbol{\mu}$, and $\boldsymbol{\lambda}$ is mathematically equivalent to solving *LP*. System (2.13) will be used later in our study of network programming in the development of an algorithm (the out-of-kilter algorithm) for the minimal cost network flow problem.

EXERCISES

2.1 Let $\mathbf{x}$ be an extreme point of some polytope. Show that $\{\mathbf{x}\}$ is an extreme subset.

2.2 Let $\Lambda=\{\mathbf{x}\}$ be an extreme subset of some polytope. Show that $\mathbf{x}$ is an extreme point.

2.3 Show that the convex hull of a set $X\subset R^n$ is a convex set.

2.4 Explain why the word *maximal* is used in the definitions of finite and infinite edge.

2.5 Show that a closed half-space is a closed set.

2.6 Let $(\mathbf{x}^*,\boldsymbol{\pi}^*,\boldsymbol{\mu}^*,\boldsymbol{\lambda}^*)$ be a solution to (2.13). Show that $\mathbf{c}\mathbf{x}^*=\mathbf{b}\boldsymbol{\pi}^*-\mathbf{u}\boldsymbol{\mu}^*$.

2.7 Show that, if the system

$$\left\{\begin{array}{l} \mathbf{Ax}=\mathbf{b} \\ \mathbf{0}\leqslant\mathbf{x}\leqslant\mathbf{u} \\ \boldsymbol{\pi}\mathbf{A}-\boldsymbol{\mu}\leqslant\mathbf{c} \\ \boldsymbol{\mu}\geqslant\mathbf{0} \end{array}\right\}$$

has a solution, $\boldsymbol{\pi}\mathbf{b}-\boldsymbol{\mu}\mathbf{u}\leqslant\mathbf{cx}$.

2.8 Solve the three-node minimal cost network flow problem in Table E2.8 using ALG 2.2. Begin with an initial solution of $[x_1 \;\; x_2 \;\; x_3 \;\; x_4 \;\; x_5]=[0 \;\; 1 \;\; 0 \;\; 0 \;\; 9]$.

Table E2.8

Node Number	Supply	Demand
1	0	1
2	9	0
3	0	8

Arc Number	Directed Away From	Directed Toward	Unit Cost	Arc Capacity
1	2	1	0	8
2	3	1	100	5
3	1	3	1	9
4	2	3	2	10
5	2	3	100	10

Give the corresponding dual problem, the dual solution, and the solution to (2.13). (*Note*: The constraint matrix will have a rank of 2. Append an artificial variable, say x_6, having $c_6=u_6=0$ to row 1. Then the constraint matrix $[\mathbf{A}\,|\,\mathbf{e}^1]$ will have full row rank.) (Optimal cost equals 9.)

2.9 Solve the four-node minimal cost network flow problem in Table E2.9 using ALG 2.2. Begin with an initial solution of $[x_1 \;\; x_2 \;\; x_3 \;\; x_4 \;\; x_5]=[1 \;\; 0 \;\; 1 \;\; 1 \;\; 2]$.

Give the corresponding dual problem, the dual solution, and the solution to (2.13). (*Note*: The constraint matrix **A** will have a rank of 3. Append an artificial variable, say x_6, having $c_6=u_6=0$ to row 1. Then the constraint matrix $[\mathbf{A}\,|\,\mathbf{e}^1]$ will have full row rank.) (Optimal cost equals 0.)

Table E2.9

Node Number	Supply	Demand
1	11	10
2	0	1
3	3	0
4	0	3

Arc Number	Directed Away From	Directed Toward	Unit Cost	Arc Capacity
1	1	2	10	1
2	1	3	1	2
3	3	2	−3	4
4	2	4	3	2
5	3	4	2	2

2.10 Solve the two-source, three-destination transportation problem in Table E2.10 using ALG 2.2. Begin with an initial solution of $[x_1 \quad x_2 \quad x_3 \quad x_4 \quad x_5 \quad x_6] = [8 \quad 2 \quad 0 \quad 0 \quad 6 \quad 8]$.

Table E2.10

Node Number	Supply	Demand
1	10	0
2	14	0
3	0	8
4	0	8
5	0	8

Arc Number	Directed Away From	Directed Toward	Unit Cost
1	1	3	10
2	1	4	11
3	1	5	0
4	2	3	0
5	2	4	12
6	2	5	12

Give the corresponding solution to the dual problem. (Optimal cost equals 94.)

2.11 Solve the following 3×3 assignment problem using ALG 2.2. The cost matrix is as follows:

$$\mathbf{C}=\begin{bmatrix} 10 & 5 & 0 \\ 8 & 13 & 2 \\ 13 & 11 & 17 \end{bmatrix},$$

where C_{ij} denotes the cost of assigning man i to job j. Begin with the initial solution given by assigning man i to job i for $i=1,2,3$. (Optimal cost equals 19.)

2.12 Prove the following: If $z(\boldsymbol{\alpha})=\min\{\mathbf{cx}:\mathbf{Ax}=\mathbf{b}+\boldsymbol{\alpha},\mathbf{0}\leqslant\mathbf{x}\leqslant\mathbf{u}\}$ has a feasible solution over $\mathbf{0}\leqslant\boldsymbol{\alpha}\leqslant\boldsymbol{\beta}$, then $z(\boldsymbol{\alpha})$ is convex over $\mathbf{0}\leqslant\boldsymbol{\alpha}\leqslant\boldsymbol{\beta}$.

2.13 Let S be a convex set, and let $\mathbf{x}\notin S$. Consider the problem $\min\{\|\mathbf{y}-\mathbf{x}\|:\mathbf{y}\in S\}$. Does this problem always have a solution? Show that, if a solution exists, it is unique.

NOTES AND REFERENCES

Section 2.1

Some of the most popular textbooks devoted primarily to the study of linear programming are as follows: Bazaraa and Jarvis [20], Cooper and Steinberg [36], Dantzig [47], Gass [71], Hadley [91], Loomba [138], Orchard-Hays [148], Simmons [168], Simonnard [169], Smythe and Johnson [170], and Zionts [188].

Section 2.2

Proofs of the propositions in this section can be found in Bartle [19], Berge [23], and Rockafellar [157].

Section 2.3

The proof of Proposition 2.8 follows that of Hadley [91].

Section 2.4

George Dantzig is credited with the discovery of the simplex method. Convergence proofs for this method were first provided by Charnes [28] and later by Dantzig, Orden, and Wolfe [45]. The general feasible direction approach was first investigated by Zoutendijk [189]. Discussions of cycling and a convergence proof of the simplex method can be found in Dantzig [47], Hadley [91], Bazaraa and Jarvis [20], and Kotiah and Steinberg [127].

Section 2.5

The famous mathematician John Von Neumann is credited with having first postulated the existence of a dual linear program.

Section 2.6

Proposition 2.11 is a special case of results presented in Kuhn and Tucker [128].

CHAPTER 3

The Simplex Method for the Network Program

In this chapter we specialize George Dantzig's primal simplex algorithm for NP. This specialization is of great importance in that it completely eliminates the need for carrying and updating the basis inverse. In fact, when this specialization is used, the primal simplex method can be performed directly on a network diagram. Hence we call this procedure the *simplex on a graph algorithm.*

For the development of this chapter, we assume that the network does not contain multiple arcs (i.e., there are no arcs e_j and e_k such that the corresponding columns of the node-arc incidence matrix are identical). This assumption is for notational convenience only and poses no mathematical difficulties. If there are two arcs from node i to node j, one of these may be replaced by the two arcs (i,k) and (k,j), where k is a new node. Such a transformation of the network of Figure 1.1 is illustrated in Figure 3.1. The corresponding node-arc incidence matrix is as follows:

$$\mathbf{A}=\overset{\text{arcs}}{\overbrace{\begin{array}{cccccccc} 1 & 2 & 3 & 4 & 5 & 6 & 7 & 8 \end{array}}}\ \text{nodes}$$

$$\mathbf{A}=\begin{bmatrix} 1 & 1 & & & & & -1 & \\ & -1 & 1 & 1 & & & & -1 \\ & & & -1 & -1 & 1 & & \\ & & -1 & & 1 & -1 & 1 & \\ -1 & & & & & & & 1 \end{bmatrix}\begin{array}{c} 1 \\ 2 \\ 3 \\ 4 \\ 5 \end{array}$$

Note that the columns of $\mathbf{A}$ are distinct.

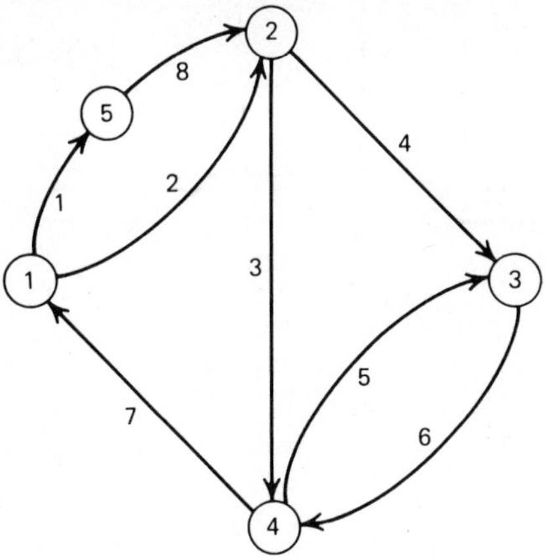

FIGURE 3.1 Network having no multiple arcs.

3.1 RESULTS FROM GRAPH THEORY

In this section we develop results from graph theory that will be used in our specialization of the primal simplex algorithm for NP. Consider a network having $\bar{I}$ nodes and $\bar{J}$ arcs which have been placed into one-to-one correspondence with the integers $1,\ldots,\bar{I}$ and $1,\ldots,\bar{J}$, respectively. Let **A** denote the node-arc incidence matrix associated with this network. For each arc j, let $F(j)=i$, where $A_{ij}=1$, and let $T(j)=k$, where $A_{kj}=-1$. The $F(j)$ and $T(j)$ functions for the network of Figure 3.1 are illustrated in Table 3.1. Then arc j may be formally described as the pair $(F(j),T(j))$. We require that $F(j)\neq T(j)$.

Let the set $\mathscr{A}$, called the *arc set*, be denoted by $\mathscr{A}=\{e_1,e_2,\ldots,e_{\bar{J}}\}$, where $e_j=(F(j),T(j))$. Hence, for the network illustrated in Figure 3.1,

Table 3.1 "From" and "To" Functions for the Network of Figure 3.1

Arc Number, j	"From" Node, $F(j)$	"To" Node, $T(j)$
1	1	5
2	1	2
3	2	4
4	2	3
5	4	3
6	3	4
7	4	1
8	5	2

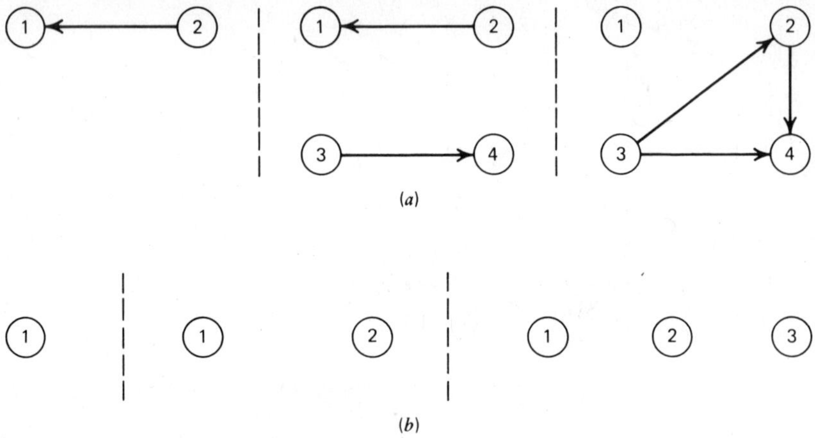

FIGURE 3.2 Illustrations of proper graphs and graphs that are not proper. (*a*) Proper graphs. (*b*) Graphs that are not proper.

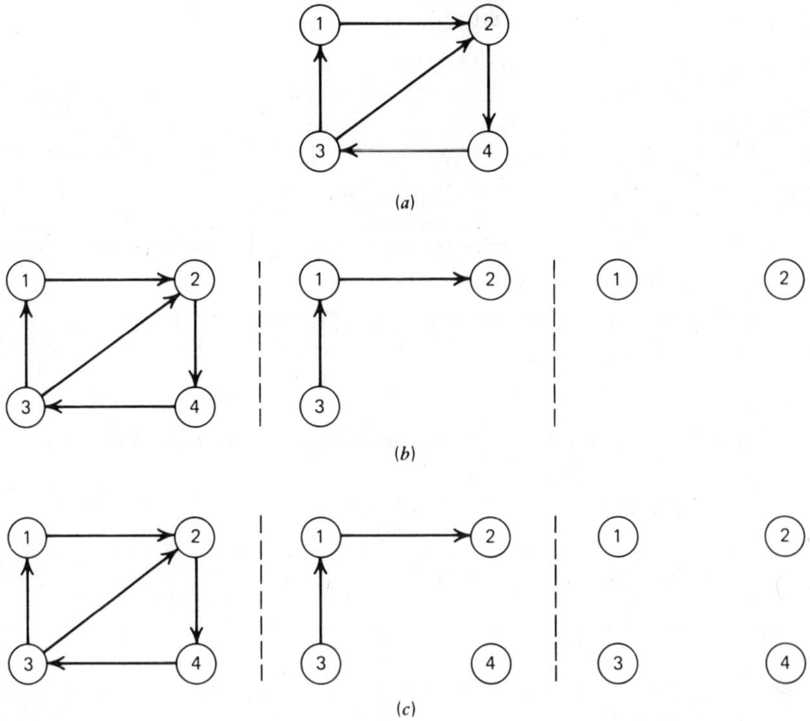

FIGURE 3.3 Illustrations of subgraphs. (*a*) A proper graph, $\mathcal{G}$. (*b*) Subgraphs of $\mathcal{G}$. (*c*) Spanning subgraphs of $\mathcal{G}$.

$\mathcal{A} = \{e_1 = (1,5), e_2 = (1,2), e_3 = (2,4), e_4 = (2,3), e_5 = (4,3), e_6 = (3,4), e_7 = (4,1), e_8 = (5,2)\}$. Let the set $\mathcal{N}$, called the *node set*, be denoted by $\mathcal{N} = \{1,\ldots,\bar{I}\}$. Then the *graph* $\mathcal{G}$ consists of the sets $[\mathcal{N}, \mathcal{A}]$. We say that $\mathcal{G}$ is a *proper graph* if $\bar{I} \geqslant 2$ and $\bar{J} \geqslant 1$. A graph $\hat{\mathcal{G}} = [\hat{\mathcal{N}}, \hat{\mathcal{A}}]$ is said to be a *subgraph* of $\mathcal{G}$ if $\hat{\mathcal{N}} \subset \mathcal{N}$ and $\hat{\mathcal{A}} \subset \mathcal{A}$. If $\hat{\mathcal{N}} = \mathcal{N}$, we say that $\hat{\mathcal{G}}$ *spans* $\mathcal{G}$ *or that* $\hat{\mathcal{G}}$ *is a spanning subgraph* for $\mathcal{G}$. These definitions are illustrated in Figures 3.2 and 3.3.

Given the formal definition of a graph, we now present other notions that are required to describe the results we seek. A finite sequence $P = \{s_1, e_{j_1}, s_2, e_{j_2}, s_3, e_{j_3}, \ldots, s_n, e_{j_n}, s_{n+1}\}$ having at least one arc is defined to be a *path* in a graph $\mathcal{G}$ if its odd elements are distinct nodes of $\mathcal{G}$, its even elements are arcs of $\mathcal{G}$, and, for each arc, $e_{j_i} \varepsilon \{(s_i, s_{i+1}), (s_{i+1}, s_i)\}$. An example of a path in the graph illustrated in Figure 3.1 is $P = \{5, e_1, 1, e_7, 4, e_5, 3\} = \{5, (1,5), 1, (4,1), 4, (4,3), 3\}$ with the corresponding representation illustrated in Figure 3.4. A path is said to *link* its first element to its last element.

Proposition 3.1

The arcs of a path are distinct.

Proof. If two arcs of a path were identical, the nodes of the path would not be distinct. ■

A finite sequence $C = \{s_1, e_{j_1}, s_2, e_{j_2}, \ldots, s_n, e_{j_n}, s_{n+1}\}$ having at least two arcs is called a *cycle* in $\mathcal{G}$ if the subsequence $\{s_1, e_{j_1}, s_2, e_{j_2}, \ldots, s_n\}$ is a path in $\mathcal{G}$, $e_{j_n} \varepsilon \{(s_n, s_{n+1}), (s_{n+1}, s_n)\}, s_1 = s_{n+1}$, and $e_{j_n} \neq e_{j_1}$. An example of a cycle in the graph represented by Figure 3.1 is $C = \{2, e_8, 5, e_1, 1, e_2, 2\}$.

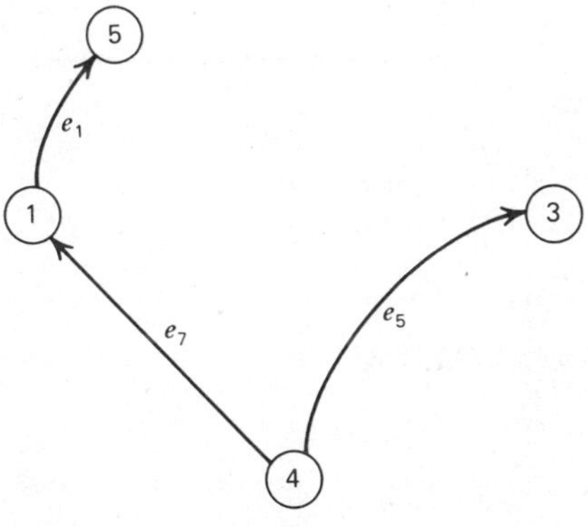

FIGURE 3.4 Illustration of a path.

Proposition 3.2

The arcs of a cycle are distinct.

Proof. Suppose the cycle has n arcs. Since the subsequence $\{s_1, e_{j_1}, s_2, e_{j_2}, \ldots, s_{n-1}, e_{j_{n-1}}, s_n\}$ is a path, the first $n-1$ arcs must be distinct by Proposition 3.1. The subsequence $\{s_2, e_{j_2}, s_3, e_{j_3}, \ldots, s_n, e_{j_n}, s_{n+1}\}$ also satisfies the definition of a path, so the last $n-1$ arcs are distinct. Since C has n arcs, the only possibility for replication is that the last arc (e_{j_n}) equals the first arc (e_{j_1}), and this is expressly prohibited by the definition of a cycle.

The *length* of a path or cycle is taken to be the number of arcs in the path or cycle. For any path or cycle P, with length n, we define the *orientation sequence* $O(P)$ having n elements as follows:

$$O_i(P) = \begin{cases} +1 & \text{if } e_{j_i} = (s_i, s_{i+1}), \\ -1 & \text{if } e_{j_i} = (s_{i+1}, s_i) \end{cases}.$$

For the path $\{5, e_1, 1, e_7, 4, e_5, 3\}$ illustrated in Figure 3.4 the orientation sequence is $\{-1, -1, 1\}$.

Proposition 3.3

Let the finite sequence $P = \{s_1, e_{j_1}, s_2, e_{j_2}, \ldots, s_n, e_{j_n}, s_{n+1}\}$ be a path or cycle in a proper graph $\mathcal{G}$ with node-arc incidence matrix $\mathbf{A}$; then

$$\sum_{i=1}^{i=n} O_i(P)\mathbf{A}(j_i) = \mathbf{e}^{s_1} - \mathbf{e}^{s_{n+1}}. \tag{3.1}$$

Proof. Let $i \varepsilon \{1, \ldots, n\}$.

Case 1 Suppose $e_{j_i} = (s_i, s_{i+1})$. Then

$$O_i(P)\mathbf{A}(j_i) = (+1)(e^{s_i} - \mathbf{e}^{s_{i+1}}) = \mathbf{e}^{s_i} - \mathbf{e}^{s_{i-1}}.$$

Case 2 Suppose $e_{j_i} = (s_{i+1}, s_i)$. Then

$$O_i(P)\mathbf{A}(j_i) = (-1)(\mathbf{e}^{s_{i+1}} - \mathbf{e}^{s_i}) = \mathbf{e}^{s_i} - \mathbf{e}^{s_{i-1}}.$$

Hence in either case $O_i(P)\mathbf{A}(j_i) = \mathbf{e}^{s_i} - \mathbf{e}^{s_{i+1}}$. Thus, because of the telescoping property of the sum, (3.1) holds. ■

Proposition 3.4

Let $C=\{s_1,e_{j_1},s_2,e_{j_2},\ldots,s_n,e_{j_n},s_{n+1}\}$ be a cycle in a proper graph $\mathcal{G}$ with node-arc incidence matrix $\mathbf{A}$; then

$$\sum_{i=1}^{i=n} O_i(C)\mathbf{A}(j_i)=\mathbf{0}. \tag{3.2}$$

Proof.

$$\sum_{i=1}^{i=n} O_i(C)\mathbf{A}(j_i)=\sum_{i=1}^{i=n-1} O_i(C)\mathbf{A}(j_i)+\mathbf{e}^{s_n}-e^{s_{n+1}}.$$

By Proposition 3.3,

$$\sum_{i=1}^{i=n-1} O_i(C)\mathbf{A}(j_i)=\mathbf{e}^{s_1}-\mathbf{e}^{s_n}.$$

Then

$$\sum_{i=1}^{i=n} O_i(C)\mathbf{A}(j_i)=\mathbf{e}^{s_1}-\mathbf{e}^{s_n}+\mathbf{e}^{s_n}-\mathbf{e}^{s_{n+1}}.$$

Since C is a cycle, $s_1=s_{n+1}$; hence

$$\sum_{i=1}^{i=n} O_i(C)\mathbf{A}(j_i)=\mathbf{0}.$$

■

Corollary 3.5

Let $C=\{s_1,e_{j_1},s_2,e_{j_2},\ldots,s_n,e_{j_n},s_{n+1}\}$ be a cycle in a proper graph $\mathcal{G}$ with node-arc incidence matrix $\mathbf{A}$; then $\{\mathbf{A}(j_i):i=1,\ldots,n\}$ is linearly dependent.

Consider the cycle illustrated in Figure 3.5. Then the set of columns

$$\left\{\begin{bmatrix} \\ \\ -1 \\ 1 \end{bmatrix},\ \begin{bmatrix} 1 \\ \\ -1 \\ \\ \end{bmatrix},\ \begin{bmatrix} 1 \\ -1 \\ \\ \\ \end{bmatrix},\ \begin{bmatrix} \\ 1 \\ \\ -1 \end{bmatrix}\right\}$$

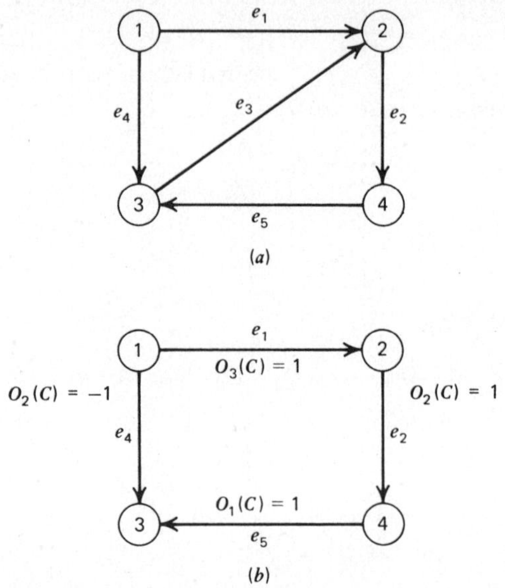

FIGURE 3.5 Example of a cycle in a proper graph. (*a*) Proper graph. (*b*) Cycle in a proper graph: $C=\{4,e_5,3,e_4,1,e_1,2,e_2,4\}$.

is linearly dependent, that is,

$$O_1(C)\begin{bmatrix} \\ \\ -1 \\ 1 \end{bmatrix} + O_2(C)\begin{bmatrix} 1 \\ \\ -1 \\ \end{bmatrix} + O_3(C)\begin{bmatrix} 1 \\ -1 \\ \\ \end{bmatrix} + O_4(C)\begin{bmatrix} \\ 1 \\ \\ -1 \end{bmatrix}$$

$$=(1)\begin{bmatrix} \\ \\ -1 \\ 1 \end{bmatrix} + (-1)\begin{bmatrix} 1 \\ \\ -1 \\ \end{bmatrix} + (1)\begin{bmatrix} 1 \\ -1 \\ \\ \end{bmatrix} + (1)\begin{bmatrix} \\ 1 \\ \\ -1 \end{bmatrix} = \begin{bmatrix} 0 \\ 0 \\ 0 \\ 0 \end{bmatrix}.$$

A graph $\mathcal{G}=[\mathfrak{N},\mathcal{A}]$ is said to be *acyclic* if no cycles can be formed from $\mathfrak{N}$ and $\mathcal{A}$. A graph $\mathcal{G}=[\mathfrak{N},\mathcal{A}]$ is said to be *connected* if, for every distinct pair of nodes (i,j) from $\mathfrak{N}$, a path can be formed from $\mathfrak{N}$ and $\mathcal{A}$ that links i to j. A *tree* is a connected acyclic graph. A tree that is a spanning subgraph of a graph $\mathcal{G}$ is called a *spanning tree* for $\mathcal{G}$. These definitions are illustrated in Figure 3.6.

The important proposition to follow gives other characterizations of a tree. Because of the length of the proof of this proposition, we state the proposition here and refer the reader to Appendix A for the proof.

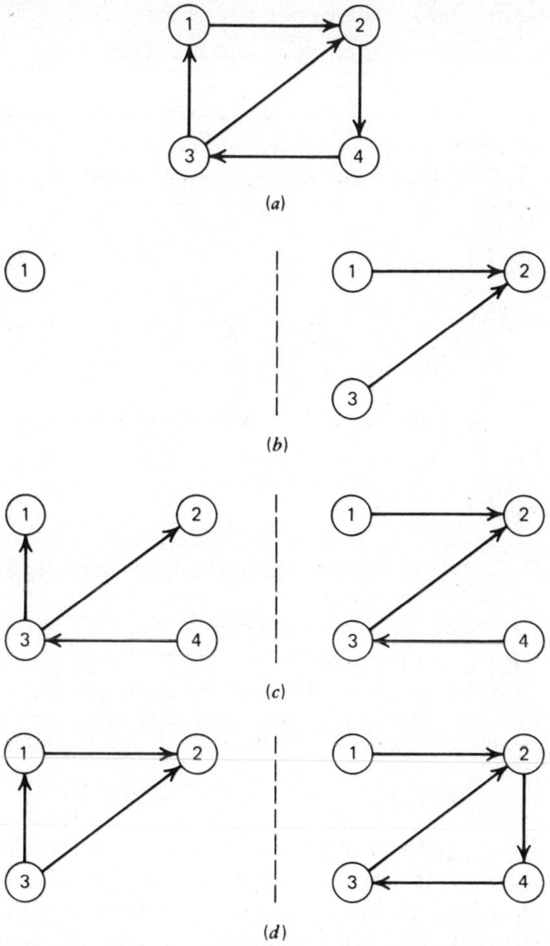

FIGURE 3.6 Examples of trees and spanning trees. (*a*) Graph $\mathcal{G}$. (*b*) Subgraphs of $\mathcal{G}$ that are trees but are not spanning trees. (*c*) Subgraphs of $\mathcal{G}$ that are spanning trees. (*d*) Subgraphs of $\mathcal{G}$ that are not trees.

Proposition 3.6

For a graph $\mathcal{T}=[\mathcal{N},\mathcal{Q}]$ having at least one node, the following are equivalent:

1 $\mathcal{T}$ is a tree.

2 For every distinct pair of nodes (p,q) of $\mathcal{N}$, there is a unique path in $\mathcal{T}$ that links p to q.

3 $\mathcal{T}$ has one less arc than node and is connected.

4 $\mathcal{T}$ has one less arc than node and is acyclic.

The *degree* of a node i of a graph $\mathcal{G}$, denoted by $D(i)$, is the number of arcs of $\mathcal{G}$ incident on node i. The degree of node 3 of the graph of Figure 3.6*a* is 3. A node in a tree that has a degree equal to 1 is called an *endpoint* of the tree.

Proposition 3.7

The sum of the degrees of all nodes of a graph is twice the number of arcs of the graph.

Proof. Every arc is incident on two nodes, thus contributing 2 to the sum of the degrees. ■

Proposition 3.8

Every tree with at least two nodes has at least two endpoints.

Proof. Let $\mathcal{T}=[\mathcal{N},\mathcal{A}]$ be a tree having $m\geqslant 2$ nodes. Every node of $\mathcal{T}$ is incident on at least one arc, since otherwise $\mathcal{T}$ would not be connected. Thus the degree of each node is at least 1. By Proposition 3.6, $\mathcal{T}$ has $m-1$ arcs. We assume that $\mathcal{T}$ has less than two endpoints and derive a contradiction.

Case 1 Suppose $\mathcal{T}$ has no endpoints. Then, for each $i\varepsilon\mathcal{N}, D(i)\geqslant 2$. Thus $\Sigma_{i\varepsilon\mathcal{N}}D(i)\geqslant 2m>2(m-1)$, contradicting Proposition 3.7.

Case 2 Suppose $\mathcal{T}$ has one endpoint. Let the endpoint be j. Then $D(j)=1$, and, for each $i\varepsilon\mathcal{N}-\{j\}, D(i)\geqslant 2$. Thus $\Sigma_{i\varepsilon\mathcal{N}}D(i)= D(j)+\Sigma_{i\varepsilon\mathcal{N}-\{j\}}D(i)=1+2(m-1)\neq 2(m-1)$, contradicting Proposition 3.7.

Hence $\mathcal{T}$ has at least two endpoints. ■

It should be noted that trees may have more than two endpoints. The two trees illustrated in Figure 3.6*c* have three and two endpoints.

We now show that removal of an endpoint and its incident arc from a tree results in a subgraph that is also a tree.

Proposition 3.9

Let $\mathcal{T}=[\mathcal{N},\mathcal{A}]$ be a tree having $m\geqslant 2$ nodes. Let i be an endpoint of $\mathcal{T}$, and let e_j be the arc incident on i in $\mathcal{T}$. If $\hat{\mathcal{N}}=\mathcal{N}-\{i\}$ and $\hat{\mathcal{A}}=\mathcal{A}-\{e_j\}$, then $\hat{\mathcal{T}}=[\hat{\mathcal{N}},\hat{\mathcal{A}}]$ is a tree.

Proof. Since $\hat{\mathcal{T}}$ is a subgraph of a tree, by Proposition 3.6 $\hat{\mathcal{T}}$ is acyclic. By construction, $\hat{\mathcal{T}}$ has one less arc than node. Hence, by Proposition 3.6, $\mathcal{T}$ is a tree. ■

We are now ready to present an important result concerning the arcs of a tree and the corresponding columns in the node-arc incidence matrix. Recall that we assumed that the network does not contain multiple arcs. Therefore there is a one-to-one correspondence between the arcs of a graph and the columns of the associated node-arc incidence matrix. Furthermore, arc e_j from $\mathcal{A}$ corresponds to the jth column of the node-arc incidence matrix.

Proposition 3.10

Let $\mathbf{A}$ be the node-arc incidence matrix for a proper graph $\mathcal{G}$. Let $\mathcal{T}=[\mathcal{N},\mathcal{A}]$ be a subgraph of $\mathcal{G}$ that is a tree having at least two nodes. Then $\{\mathbf{A}(j): e_j \varepsilon \mathcal{A}\}$ is linearly independent.

Proof. Let m denote the number of nodes in $\mathcal{T}$. Suppose $m=2$. Then, by Proposition 3.6, $\mathcal{A}$ contains one arc, say e_j. Since $\mathbf{A}(j)=\mathbf{e}^{F(j)}-\mathbf{e}^{T(j)}\neq\mathbf{0}$, the proposition holds for $m=2$.

Suppose $m>2$. Let n be the largest integer such that, for all subgraphs $\hat{\mathcal{T}}=[\hat{\mathcal{N}},\hat{\mathcal{A}}]$ of $\mathcal{G}$ that are trees having k nodes, where $2\leqslant k\leqslant n$, $\{\mathbf{A}(j): e_j \varepsilon \hat{\mathcal{A}}\}$ is linearly independent. From the above, we know that $n\geqslant 2$. We now assume that $n<m$ and derive a contradiction. By the definition of n, there must be a tree $\hat{\mathcal{T}}=[\hat{\mathcal{N}},\hat{\mathcal{A}}]$ having $n+1$ nodes for which $\{\mathbf{A}(j): e_j \varepsilon \hat{\mathcal{A}}\}$ is linearly dependent. Thus there is a set of constants $\{v_j : e_j \varepsilon \hat{\mathcal{A}}\}$, not all zero, such that $\Sigma_{e_j \varepsilon \hat{\mathcal{A}}} v_j \mathbf{A}(j)=\mathbf{0}$. Let $\mathcal{L}$ denote the set of endpoints for $\hat{\mathcal{T}}$. By Proposition 3.8 $\mathcal{L}\neq\phi$. Let p be any member of $\mathcal{L}$, and let e_w be the arc incident on p. Let $\bar{\mathcal{T}}=[\bar{\mathcal{N}},\bar{\mathcal{A}}]$, where $\bar{\mathcal{N}}=\hat{\mathcal{N}}-\{p\}$ and $\bar{\mathcal{A}}=\hat{\mathcal{A}}-\{e_w\}$. By Proposition 3.9 $\bar{\mathcal{T}}$ is a tree. Since node p is incident only on arc e_w of $\hat{\mathcal{A}}$, $A_{p,j}=0$ for all $e_j \varepsilon \bar{\mathcal{A}}$. Thus for the pth component

$$\sum_{e_j \varepsilon \hat{\mathcal{A}}} v_j A_{p,j} = v_w A_{p,w} + \sum_{e_j \varepsilon \bar{\mathcal{A}}} v_j A_{p,j} = v_w A_{p,w} = 0.$$

Since $A_{p,w} \varepsilon \{+1,-1\}, v_w=0$. But then $\Sigma_{e_j \varepsilon \bar{\mathcal{A}}} v_j \mathbf{A}(j)=\mathbf{0}$ and $\{v_j : e_j \varepsilon \bar{\mathcal{A}}\}$ are not all zero. So $\{\mathbf{A}(j): e_j \varepsilon \bar{\mathcal{A}}\}$ is linearly dependent. Since $\bar{\mathcal{T}}$ is a tree with n nodes, this contradicts the definition of n. Hence $n=m$, and all subgraphs of $\mathcal{G}$ that are trees with two or more nodes have the specified property. ■

We now show that, if a graph $\mathcal{G}$ is connected, there exists a spanning tree for $\mathcal{G}$.

Proposition 3.11

Every connected graph $\mathcal{G}=[\mathcal{N},\mathcal{A}]$ with $\mathcal{N}\neq\phi$ has a subgraph that is a spanning tree.

Proof. Let $\mathcal{G}$ be connected, and let $n \geqslant 1$ denote the number of nodes of $\mathcal{G}$. Suppose $n=1$. Then $\mathcal{A}=\phi$, and it follows that $\mathcal{G}$ is a tree and thus is a spanning tree for itself. Suppose $n \geqslant 2$. Let m denote the largest positive integer such that there is a subgraph of $\mathcal{G}$ that is a tree and has m nodes. Any subgraph of $\mathcal{G}$ with one node is a tree, so $m \geqslant 1$. We assume $m<n$ and derive a contradiction. By the definition of m, there exists a subgraph of $\mathcal{G}$, say $\hat{\mathcal{T}}=[\hat{\mathcal{N}},\hat{\mathcal{A}}]$, that is a tree and has m nodes. Since $\hat{\mathcal{N}}$ has $m<n$ nodes, $\mathcal{N}-\hat{\mathcal{N}}\neq\phi$. Let $\mathcal{L}$ denote the set of arcs of $\mathcal{A}$ that are incident on a node in $\hat{\mathcal{N}}$ and a node in $\mathcal{N}-\hat{\mathcal{N}}$. The set $\mathcal{L}\neq\phi$ since otherwise $\mathcal{G}$ would not be connected. Choose any arc $e_i \varepsilon \mathcal{L}$. Then $e_i \not\varepsilon \hat{\mathcal{A}}$. Let j be the node of $\{F(i), T(i)\}$ that is in $\mathcal{N}-\hat{\mathcal{N}}$. Let $\bar{\mathcal{T}}=[\bar{\mathcal{N}},\bar{\mathcal{A}}]$, where $\bar{\mathcal{N}}=\hat{\mathcal{N}}\cup\{j\}$ and $\bar{\mathcal{A}}=\hat{\mathcal{A}}\cup\{e_i\}$. By construction, $\bar{\mathcal{T}}$ is a subgraph of $\mathcal{G}$ that is a tree and has $m+1$ nodes. This contradicts the definition of m. Hence $m=n$, and there is a spanning tree for $\mathcal{G}$. ■

Propositions 3.10 and 3.11 taken together show that for a given graph $\mathcal{G}$, with corresponding node-arc incidence matrix **A**, there exists a spanning tree from $\mathcal{G}$, say $\mathcal{T}$, such that the columns of **A** corresponding to arcs in $\mathcal{T}$ are linearly independent. We now show (1) that the rank of **A** is one less than the number of rows (nodes), and (2) that a maximal set of linearly independent columns from **A** corresponds to a spanning tree.

Proposition 3.12

Let **A** be the node-arc incidence matrix for a proper graph $\mathcal{G}$ that is connected and has n nodes. Then the rank of **A** is $n-1$.

Proof. By Proposition 3.11, $\mathcal{G}$ has a spanning tree $\mathcal{T}$. By Proposition 3.6, $\mathcal{T}$ has $n-1$ arcs. By Proposition 3.10, $\{\mathbf{A}(j): e_j \varepsilon \mathcal{T}\}$ are linearly independent. Thus the rank of **A** is at least $n-1$. Since $\mathbf{1A}=\mathbf{0}$, the rank of **A** is less than n. Hence the rank of **A** is $n-1$. ■

Proposition 3.13

Let **A** be the node-arc incidence matrix for a proper graph $\mathcal{G}=[\mathcal{N},\mathcal{A}]$, where $\mathcal{G}$ has n nodes. Let $\hat{\mathcal{A}}$ be a subset of $\mathcal{A}$ such that $\{\mathbf{A}(j): e_j \varepsilon \hat{\mathcal{A}}\}$ is linearly independent and $\hat{\mathcal{A}}$ has $n-1$ arcs. Then $\mathcal{T}=[\mathcal{N},\hat{\mathcal{A}}]$ is a tree.

Proof. Corollary 3.5 implies that a cycle cannot be formed from $\mathcal{T}$, since $\{\mathbf{A}(j): e_j \varepsilon \hat{\mathcal{Q}}\}$ is linearly independent. Since $\mathcal{T}$ has one less arc than node and is acyclic, $\mathcal{T}$ is a tree by Proposition 3.6. ■

3.2 NETWORK BASIS CHARACTERIZATION

Recall that in our definition of LP (see Chapter 2) we assumed that the constraint matrix **A** has full row rank. However, we see by Proposition 3.12 that the constraint matrix for NP has a rank one less than the number of rows. There are two obvious ways in which NP may be reformulated so that the constraint matrix has full row rank: we can drop one row of the node-arc incidence matrix, or we can add a sufficient number of linearly independent columns (i.e., variables), each having upper bound of zero, so that the constraint matrix has full row rank. In this presentation we have selected the latter approach. Consider the following formulation:

$$\left.\begin{array}{ll}\min & \mathbf{cx}\\ \text{s.t.} & \mathbf{Ax}+a\mathbf{e}^l=\mathbf{r}\\ & \mathbf{0}\leqslant\mathbf{x}\leqslant\mathbf{u}\\ & 0\leqslant a\leqslant 0\end{array}\right\}\mathcal{NP},$$

where **A** is a node-arc incidence matrix and l is any positive integer no larger than the number of nodes, n. Since a is restricted to be zero, any optimum for $\mathcal{NP}$ will be an optimum for NP. We now show that there exists a set of n columns from $[\mathbf{A}\vdots\mathbf{e}^l]$ that spans E^n.

Proposition 3.14

Let **A** be the node-arc incidence matrix for a connected proper graph $\mathcal{G}=[\mathcal{N},\mathcal{Q}]$ having n nodes. Let $\mathcal{T}=[\mathcal{N},\hat{\mathcal{Q}}]$ be a spanning tree for $\mathcal{G}$. Then $\Omega=\{\mathbf{A}(j): e_j\varepsilon\hat{\mathcal{Q}}\}\cup\{\mathbf{e}^l\}$ spans E^n.

Proof. We must show that $\{\mathbf{e}^i: i=1,\ldots,n\}$ are linear combinations of the vectors of Ω. The vector $\mathbf{e}^l$ is certainly such a combination. Let $p\varepsilon\{1,\ldots,n\}$ and $p\neq l$. Since $\mathcal{T}$ is a spanning tree for $\mathcal{G}$ by Proposition 3.6, there is a unique path $P=\{s_1,e_{j_1},s_2,e_{j_2},\ldots,s_q,e_{j_q},s_{q+1}\}$ in $\mathcal{T}$ which links nodes p and l. By Proposition 3.3

$$\sum_{i=1}^{q} O_i(P)\mathbf{A}(j_i)=\mathbf{e}^p-\mathbf{e}^l.$$

Thus the linear combination of vectors of Ω, $\mathbf{e}^l+\sum_{i=1}^q O_i(P)\mathbf{A}(j_i)=\mathbf{e}^p$. Hence Ω spans E^n. ■

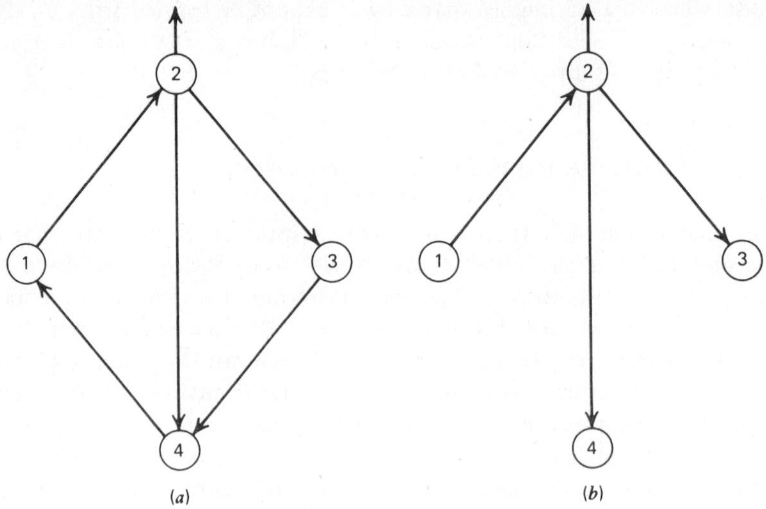

FIGURE 3.7 Networks having roots. (*a*) Rooted graph. (*b*) Rooted tree.

When we refer to the graph associated with $\mathcal{NP}$, we call a the *root arc* and node l the *root node*. Furthermore, the corresponding graph is called a *rooted graph*, and a rooted graph that is a tree is termed a *rooted tree*. We represent the root arc in our drawing by an arc incident on a single node. Figure 3.7 illustrates both a rooted graph and a rooted tree.

Proposition 3.15

Let $\mathbf{A}$ be the node-arc incidence matrix for a proper rooted graph $\mathcal{G} = [\mathcal{N}, \mathcal{A}]$ with root node l. If Ω is a basis for $[\mathbf{A} \vdots \mathbf{e}^l]$, then $\mathbf{e}^l \varepsilon \Omega$ and $\mathcal{T} = [\mathcal{N}, \hat{\mathcal{A}}]$ is a spanning tree for $\mathcal{G}$, where $\hat{\mathcal{A}} = \{e_j : \mathbf{A}(j) \varepsilon \Omega\}$.

Proof. Let Ω be any basis for $[\mathbf{A} \vdots \mathbf{e}^l]$. By Proposition 3.14, $[\mathbf{A} \vdots \mathbf{e}^l]$ has full row rank. By Proposition 3.12, $\mathbf{A}$ does not have full row rank. Hence $\mathbf{e}^l \varepsilon \Omega$. Let n denote the number of nodes in $\mathcal{G}$. Then $\hat{\mathcal{A}}$ must correspond to $n-1$ linearly independent columns from $\mathbf{A}$. Then, by Proposition 3.13, $\mathcal{T}$ is a spanning tree for $\mathcal{G}$. ■

We are now ready to characterize bases of $\mathcal{NP}$.

Proposition 3.16

Let $\mathbf{A}$ be a node-arc incidence matrix for a proper rooted graph with root node l which is connected. The only bases for $[\mathbf{A} \vdots \mathbf{e}^l]$ are $\mathbf{e}^l$ along with a set of columns from $\mathbf{A}$ corresponding to a spanning tree for $\mathcal{G}$.

Proof. The proof follows directly from Propositions 3.14 and 3.15. ■

We now have a graph-theoretic characterization of a basis for $\mathcal{NP}$. Next we develop an important algebraic property for these bases. These two properties taken together provide the theoretical justification for our specialization of the primal simplex algorithm.

A nonsingular *triangular* matrix is a square matrix with nonzeros on the main diagonal and all zeros above the main diagonal, or one that can be brought to such a form by row or column interchanges. We now show that bases for any $\mathcal{NP}$ defined on a proper graph are triangular.

Proposition 3.17

Let $\mathbf{A}$ be a node-arc incidence matrix for a proper rooted graph $\mathcal{G}$ with root node l which is connected. Let $\mathbf{B}$ be any basis from $[\mathbf{A} \,\vdots\, \mathbf{e}^l]$. Then $\mathbf{B}$ is triangular.

Proof. Let $\mathcal{T} = [\mathcal{N}, \mathcal{Q}]$ denote the rooted spanning tree associated with $\mathbf{B}$, and let n be the number of nodes of $\mathcal{T}$. Since $\mathcal{G}$ is proper, $n \geqslant 2$. By Proposition 3.8, $\mathcal{T}$ has at least two endpoints. Let n_1 be an endpoint of $\mathcal{T}$, not the root, and let e_{j_1} be the arc in $\mathcal{T}$ incident on n_1. Then row n_1 of $\mathbf{B}$ has one nonzero element. Thus, by letting the first row and column correspond to node n_1 and arc e_{j_1}, we may display $\mathbf{B}$ as follows:

$$\left[\begin{array}{c|c} \overbrace{\pm 1}^{e_{j_1}} & \\ \hline \pm \mathbf{e}^s & \mathbf{B}^1 \end{array}\right] \} \text{ node } n_1.$$

Let $\mathcal{T}^1 = [\mathcal{N}, -\{n_1\}, \mathcal{Q} - \{e_{j_1}\}]$. By Proposition 3.9, $\mathcal{T}^1$ is a tree with $n-1$ arcs. If $n-1=1$, then $\mathbf{B}^1$ is 1×1 and $\mathbf{B}$ is triangular. Suppose $n-1>1$. Then, by Proposition 3.8, $\mathcal{T}^1$ has at least two endpoints. Let n_2 be an endpoint of $\mathcal{T}^1$, not the root, and let e_{j_2} be the arc in $\mathcal{T}^1$ incident on n_2. Then row n_2 of $\mathbf{B}^1$ has one nonzero element. Thus, by letting the first row and column of $\mathbf{B}^1$ correspond to node n_2 and arc e_{j_2}, we may display $\mathbf{B}$ as follows:

$$\begin{array}{ccc} e_{j_1} & e_{j_2} & \\ \end{array}$$

$$\left[\begin{array}{c|c|c} \pm 1 & & \\ \hline & \pm 1 & \\ \pm \mathbf{e}^s & \pm \mathbf{e}^t & \mathbf{B}^2 \end{array}\right] \begin{array}{l} \text{node } n_1 \\ \text{node } n_2 . \\ \\ \end{array}$$

We may repeat this process exactly n times. Hence $\mathbf{B}$ is triangular. ■

Using the strategy of the proof of Proposition 3.17, we now develop an algorithm for determining the row and column order that makes it possible to display any network basis in lower triangular form.

ALG 3.1 ALGORITHM FOR TRIANGULARIZING A NETWORK BASIS

0 *Initialization* Let **B** be any basis for $\mathfrak{N}\mathfrak{P}$, and let $\mathfrak{T}=[\mathfrak{N},\mathfrak{A}]$ be the associated rooted spanning tree with root node l. Let n be the number of nodes in $\mathfrak{T}$, and set $i \leftarrow 1$.

1 *Find Endpoint Not Root Node.* Let $r \neq l$ be any endpoint in $\mathfrak{T}$, and let e_s be the arc in $\mathfrak{T}$ incident on r.

2 *Insert* ith *Row and Column.* Let the ith row of $\hat{\mathbf{B}}$ correspond to node r, and let the ith column of $\hat{\mathbf{B}}$ correspond to arc e_s.

3 *Reduce Tree.* If $i=n-1$, go to Step 4; otherwise $\mathfrak{T} \leftarrow [\mathfrak{N}-\{r\}, \mathfrak{A}-\{e_s\}]$, $i \leftarrow i+1$, and go to Step 1.

4 *Insert Root Node and Arc.* Let the nth row of $\hat{\mathbf{B}}$ correspond to node l, and let the nth column of $\hat{\mathbf{B}}$ be $\mathbf{e}^n$.

We return to Example 2.1 (see Section 2.4) and examine the spanning trees corresponding to the various bases developed. The network associated with Example 2.1 is illustrated in Figure 3.8, while the rooted spanning trees and triangularized bases can be found in Figure 3.9. Therefore we see that the rooted spanning trees

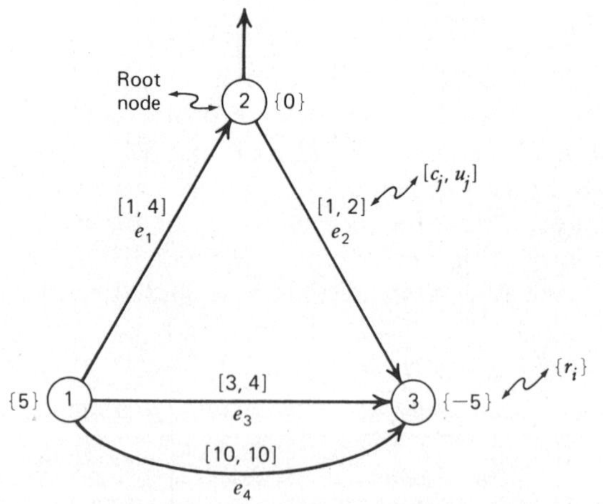

FIGURE 3.8 Network for Example 2.1.

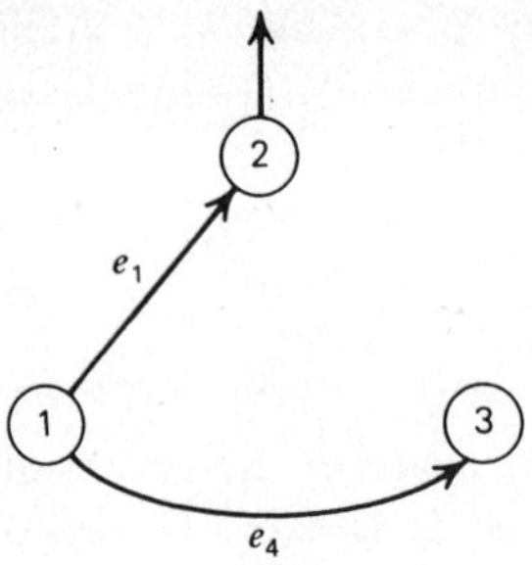

(a)

$$
\begin{array}{c}
\begin{matrix} e_4 & e_1 & \end{matrix} \\
\begin{bmatrix} -1 & 0 & 0 \\ 1 & 1 & 0 \\ 0 & -1 & 1 \end{bmatrix}
\end{array}
\begin{matrix} \text{node} \\ 3 \\ 1 \\ 2 \end{matrix}
$$

(b)

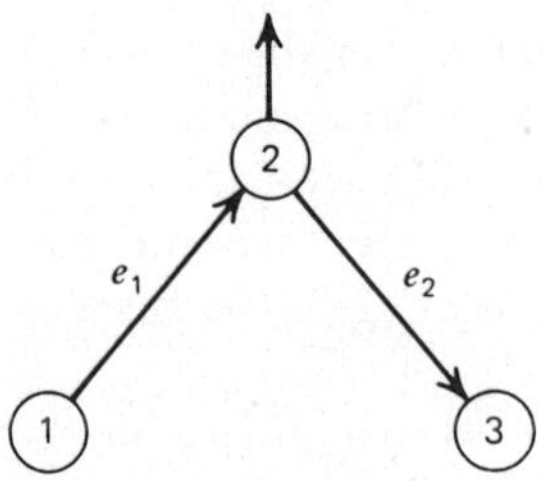

(c)

$$
\begin{array}{c}
\begin{matrix} e_1 & e_2 & \end{matrix} \\
\begin{bmatrix} 1 & 0 & 0 \\ 0 & -1 & 0 \\ -1 & 1 & 1 \end{bmatrix}
\end{array}
\begin{matrix} \text{node} \\ 1 \\ 3 \\ 2 \end{matrix}
$$

(d)

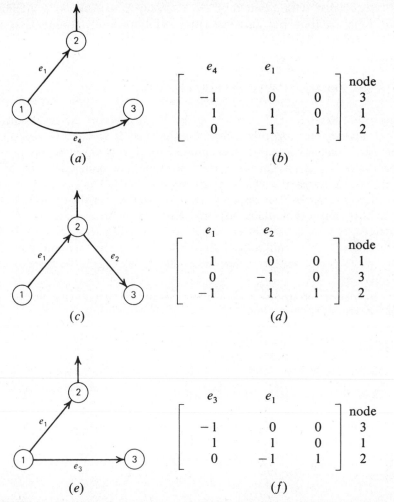

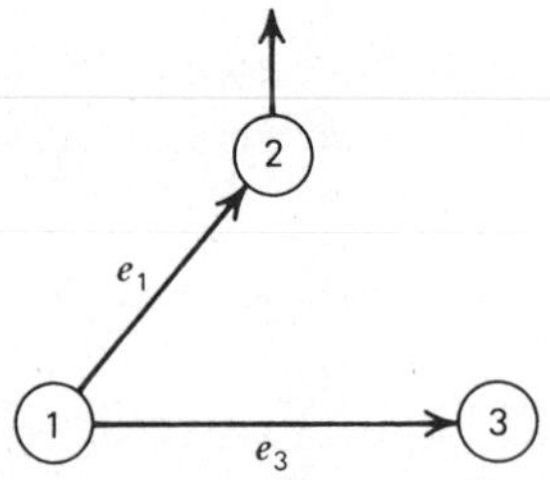

(e)

$$
\begin{array}{c}
\begin{matrix} e_3 & e_1 & \end{matrix} \\
\begin{bmatrix} -1 & 0 & 0 \\ 1 & 1 & 0 \\ 0 & -1 & 1 \end{bmatrix}
\end{array}
\begin{matrix} \text{node} \\ 3 \\ 1 \\ 2 \end{matrix}
$$

(f)

FIGURE 3.9 Rooted spanning trees corresponding to bases for Example 2.1. (*a*) Basis tree for iteration 1. (*b*) Triangularized basis for iteration 1. (*c*) Basis tree for iteration 2. (*d*) Triangularized basis for iteration 2. (*e*) Basis tree for iteration 3. (*f*) Triangularized basis for iteration 3.

associated with bases of $\mathcal{NP}$ not only provide a nice graphical representation but also make it possible to easily triangularize the corresponding basis.

3.3 PRIMAL SIMPLEX SPECIALIZATION

Given that bases for $\mathcal{NP}$ are triangular, let us examine the operations of the primal simplex algorithm involving the basis $\mathbf{B}$ or $\mathbf{B}^{-1}$. To determine whether a nonbasic variable is a candidate for flow change (i.e., the pricing operation), we must calculate the corresponding component of $[\mathbf{c}^N - \mathbf{c}^B\mathbf{B}^{-1}\mathbf{N}]$. Consider only $\mathbf{c}^B\mathbf{B}^{-1}\mathbf{N}$, and let $\boldsymbol{\pi}=\mathbf{c}^B\mathbf{B}^{-1}$. The π_i's are generally called *dual variables* or *node potentials*. But $\boldsymbol{\pi}=\mathbf{c}^B\mathbf{B}^{-1}$ is the solution to the system of linear equations $\boldsymbol{\pi}\mathbf{B}=\mathbf{c}^B$. Since $\mathbf{B}$ is triangular, $\boldsymbol{\pi}$ may be obtained by solving for the last component first and backward substituting to iteratively obtain all components. Furthermore, we indicate how one may make use of the tree associated with $\mathbf{B}$, say $\mathcal{T}_B$, in performing this calculation.

Consider the rooted spanning tree illustrated in Figure 3.10 with the corresponding basis;

$$\mathbf{B}=\begin{array}{c} \\ \end{array}\begin{bmatrix} 1 & & & & \\ & -1 & & & \\ & & -1 & & \\ -1 & & 1 & -1 & \\ & 1 & & 1 & 1 \end{bmatrix}$$

e_1	e_4	e_2	e_3	a	nodes
1					1
	−1				4
		−1			2
−1		1	−1		3
	1		1	1	5

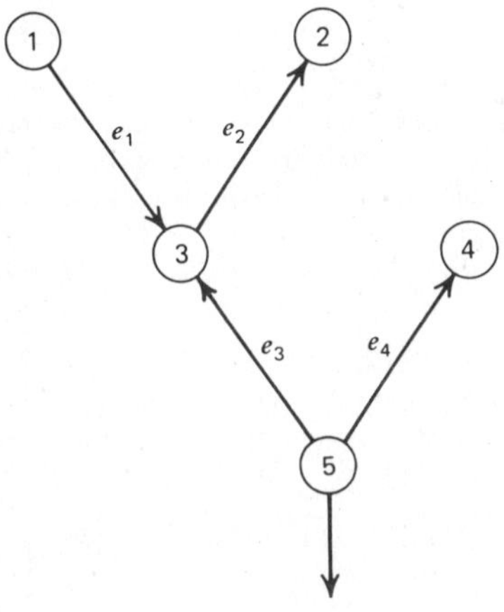

FIGURE 3.10 Rooted spanning tree.

Then $\boldsymbol{\pi}\mathbf{B}=\mathbf{c}^B$ reduces to the triangular system

$$\begin{array}{ccccccl}
\pi_1 & & & -\pi_3 & & & =c_1^B \\
 & & -\pi_4 & & & +\pi_5 & =c_4^B \\
 & -\pi_2 & & +\pi_3 & & & =c_2^B \\
 & & & -\pi_3 & & +\pi_5 & =c_3^B \\
 & & & & & \pi_5 & =0
\end{array}.$$

Clearly, this system may be solved by backward substitution, beginning with $\pi_5=0$. In general, for a basis $\mathbf{B}$ and basis tree $\mathcal{T}_B$ with root node $l, \boldsymbol{\pi}\mathbf{B}=\mathbf{c}^B$ reduces to

$$\left\{\begin{array}{lll}\pi_l & =0 & \\ \pi_{F(j)}-\pi_{T(j)} & =c_j & \text{for } e_j \varepsilon \mathcal{T}_B\end{array}\right\},$$

which is a triangular system since $\mathbf{B}$ is triangular. Therefore the following algorithm may be used to iteratively determine $\boldsymbol{\pi}$.

ALG 3.2 DETERMINATION OF DUAL VARIABLES

0 *Initialization.* Let $\mathcal{T}=[\mathcal{N},\mathcal{A}]$ denote the basis tree with root node l. Set $\pi_l \leftarrow 0$, $\mathcal{N}^L \leftarrow \{l\}$, and $\mathcal{N}^U \leftarrow \mathcal{N}-\{l\}$.

1 *Find an Arc with "To" Node Labeled.* Let $e_j \varepsilon \mathcal{A}$ be such that $F(j) \varepsilon \mathcal{N}^U$ and $T(j) \varepsilon \mathcal{N}^L$. If no such arc exists, go to Step 3.

2 *Label "From" Node.* Set $\pi_{F(j)} \leftarrow c_j+\pi_{T(j)}$, $\mathcal{N}^L \leftarrow \mathcal{N}^L \cup \{F(j)\}$, $\mathcal{N}^U \leftarrow \mathcal{N}^U-\{F(j)\}$, and go to Step 1.

3 *Find an Arc with "From" Node Labeled.* Let $e_j \varepsilon \mathcal{A}$ be such that $T(j) \varepsilon \mathcal{N}^U$ and $F(j) \varepsilon \mathcal{N}^L$. If no such arc exists, terminate.

4 *Label "To" Node.* Set $\pi_{T(j)} \leftarrow -c_j+\pi_{F(j)}$, $\mathcal{N}^L \leftarrow \mathcal{N}^L \cup \{T(j)\}$, $\mathcal{N}^U \leftarrow \mathcal{N}^U-\{T(j)\}$, and go to Step 1.

All of the dual variables are uniquely determined iteratively because a rooted tree is connected and has no cycles. When $\boldsymbol{\pi}=\mathbf{c}^B\mathbf{B}^{-1}$ is used, the candidate sets used in the primal simplex algorithm (see section 2.4) specialize to

$$\psi_1=\{e_j : x_j=0 \text{ and } \pi_{F(j)}-\pi_{T(j)}-c_j>0\}$$

and

$$\psi_2=\{e_j : x_j=u_j \text{ and } \pi_{F(j)}-\pi_{T(j)}-c_j<0\}.$$

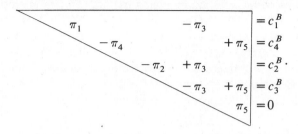

The components of $\mathbf{c}^B\mathbf{B}^{-1}\mathbf{N}$ can also be determined in another way. Let $\mathbf{B}$ be the basis with corresponding basis tree $\mathcal{T}_B$. The ith component of $\mathbf{c}^B\mathbf{B}^{-1}\mathbf{N}$ is given by $\mathbf{c}^B\mathbf{B}^{-1}\mathbf{N}(i)$. Suppose $\mathbf{A}(k)=\mathbf{N}(i)$. Then one may calculate $\mathbf{y}=\mathbf{B}^{-1}\mathbf{A}(k)$ and then calculate $\mathbf{c}^B\mathbf{y}$, where $e_k \notin \mathcal{T}_B$. But this is equivalent to solving $\mathbf{B}\mathbf{y}=\mathbf{A}(k)=\mathbf{e}^{F(k)}-\mathbf{e}^{T(k)}$. Since $\mathbf{B}$ is triangular, $\mathbf{y}$ may be obtained directly and hence $\mathbf{B}^{-1}$ is not required. Again we make use of $\mathcal{T}_B$ to solve this triangular system. Let $P=\{s_1, e_{j_1}, s_2, e_{j_2}, \ldots, s_n, e_{j_n}, s_{n+1}\}$ be the unique path in $\mathcal{T}_B$ linking $F(k)$ to $T(k)$. Then by Proposition 3.3

$$\sum_{i=1}^{i=n} O_i(P)\mathbf{A}(j_i)=\mathbf{e}^{F(k)}-\mathbf{e}^{T(k)}. \tag{3.3}$$

Then $\mathbf{c}^B\mathbf{B}^{-1}\mathbf{N}(i)=\mathbf{c}^B\mathbf{y}$ is simply

$$\sum_{i=1}^{i=n} c_{j_i} O_i(P). \tag{3.4}$$

Either of these approaches may be used for the pricing routine, but the former has been adopted by most groups developing network codes.

Equation 3.3 also indicates how one may specialize the ratio test that requires the calculation of $\mathbf{y}=\mathbf{B}^{-1}\mathbf{A}(k)$. Let the arcs in $\mathcal{T}_B$ be ordered as follows: $e_{k_1}, e_{k_2}, \ldots, e_{k_{\bar{I}}}$, corresponding to the $\bar{I}$ columns of $\mathbf{B}$. Then the $\bar{I}$ components of y may be determined by the orientation sequence;

$$y_n=\begin{cases} O_i(P) & \text{if } e_{k_n}=e_{j_i}\varepsilon P, \\ 0 & \text{otherwise} \end{cases}. \tag{3.5}$$

Therefore the calculation of Δ_1 and Δ_2 in the ratio test may be specialized to

$$\Delta_1 \leftarrow \min_{O_i(P)=\delta} \{x_{j_i}, \infty\} \qquad \text{and} \qquad \Delta_2 \leftarrow \min_{-O_i(P)=\delta} \{u_{j_i}-x_{j_i}, \infty\}.$$

Consider the rooted spanning tree of Figure 3.10. Suppose the out-of-tree arc is $e_k=(2,4)$. Then $P=\{2, e_2, 3, e_3, 5, e_4, 4\}$ and $O(P)=\{-1, -1, 1\}$. The cycle formed and the calculation of $\mathbf{y}$ are illustrated in Figure 3.11.

We now present the primal simplex algorithm on a graph.

ALG 3.3 PRIMAL SIMPLEX METHOD ON A GRAPH

0 *Initialization.* Let $[\mathbf{x}^B \,\vdots\, \mathbf{x}^N]$ be a basic feasible solution with basis tree $\mathcal{T}_B$. Calculate $\boldsymbol{\pi}$ using ALG 3.2.

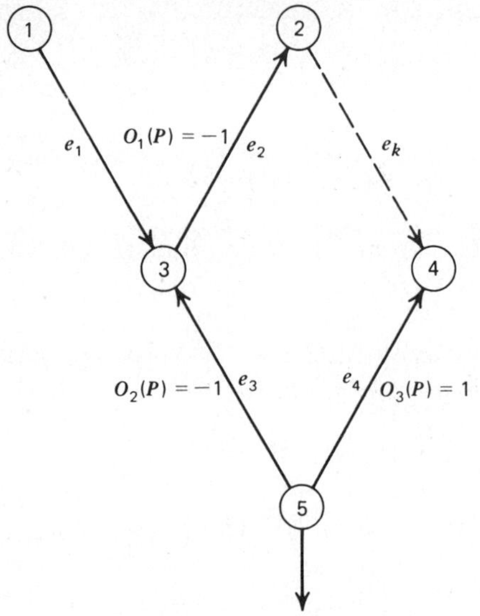

FIGURE 3.11 Cycle formed by (2, 4).

1 *Pricing.* Let

$$\psi_1 = \{e_j : x_j = 0 \text{ and } \pi_{F(j)} - \pi_{T(j)} - c_j > 0\}$$

and

$$\psi_2 = \{e_j : x_j = u_j \text{ and } \pi_{F(j)} - \pi_{T(j)} - c_j < 0\}.$$

If $\psi_1 \cup \psi_2 = \phi$, terminate with $[\mathbf{x}^B \mid \mathbf{x}^N]$ an optimum; otherwise select $e_k \varepsilon \psi_1 \cup \psi_2$ and set

$$\delta \leftarrow \begin{cases} +1 & \text{if } e_k \varepsilon \psi_1, \\ -1 & \text{if } e_k \varepsilon \psi_2. \end{cases}$$

2 *Ratio Test.* Let $P = \{s_1, e_{j_1}, s_2, e_{j_2}, \ldots, s_n, e_{j_n}, s_{n+1}\}$ denote the path in $\mathfrak{T}_B$ linking $F(k)$ to $T(k)$. Set

$$\Delta_1 \leftarrow \min_{O_i(P) = \delta} \{x_{j_i}, \infty\},$$

$$\Delta_2 \leftarrow \min_{-O_i(P) = \delta} \{u_{j_i} - x_{j_i}, \infty\},$$

$$\Delta \leftarrow \min\{\Delta_1, \Delta_2, u_k\}.$$

3 *Update Flows.* Set $x_k \leftarrow x_k + \Delta\delta$. For all $e_j \varepsilon P$ set $x_{j_i} \leftarrow x_{j_i} - \Delta\delta O_i(P)$. If $\Delta = u_k$, go to Step 1.

4 *Update Tree and Duals.* Let

$$\psi_3 = \{ e_{j_i} : x_{j_i} = 0, \text{ where } O_i(P) = \delta \}$$

and

$$\psi_4 = \{ e_{j_i} : x_{j_i} = u_{j_i}, \text{ where } -O_i(P) = \delta \}.$$

Select any $e_m \varepsilon \psi_3 \cup \psi_4$. Replace e_m in $\mathcal{T}_B$ with e_k, update the dual variables, and return to Step 1.

We now solve Example 2.1 using ALG 3.3. The basis trees and ratio test cycles are given in Figure 3.12.

0 (Initialization). Let

$$[x_1 \quad x_2 \quad x_3 \quad x_4 \quad x_5] = [0 \quad 0 \quad 4 \quad 1 \quad 0], \mathbf{x}^B = [x_1 \quad x_4 \quad x_5],$$
$$\text{and} \quad \mathbf{x}^N = [x_2 \quad x_3].$$

1 (Pricing at Iteration 1)
For e_2: $\pi_2 - \pi_3 - c_2 = 0 - (-9) - 1 = 8 \Rightarrow e_2 \varepsilon \psi_1$.
For e_3: $\pi_1 - \pi_3 - c_3 = 1 - (-9) - 3 = 7 \Rightarrow e_3 \not\varepsilon \psi_1$ or ψ_2.
Also, $\psi_1 = \{e_2\}, \psi_2 = \phi$. Then $e_k = e_2$ and $\delta = 1$.

2 (Ratio Test) $P = \{2, e_1, 1, e_4, 3\}$

$$\Delta_1 \leftarrow \min\{x_4, \infty\} = \min\{1, \infty\} = 1,$$
$$\Delta_2 \leftarrow \min\{u_1 - x_1, \infty\} = \min\{4 - 0, \infty\} = 4,$$
$$\Delta \leftarrow \min\{1, 4, 2\} = 1.$$

3 (Update Flows)

$$x_2 \leftarrow 0 + 1(1) = 1,$$
$$x_1 \leftarrow 0 - 1(1)(-1) = 1,$$
$$x_4 \leftarrow 1 - 1(1)(1) = 0.$$

4 (Update Tree and Duals). $\psi_3 = \{e_4\}, \psi_4 = \phi$. Then $e_m = e_4$, $\mathbf{x}^B = [x_1 \quad x_2 \quad x_5]$, and $\mathbf{x}^N = [x_3 \quad x_4]$.

1 (Pricing at Iteration 2 with $[x_1 \quad x_2 \quad x_3 \quad x_4 \quad x_5] = [1 \quad 1 \quad 4 \quad 0 \quad 0]$)
For e_3: $\pi_1 - \pi_3 - c_3 = 1 - (-1) - 3 = -1 \Rightarrow e_3 \varepsilon \psi_2$.
For e_4: $\pi_1 - \pi_3 - c_4 = 1 - (-1) - 10 = -8 \Rightarrow e_4 \not\varepsilon \psi_1$ or ψ_2.
Then $e_k = e_3$ and $\delta = -1$.

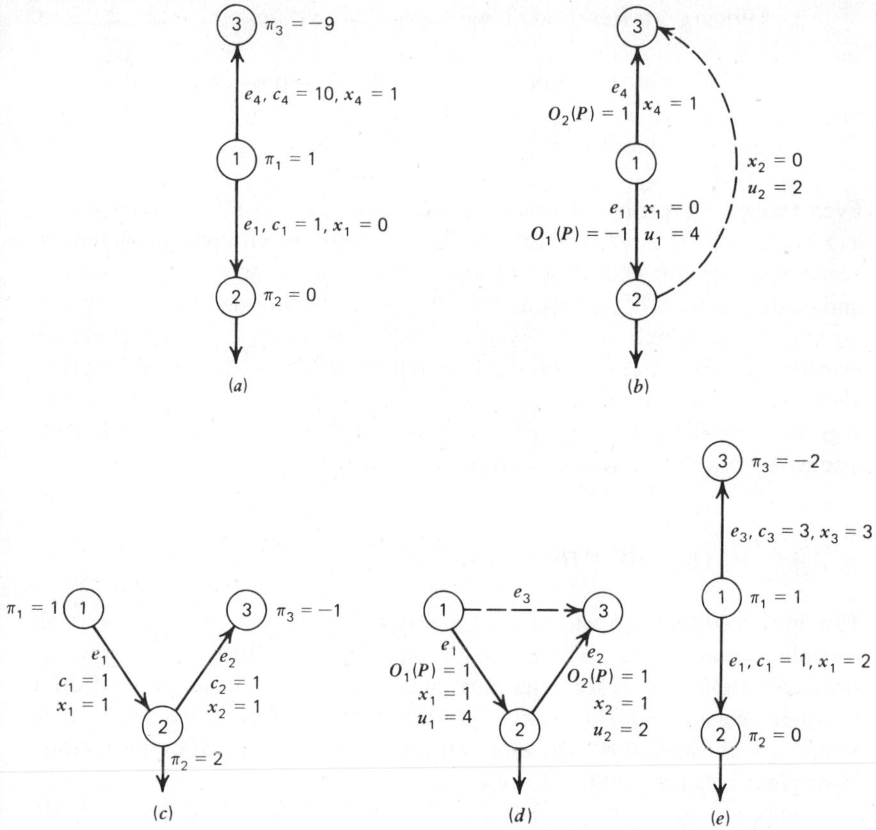

FIGURE 3.12 Basis trees and ratio test cycles for Example 2.1. (*a*) Basis tree for iteration 1. (*b*) Ratio test cycle for iteration 1. (*c*) Basis tree for iteration 2. (*d*) Ratio test cycle for iteration 2. (*e*) Basis tree for iteration 3.

2 (Ratio Test) $P=\{1,e_1,2,e_2,3\}$

$$\Delta_1 \leftarrow \min\{\infty\}=\infty, \Delta_2 \leftarrow \min\{4-1,2-1,\infty\}=1,$$
$$\Delta \leftarrow \min\{\infty,1,4\}=1.$$

3 (Update Flows)

$$x_3 \leftarrow 4+1(-1)=3,$$
$$x_1 \leftarrow 1-1(-1)(1)=2,$$
$$x_2 \leftarrow 1-1(-1)(1)=2.$$

4 (Update Tree and Duals). $\psi_3=\phi, \psi_4=\{e_2\}$. Then $e_m=e_2$, $\mathbf{x}^B=[x_1 \quad x_3 \quad x_5]$, and $\mathbf{x}^N=[x_2 \quad x_4]$.

1 (Pricing at Iteration 3 with $[x_1 \quad x_2 \quad x_3 \quad x_4 \quad x_5]=[2 \quad 2 \quad 3 \quad 0 \quad 0]$)

For e_2: $\pi_2-\pi_3-c_2=0-(-2)-1=1 \Rightarrow e_2 \notin \psi_1$ or ψ_2.

For e_4: $\pi_1-\pi_3-c_4=1-(-2)-10=-7 \Rightarrow e_4 \notin \psi_1$ or ψ_2. Optimum attained!

Even though the primal simplex method on a graph is straightforward for hand calculation, there is much to be said about an efficient implementation (i.e., computer code) for this algorithm. In fact, there exists a body of knowledge concerning efficient computational techniques for implementing ALG 3.3, which we call *implementation technology*. This technology involves the development of data structures along with the appropriate algorithms for executing the various steps of the primal simplex method on a graph. The interested reader is referred to Appendix B for a complete discussion of data structures and related algorithms.

3.4 INITIAL SOLUTION

The initialization step of ALG 3.3 assumes that one can find a basic feasible solution with which to initiate the algorithm, and numerous strategies are available for obtaining such a solution. The simplest of these is called the *all-artificial start*. For this method only extra arcs called artificial arcs carry flow. All other strategies allow flow on a combination of original arcs and artificial arcs.

Suppose the network associated with $\mathcal{NP}$ has $\bar{I}$ nodes. The all-artificial start involves enlarging the network by first appending a new node, $\bar{I}+1$. For each source node i with supply r_i, we append a slack arc from i to $\bar{I}+1$ having cost equal to zero and bound equal to infinity. The flow on this arc is set to r_i, thus satisfying flow conservation at node i. For each demand node k, with demand $|r_k|$, we append an artificial arc from $\bar{I}+1$ to k having cost and bound equal to infinity. The flow on this arc is also set to $|r_k|$, thus satisfying flow conservation at node k. Since the costs associated with the artificial arcs are infinite, any solution without flow on artificial arcs will dominate any solution having flow on these arcs. This formulation corresponds to what is traditionally called the Big-M method in the linear programming literature. The basis tree is completed by using original arcs, each with a flow of zero.

An example of the all-artificial start procedure is illustrated in Figure 3.13. Note that node 5 also serves as a dummy destination with a demand of 2. A variation of the all-artificial start is implemented in the NETFLO computer code (see Appendix F).

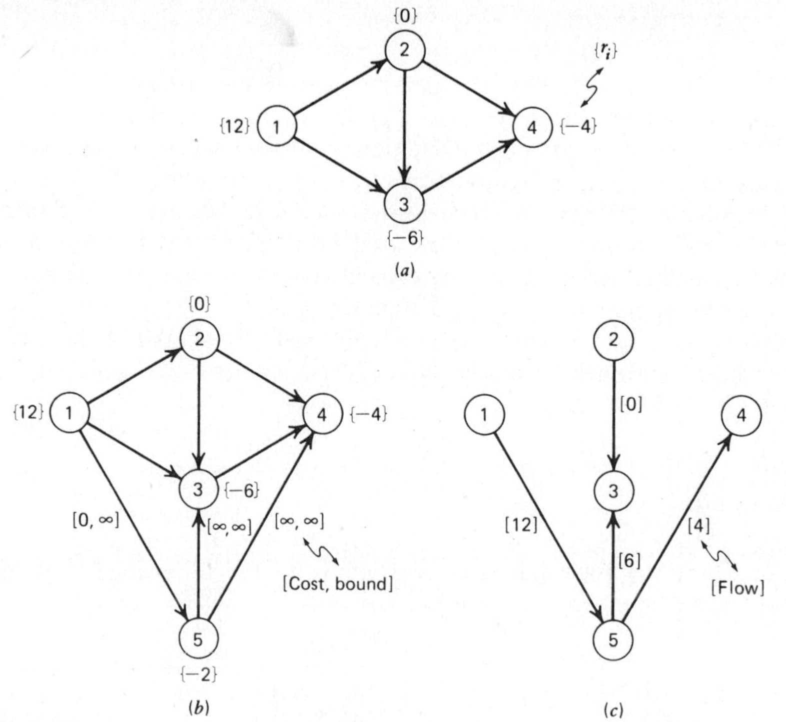

FIGURE 3.13 Example of all-artificial start. (*a*) Original network. (*b*) Transformed network. (*c*) Basis tree.

3.5 PRICING STRATEGIES

We say that an arc e_j is a candidate for flow change if $e_j \varepsilon \psi_1 \cup \psi_2$ (see Step 1 of ALG 3.3). The pricing strategy is the algorithm that selects, from among the arcs in $\psi_1 \cup \psi_2$, the particular nonbasic arc for which flow change will be allowed. Pricing strategies range from selection of the first candidate arc found to selection of the candidate $e_j \varepsilon \psi_1 \cup \psi_2$ having the largest $|\pi_{F(j)} - \pi_{T(j)} - c_j|$. Most codes incorporate some strategy lying between these two extremes.

For most pricing strategies the arc file is managed as a wrap-around stack, that is, if in a previous search the last element scanned resides in position k, the next search commences at position $k+1$. Whenever the end of the file is reached, scanning continues at position 1. The simplex method terminates when a scan through the entire arc file fails to find a candidate arc.

3.6 SUMMARY

This chapter has presented the development of the primal simplex algorithm on a graph. The specialized algorithm was developed in Section 3, and data structures for implementation are presented in Appendix B. A means for obtaining an initial feasible solution was given in Section 4, while pricing strategies were discussed briefly in Section 5. The primal simplex on a graph algorithm has itself been specialized for the uncapitated transshipment problem, the capacitated transportation problem, the transportation problem, and the assignment problem. These specializations are straightforward and are not presented in this book. All of these ideas have been implemented in the NETFLO computer code, which can be found in Appendix F.

EXERCISES

3.1 Draw the rooted tree corresponding to the following bases, and apply ALG 3.1 to display these bases in lower triangular form:

$$
\text{(a)}\quad \begin{bmatrix}
 & & & 1 & & & \\
 & & & -1 & & & -1 \\
1 & -1 & & & & -1 & \\
 & 1 & & & & & \\
 & & & & -1 & & \\
-1 & & 1 & & 1 & & 1 \\
 & & & & & 1 &
\end{bmatrix}
\begin{array}{l} 1 \text{ nodes} \\ 2 \\ 3 \\ 4 \\ 5 \\ 6 \\ 7 \end{array};
$$

$$
\text{(b)}\quad \begin{bmatrix}
1 & 1 & 1 & 1 & & \\
 & -1 & & & 1 & -1 \\
 & & -1 & & & \\
 & & & -1 & & \\
 & & & & -1 & \\
 & & & & & 1
\end{bmatrix}
\begin{array}{l} 1 \text{ nodes} \\ 2 \\ 3 \\ 4 \\ 5 \\ 6 \end{array};
$$

$$
\text{(c)}\quad \begin{bmatrix}
 & 1 & & & 1 & 1 & -1 & & \\
 & & & & & & 1 & & \\
 & & & & & -1 & & & \\
1 & -1 & & -1 & & & & 1 & \\
 & & & & -1 & & & & 1 \\
 & & & & & & & -1 & \\
 & & & & & & & & -1 \\
 & & 1 & 1 & & & & & \\
 & & -1 & & & & & &
\end{bmatrix}
\begin{array}{l} 1 \text{ nodes} \\ 2 \\ 3 \\ 4 \\ 5 \\ 6 \\ 7 \\ 8 \\ 9 \end{array}.
$$

3.2 Using Alg 3.1, display the bases (i.e., matrices) corresponding to the trees of Figure 3.14 in lower triangular form.

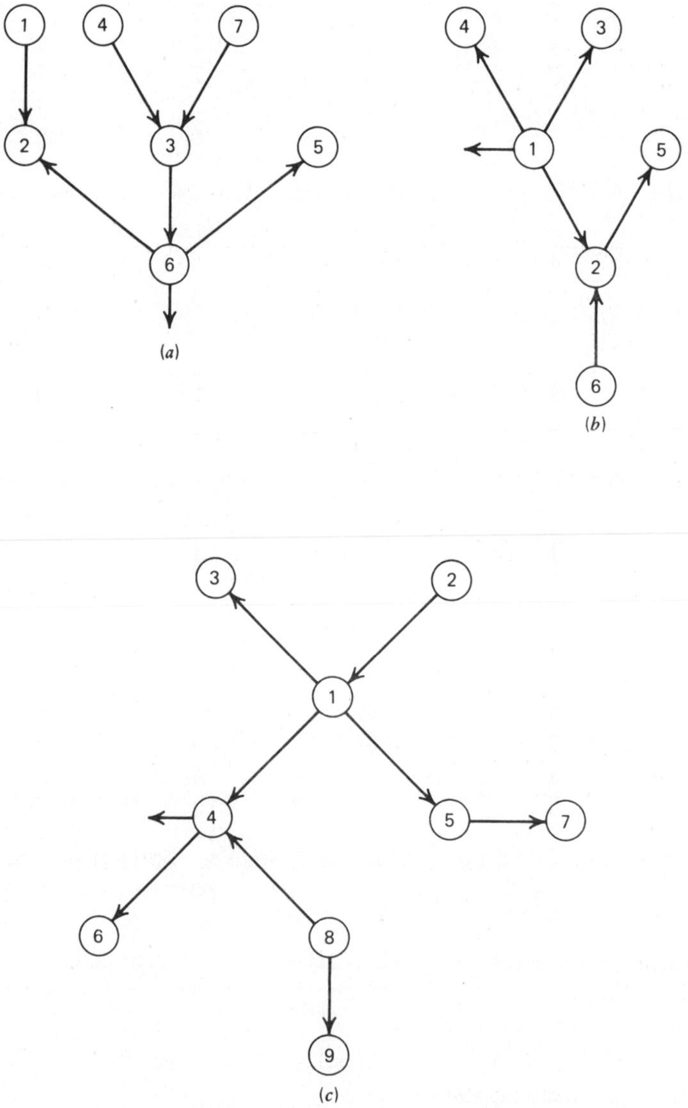

FIGURE 3.14 Sample basis trees.

3.3 Using ALG 3.2, determine the dual variables for each of the following bases:

(a)

Arc Number, j	"From" Node, $F(j)$	"To" Node, $T(j)$	Unit Cost, c_j
1	6	5	10
2	3	6	8
3	7	3	6
4	4	3	9
5	6	2	7
6	1	2	3
7	6	0	0

(b)

Arc Number, j	"From" Node, $F(j)$	"To" Node, $T(j)$	Unit Cost, c_j
1	1	0	0
2	1	3	20
3	1	4	5
4	1	2	15
5	2	5	10
6	6	2	0

(c)

Arc Number, j	"From" Node, $F(j)$	"To" Node, $T(j)$	Unit Cost, c_j
1	1	3	70
2	2	1	30
3	1	4	90
4	1	5	50
5	5	7	100
6	8	4	40
7	8	9	80
8	4	6	60
9	4	0	0

3.4 Solve Exercise 2.8 using ALG 3.3. Obtain the optimal dual variables, and show that they are feasible for the dual problem. (See Proposition 2.10.)

3.5 Solve Exercise 2.9 using ALG 3.3 with an all-artificial start. Obtain the optimal dual variables, and show that they are feasible for the dual problem. (See Proposition 2.10.)

3.6 Solve the transportation problem given in Exercise 2.10 using ALG 3.3 with an all-artificial start.

3.7 Solve the assignment problem given in Exercise 2.11 using ALG 3.3.

3.8 Show that a column of the basis inverse may be easily generated from the information available in the rooted spanning tree. Generate $\mathbf{B}^{-1}$, where

$$\mathbf{B}=\begin{bmatrix} -1 & & & & & \\ 1 & 1 & & & 1 & \\ & & & 1 & & \\ & & -1 & -1 & & \\ & & & & -1 & \\ & -1 & 1 & & & 1 \end{bmatrix}.$$

(*Hint:* Use Proposition 3.3.)

3.9 To efficiently execute a dual simplex algorithm for the minimal cost network flow problem, one needs an efficient means for generating a row of the basis inverse. Develop an algorithm for determining the ith row of $\mathbf{B}^{-1}$ via operations on $\mathcal{T}_B$. (*Hint:* Let $\boldsymbol{\beta}_i$ denote the ith row of $\mathbf{B}^{-1}$. Then the system of equations to be solved is $\boldsymbol{\beta}_i\mathbf{B}=\mathbf{e}^i$.)

3.10 A reasonable implementation of ALG 3.3 would use five node-length arrays and four arc-length arrays for both the original data and the working files. Using Appendix B as a guide, indicate the information that would be stored in each of these nine arrays. Display these arrays for each step of Exercise 3.4.

3.11 By sorting the arc data by either "from" node $F(j)$ or "to" node $T(j)$, one can execute ALG 3.3 with only three arc-length arrays. Indicate how this can be done, and discuss the advantages and disadvantages.

3.12 Develop a labeling algorithm to solve the triangular system $\mathbf{By}=\mathbf{d}$, where $\mathbf{d}$ is an arbitrary vector and $\mathcal{T}_B$ is a rooted tree corresponding to $\mathbf{B}$. Your algorithm can make no more than one scan through $\mathcal{T}_B$ (i.e., each node can be examined at most once). (*Hint:* An inverse thread or reverse thread label—see Appendix B—may be used.)

3.13 Indicate properties of the transportation problem and the assignment problem that may be exploited in a specialization of the primal simplex on a graph algorithm.

3.14 Develop an all-artificial start for Exercise 1.9. Display your answer as a rooted spanning tree.

3.15 Solve the following problem using ALG 3.3 with an artificial start:

$$\min \quad 2x_1 \; +4x_2 \; +x_3 \; +x_4 \; +9x_5 \; -10x_6$$

$$\text{s.t.} \quad \begin{bmatrix} 1 & 1 & & & & \\ -1 & & -1 & 1 & 1 & \\ & -1 & 1 & -1 & & 1 \\ & & & & -1 & -1 \end{bmatrix} \mathbf{x} \leqslant \begin{bmatrix} 6 \\ 6 \\ -3 \\ -5 \end{bmatrix}$$

$$1 \leqslant x_1 \leqslant 3, 2 \leqslant x_2 \leqslant 3, 2 \leqslant x_3 \leqslant 5, 0 \leqslant x_4 \leqslant 3, 1 \leqslant x_5 \leqslant 3, 0 \leqslant x_6 \leqslant 3.$$

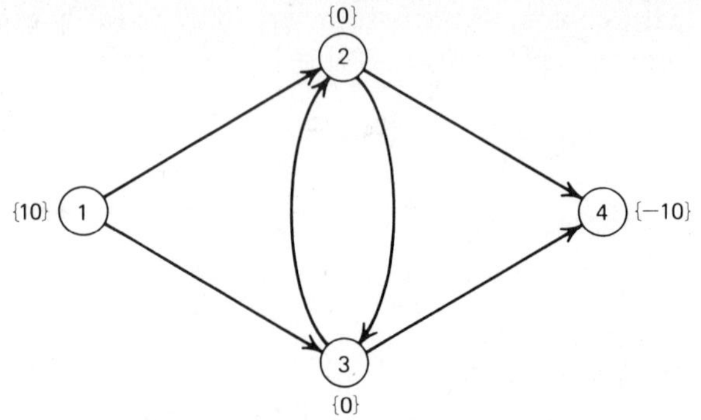

FIGURE 3.15 Minimal cost network flow problem.

(*Hint:* Add slack variables and then append a new constraint that is the negative of the sum of the other constraints. Next make variable substitutions, such as $y_1+1=x_1$, to convert to the prescribed form.)

3.16 Consider the minimal cost network flow problem of Figure 3.15. Suppose we require that the flow through node 3 be at least 3 and no greater than 5. By appending new arcs and/or new nodes, reformulate this problem as an NP in which this node capacity constraint will be enforced.

3.17 Place the following problem in the NP form (this may be viewed as a network program having undirected arcs):

$$\min \quad x_1 + x_2 + 3x_3 + x_4 + 10x_5$$

$$\text{s.t.} \quad \begin{bmatrix} -1 & & -1 & & \\ 1 & 1 & & & -1 \\ & -1 & 1 & 1 & \\ & & & -1 & 1 \end{bmatrix} \mathbf{x} = \begin{bmatrix} 10 \\ 0 \\ 0 \\ -10 \end{bmatrix}$$

$$-7 \leqslant x_i \leqslant 7, \qquad i=1,\ldots,5.$$

3.18 It can be shown that any minimal cost network flow problem can be transformed into an uncapacitated transshipment problem via substitutions of the type illustrated in Figure 3.16. If this were done for each arc, what simplifications in ALG 3.3 could be made? Compare the core storage requirements for the two approaches.

3.19 Suppose the rooted spanning tree of Figure 3.17 corresponds to a basis of a 3×3 transportation problem. (*a*) Give the corresponding basis and its inverse. (*b*) If the cost for arc (i, j) is $i+j$, determine the dual

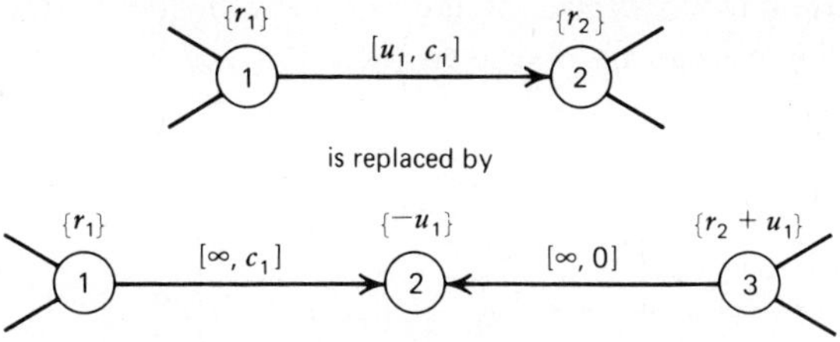

FIGURE 3.16 Equivalent network substitution.

variables corresponding to this basis. (*c*) Give the reduced cost for the nonbasic arc (1,6).

3.20 Solve the Detroit Motors production-distribution problem described in Section 1.4.1 using NETFLO. (See Appendix F.)

3.21 Develop an algorithm that takes a given basis (rooted spanning tree) and a given requirements vector and obtains the flows on the arcs in the tree. Describe a situation in which this algorithm would be needed.

3.22 Show that the columns of the node-arc incidence matrix **A** corresponding to a path can be placed in the bidiagonal triangular form shown below:

$$\left[\begin{array}{c:c:c:c:c} a_1 & & & & \\ b_1 & a_2 & & & \\ & b_2 & a_3 & & \\ & & & \ddots & \\ & & & & a_m \\ & & & & b_m \end{array}\right].$$

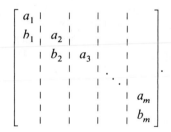

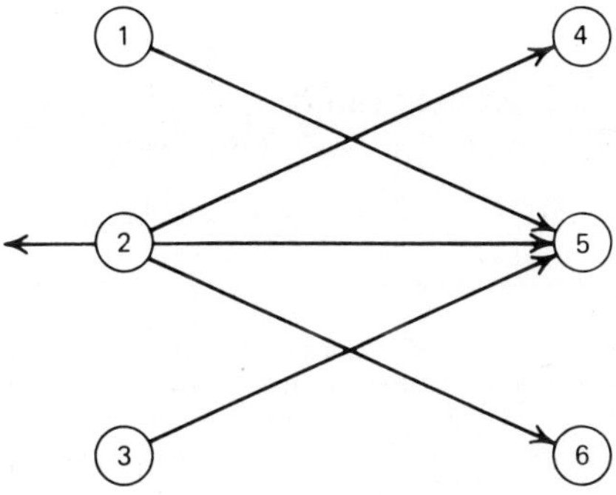

FIGURE 3.17 Rooted spanning tree.

3.23 Show that the columns of the node-arc incidence matrix **A** corresponding to a cycle can be placed in the following form:

$$\left[\begin{array}{c:c:c:c:c} a_1 & & & & b_m \\ b_1 & a_2 & & & \\ & b_2 & & & \\ & & a_3 & & \\ & & & \ddots & \\ & & & & a_m \end{array}\right].$$

NOTES AND REFERENCES

Sections 3.1, 3.2, and 3.3

The results of these sections may be traced to the pioneering work of Koopmans [125]. These results were further examined by Dantzig [47], Johnson [112], Bazaraa and Jarvis [20], and Helgason [102]. The first data structure suggested for implementation can be found in Johnson [113]. The first implementations of these ideas were by Srinivasan and Thompson [172] and by Glover, Karney, Klingman, and Napier [77]. The code of Glover, Hultz, and Klingman [81] was found to be over 100 times faster than a general linear programming code. A new convergence result for ALG 3.3 has been presented by Cunningham [41].

Section 3.4

An algebraic presentation of the all-artificial start is given in Bazaraa and Jarvis [20].

Section 3.5

The various pricing strategies that have been tried can be found in Glover, Karney, Klingman, and Napier [77], Srinivasan and Thompson [172], Bradley, Brown, and Graves [25], and Mulvey [145].

Section 3.6

A specialization of the algorithm for a transportation problem is given in Langley [130] and Langley, Kennington, and Shetty [131].

General Comments

Many solution techniques for more general distribution models frequently involve subprograms that are minimal cost network flow problems. Algorithms for the more general models usually require that the subprograms be solved successively with one or more bound changes, one or more cost coefficient changes, or both. Typical examples include the following: Rardin and Unger [155], Geoffrion and Graves [74], Kennington and Unger [117], Barr [13], and Barr, Glover, and, Klingman [14].

Exercises

Implementation of a dual simplex network code has been discussed by Helgason and Kennington [99] and by Armstrong, Klingman, and Whitman [5]. The basic idea for Exercise 3.11 is due to Dembo and Klincewicz [49], and the basic idea for 3.17 is due to Bazaraa and Jarvis [20].

CHAPTER 4

The Out-of-Kilter Algorithm for the Network Program

In this chapter we present another algorithm for solving the network program

$$\left.\begin{array}{ll}\min & \mathbf{cx} \\ \text{s.t.} & \mathbf{Ax}=\mathbf{r} \\ & \mathbf{0}\leqslant\mathbf{x}\leqslant\mathbf{u}\end{array}\right\}\mathbf{NP},$$

where **A** is a node-arc incidence matrix and $\mathbf{lr}=0$. For this chapter we assume that $\mathbf{u}>\mathbf{0}$. This alternative method, known as the *out-of-kilter algorithm*, was developed by Delbert Fulkerson in 1961. Unlike the primal simplex on a graph algorithm, the out-of-kilter algorithm is not a specialization of a more general method. This algorithm was developed specifically for **NP** and is unique in the mathematical programming literature. Although several computational comparisons of implementations of these two algorithms have appeared in the literature, there is still disagreement as to which algorithm is superior.

4.1 OPTIMALITY CONDITIONS

In this section we present a set of conditions that, when satisfied, allow termination of the out-of-kilter algorithm with the conclusion that an optimum has been attained. These conditions are a specialization of the Kuhn-Tucker conditions given in Section 2.6. Letting $[\mathcal{N}, \mathcal{A}]$ denote the network associated with **NP**, we can state these conditions as follows:

$$\mathbf{Ax} = \mathbf{r}, \qquad \mathbf{0} \leqslant \mathbf{x} \leqslant \mathbf{u}; \tag{4.1}$$

$$\pi_{F(j)} - \pi_{T(j)} - c_j = \mu_j - \lambda_j \qquad \text{for all } e_j \varepsilon \mathcal{A}; \tag{4.2}$$

$$(x_j - u_j)\mu_j = 0 \qquad \text{for all } e_j \varepsilon \mathcal{A}; \tag{4.3}$$

$$x_j \lambda_j = 0 \qquad \text{for all } e_j \varepsilon \mathcal{A}; \tag{4.4}$$

and

$$\mu_j \geqslant 0, \quad \lambda_j \geqslant 0 \qquad \text{for all } e_j \varepsilon \mathcal{A}. \tag{4.5}$$

In other words, if one obtains a set of vectors $(\mathbf{x}, \boldsymbol{\pi}, \boldsymbol{\mu}, \boldsymbol{\lambda})$ satisfying (4.1)–(4.5), $\mathbf{x}$ is an optimum for **NP**.

For the rest of this chapter we introduce the special notation $\bar{c}_j = \pi_{F(j)} - \pi_{T(j)} - c_j$. Using this notation, we show that the unknowns $\boldsymbol{\mu}$ and $\boldsymbol{\lambda}$ may be eliminated from the system given above by considering three cases.

Case 1 Suppose that, for some arc $e_j, \bar{c}_j < 0$. Then $\lambda_j > 0$ from (4.2) and (4.5). Also, $x_j = 0$ since otherwise (4.4) would not be satisfied. Furthermore $\lambda_j = -\bar{c}_j, \mu_j = 0, x_j = 0$ solves (4.2)–(4.5) for arc e_j.

Case 2 Suppose that, for some arc $e_j, \bar{c}_j > 0$. Then $\mu_j > 0$ from (4.2) and (4.5). Also, $x_j = u_j$ since otherwise (4.3) would not be satisfied. Furthermore, $\mu_j = \bar{c}_j, \lambda_j = 0, x_j = u_j$ solves (4.2)–(4.5) for arc e_j.

Case 3 Suppose that, for some arc $e_j, \bar{c}_j = 0$. Then $\mu_j = \lambda_j = 0$ solves (4.2)–(4.5).

These three cases taken together allow us to restate the optimality conditions as follows:

$$\mathbf{Ax} = \mathbf{r} \tag{4.6}$$

and

$$\text{for arc } j: \left\{ \begin{array}{ll} \bar{c}_j < 0 & \text{when } x_j = 0 \\ \bar{c}_j = 0 & \text{when } 0 \leqslant x_j \leqslant u_j \\ \bar{c}_j > 0 & \text{when } x_j = u_j \end{array} \right\}. \tag{4.7}$$

If $(\mathbf{x}, \boldsymbol{\pi})$ satisfies (4.6) and (4.7), $\mathbf{x}$ is an optimum for **NP**. Likewise, if $\mathbf{x}$ is an optimum for **NP**, there exists a vector $\boldsymbol{\pi}$ such that $(\mathbf{x}, \boldsymbol{\pi})$ satisfies (4.6) and (4.7).

4.2 BASIC STRATEGY

The out-of-kilter algorithm begins with any vector $\boldsymbol{\pi}$, called the *dual variables*, and a set of flows $\mathbf{x}$, satisfying $\mathbf{Ax}=\mathbf{r}, \mathbf{0} \leqslant \mathbf{x} \leqslant \mathbf{u}$. Such a flow can always be obtained via the introduction of artificial arcs (see Section 3.4). If $(\mathbf{x}, \boldsymbol{\pi})$ satisfies (4.6) and (4.7), we terminate with $\mathbf{x}$ an optimum; otherwise either the flows $\mathbf{x}$ or the dual variables $\boldsymbol{\pi}$ or both are changed in an attempt to satisfy (4.6) and (4.7).

Let $(\mathbf{x}, \boldsymbol{\pi})$ be given, and consider some arc e_j. If (4.7) is satisfied for e_j, we say that e_j is *in-kilter*; otherwise e_j is said to be *out-of-kilter*. The various conditions that may occur are given in Table 4.1. For expository purposes a total of nine conditions must be distinguished, five of which result in e_j being in-kilter. For each of these nine conditions we define a *kilter number*, as shown in Table 4.2. Note that the kilter numbers are all nonnegative and are the amounts of flow change required to convert an arc to the in-kilter state. Let $\mathcal{K}_j$ denote the kilter number of arc e_j. Then the kilter number of a solution, $(\mathbf{x}, \boldsymbol{\pi})$, is given by $\Sigma_{j\in\mathcal{Q}}\mathcal{K}_j$. Clearly, a solution whose kilter number is zero solves **NP**. The basic strategy of the out-of-kilter algorithm involves changing $(\mathbf{x}, \boldsymbol{\pi})$ in such a way that the following conditions hold:

1 The kilter number of an arc never increases.
2 At finite intervals the kilter number of some arc is reduced.

Table 4.1 The Possible Conditions of Arc e_j

	$\bar{c}_j < 0$	$\bar{c}_j = 0$	$\bar{c}_j > 0$
$x_j = u_j$	Out-of-kilter	In-kilter	In-kilter
$0 < x_j < u_j$	Out-of-kilter	In-kilter	Out-of-kilter
$x_j = 0$	In-kilter	In-kilter	Out-of-kilter

Table 4.2 Kilter Numbers For Possible Conditions of Arc e_j

	$\bar{c}_j < 0$	$\bar{c}_j = 0$	$\bar{c}_j > 0$
$x_j = u_j$	x_j	0	0
$0 < x_j < u_j$	x_j	0	$u_j - x_j$
$x_j = 0$	0	0	$u_j - x_j$

The out-of-kilter algorithm consists of two phases. The primal (the dual) phase makes changes in the flows (the dual variables) in an attempt to reduce the kilter number. The out-of-kilter algorithm iterates between these two phases, the primal and the dual, until

$$\sum_{e_j \varepsilon \mathscr{A}} \mathscr{K}_j = 0.$$

4.3 PRIMAL PHASE

During the primal phase the dual variables $\boldsymbol{\pi}$ are held fixed. The flows that change are always associated with arcs in a cycle. Let $C = \{s_1, e_{j_1}, s_2, e_{j_2}, \ldots, s_n, e_{j_n}, s_{n+1}\}$ be a cycle in $[\mathfrak{N}, \mathscr{A}]$, and let $O(C)$ denote the corresponding orientation sequence. Then from Proposition 3.4

$$\sum_{i=1}^{i=n} O_i(C)\mathbf{A}(j_i) = \mathbf{0}.$$

Letting

$$y_k = \begin{cases} O_i(C) & \text{if } e_{j_i} = e_k, \\ 0 & \text{otherwise,} \end{cases}$$

we have $\mathbf{Ay} = \mathbf{0}$. To help make these ideas concrete, consider the network

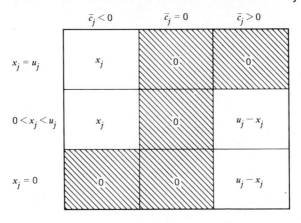

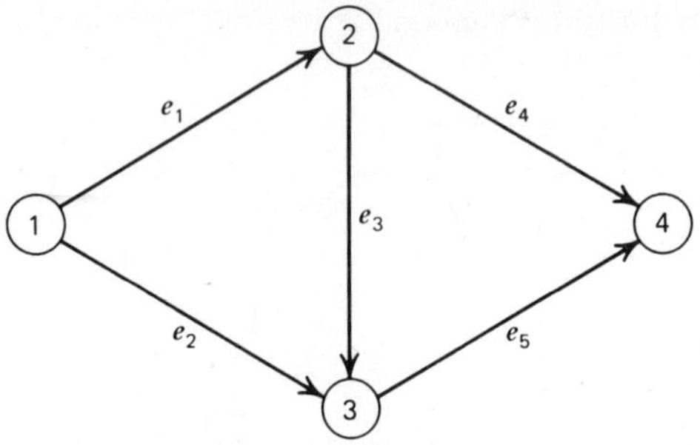

FIGURE 4.1 Sample network.

illustrated in Figure 4.1. Let $C=\{1,e_1,2,e_4,4,e_5,3,e_2,1\}$. Then $O(C)=\{1,1,-1,-1\}$, and $\mathbf{y}=[1 \quad -1 \quad 0 \quad 1 \quad -1]$, where

$$\mathbf{A}=\begin{bmatrix} 1 & 1 & & & \\ -1 & & 1 & 1 & \\ & -1 & -1 & & 1 \\ & & & -1 & -1 \end{bmatrix}.$$

Let $\mathbf{x}$ denote any set of flows such that $\mathbf{Ax}=\mathbf{r}$. Then, for any scalar Δ and any $\mathbf{y}$ corresponding to a cycle, $\hat{\mathbf{x}}=\mathbf{x}+\Delta\mathbf{y}$ satisfies $\mathbf{A}\hat{\mathbf{x}}=\mathbf{r}$. For $\Delta>0$ we say that flow is increased in the cycle C by Δ. For an arc e_k having $y_k=-1$, flow actually decreases.

Let $\mathcal{K}_k$ denote the kilter number for e_k with flow x_k, and $\hat{\mathcal{K}}_k$ denote the kilter number for e_k with flow $\hat{x}_k$. Given a set of flows $\mathbf{x}$, the primal phase attempts to find a cycle C with corresponding $\mathbf{y}$ such that, for $\hat{\mathbf{x}}=\mathbf{x}+\mathbf{y}$, $\hat{\mathcal{K}}_k \leqslant \mathcal{K}_k$ for all e_k and $\hat{\mathcal{K}}_m < \mathcal{K}_m$ for at least one arc e_m. In other words, we wish to find a cycle such that an increment of flow on this cycle will not increase any kilter number and will decrease at least one kilter number. Table 4.3 illustrates the permissible flow changes for arcs in such a cycle.

We may now state formally the problem addressed in the primal phase: *Given any out-of-kilter arc, say* e_s, *find a cycle that includes* e_s *and permits* (*by Table 4.3*) *an increment of flow*. Suppose the out-of-kilter arc selected has $\bar{c}_s>0$. Then one strategy for determining whether such a cycle exists involves the development of a special tree, say $\mathcal{T}$, rooted at $F(s)$. This tree is constructed so that, if node $i\varepsilon\mathcal{T}$, an increment of flow is permitted in the unique path in $\mathcal{T}$ from i to the root. If $T(s)\varepsilon\mathcal{T}$, the cycle is the unique path in $\mathcal{T}$ from $T(s)$ to $F(s)$ along with e_s.

We now present the algorithm for the primal phase. This algorithm is initiated with the out-of-kilter arc e_s.

Table 4.3 Permissible Flow Changes for Arc e_j

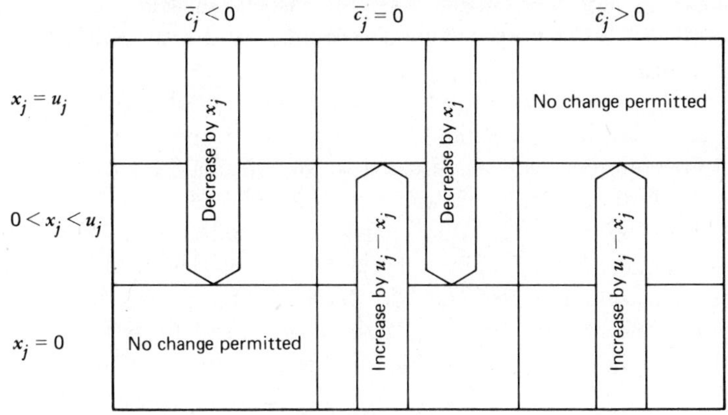

ALG 4.1 PRIMAL PHASE

0 *Initialization* If $\bar{c}_s < 0$, then $\hat{\mathfrak{N}} \leftarrow \{T(s)\}$ and $\Delta_{T(s)} \leftarrow x_s$; otherwise $\hat{\mathfrak{N}} \leftarrow \{F(s)\}$ and $\Delta_{F(s)} \leftarrow u_s - x_s$. Set $\hat{\mathcal{Q}} \leftarrow \phi$.

1 *Determine Candidates For Tree.* Let

$$\psi_1 = \{e_j : e_j \neq e_s, \bar{c}_j \geqslant 0, x_j < u_j, F(j) \not\varepsilon \hat{\mathfrak{N}}, \text{ and } T(j) \varepsilon \hat{\mathfrak{N}}\}$$

and

$$\psi_2 = \{e_j : e_j \neq e_s, \bar{c}_j \leqslant 0, x_j > 0, F(j) \varepsilon \hat{\mathfrak{N}}, \text{ and } T(j) \not\varepsilon \hat{\mathfrak{N}}\}.$$

If $\psi_1 \cup \psi_2 = \phi$, terminate with the conclusion that no cycle exists.

2 *Append New Arc to Tree.* Select $e_k \varepsilon \psi_1 \cup \psi_2$. If $e_k \varepsilon \psi_1$, set $\Delta_{F(k)} \leftarrow \min[\Delta_{T(k)}, u_k - x_k]$. If $e_k \varepsilon \psi_2$, set $\Delta_{T(k)} \leftarrow \min[\Delta_{F(k)}, x_k]$. Set $\hat{\mathfrak{N}} \leftarrow \hat{\mathfrak{N}} \cup \{F(k), T(k)\}$ and $\hat{\mathcal{Q}} \leftarrow \hat{\mathcal{Q}} \cup \{e_k\}$. If $\{F(s), T(s)\} \subset \hat{\mathfrak{N}}$, go to Step 3; otherwise go to Step 1.

3 *Breakthrough.* If $\bar{c}_s < 0$, increase the flow in the cycle by $\Delta_{F(s)}$; otherwise increase the flow in the cycle by $\Delta_{T(s)}$.

4.4 DUAL PHASE

If the primal phase terminates with the conclusion that no cycle exists, we attempt to reduce the kilter number for the solution, that is, $\Sigma_{e_j \varepsilon \mathcal{Q}} \mathcal{K}_j$, by holding the flows fixed and adjusting the dual variables.

Let $\mathcal{T}=[\hat{\mathcal{N}},\hat{\mathcal{Q}}]$ denote the tree developed in the primal phase. In the dual phase we reduce each dual variable associated with a node in $\hat{\mathcal{N}}$ by the amount $\theta>0$ (where θ will be determined by the algorithm to follow). Let $e_r \varepsilon \hat{\mathcal{Q}}$. Suppose $\hat{\pi}_{F(r)}=\pi_{F(r)}-\theta$ and $\hat{\pi}_{T(r)}=\pi_{T(r)}-\theta$. Then $\hat{\pi}_{F(r)}-\hat{\pi}_{T(r)}-c_r$ $=\pi_{F(r)}-\pi_{T(r)}-c_r$. Therefore reducing each dual variable corresponding to a node in $\mathcal{T}$ does not affect the kilter numbers for arcs in $\mathcal{T}$. The only arcs whose kilter numbers will be affected by such a change are those that are incident on both a node in $\hat{\mathcal{N}}$ and a node in $\mathcal{N}-\hat{\mathcal{N}}$.

The objective of the dual phase is to determine the permissible decrease in the dual variables associated with nodes in $\mathcal{T}$ that will not increase any kilter number and will decrease at least one kilter number. The permissible change for arc e_j with $T(j)\varepsilon\hat{\mathcal{N}}$ and $F(j)\varepsilon\mathcal{N}-\hat{\mathcal{N}}$ is given in Table 4.4, while the permissible changes for arcs e_j with $F(j)\varepsilon\hat{\mathcal{N}}$ and $T(j)\varepsilon\mathcal{N}-\hat{\mathcal{N}}$ are given in Table 4.5. The shaded areas in both tables correspond to conditions that cannot occur.

The dual phase algorithm is initiated with the tree $\mathcal{T}=[\hat{\mathcal{N}},\hat{\mathcal{Q}}]$, developed in the primal phase.

Table 4.4 Permissible Decrease in $\pi_{T(j)}$

	$\bar{c}_j<0$	$\bar{c}_j=0$	$\bar{c}_j>0$
$x_j=u_j$	$-\bar{c}_j$	∞	∞
$0<x_j<u_j$	$-\bar{c}_j$	(shaded)	(shaded)
$x_j=0$	$-\bar{c}_j$	(shaded)	(shaded)

ALG 4.2 DUAL PHASE

1 *Determine Arcs Incident on $\mathcal{T}$.* Let

$$\psi_1=\{e_j : T(j)\varepsilon\hat{\mathcal{N}}, F(j)\not\varepsilon\hat{\mathcal{N}}, \text{ and } \bar{c}_j<0\},$$

and

$$\psi_2=\{e_j : T(j)\not\varepsilon\hat{\mathcal{N}}, F(j)\varepsilon\hat{\mathcal{N}}, \text{ and } \bar{c}_j>0\}.$$

2 *Determine Maximum Permissible Change.* Set

$$\theta \leftarrow \min_{e_j \varepsilon \psi_1 \cup \psi_2} \{ |\bar{c}_j| \}.$$

3 *Reduce Duals.* Set $\pi_i \leftarrow \pi_i - \theta$ for all $i \varepsilon \hat{\mathfrak{N}}$.

Note that an arc $e_k \varepsilon \psi_1 \cup \psi_2$, which at Step 2 of the dual phase had $|\bar{c}_k| = \theta$, will have $\bar{c}_k = 0$ after Step 3. Hence, if $|\bar{c}_s| \neq \theta$ and the primal phase were repeated, the same tree as before could be constructed with the addition of e_k and the new endpoint. If $|\bar{c}_s| = \theta$, then e_s will be in-kilter.

Table 4.5 Permissible Decrease in $\pi_{F(j)}$

	$\bar{c}_j < 0$	$\bar{c}_j = 0$	$\bar{c}_j > 0$
$x_j = u_j$			$\bar{c}_j$
$0 < x_j < u_j$			$\bar{c}_j$
$x_j = 0$	∞	∞	$\bar{c}_j$

4.5 OUT-OF-KILTER ALGORITHM

The out-of-kilter algorithm may now be stated as follows.

ALG 4.3 OUT-OF-KILTER ALGORITHM

0 *Initialization.* Let $\mathbf{x}$ be any set of flows with $\mathbf{Ax} = \mathbf{r}, \mathbf{0} \leqslant \mathbf{x} \leqslant \mathbf{u}$. (A feasible solution can always be obtained via the use of artificial variables as shown in Section 3.4.) Let $\boldsymbol{\pi}$ be any vector of dual variables.

1 *Find Out-of-Kilter Arc.* Let e_s be any out-of-kilter arc. If all arcs are in-kilter, terminate with $\mathbf{x}$ an optimum.

2 *Primal Phase.* Execute ALG 4.1 with out-of-kilter arc e_s. If ALG 4.1 terminates with the conclusion that no cycle exists, go to Step 3; otherwise return to Step 1.

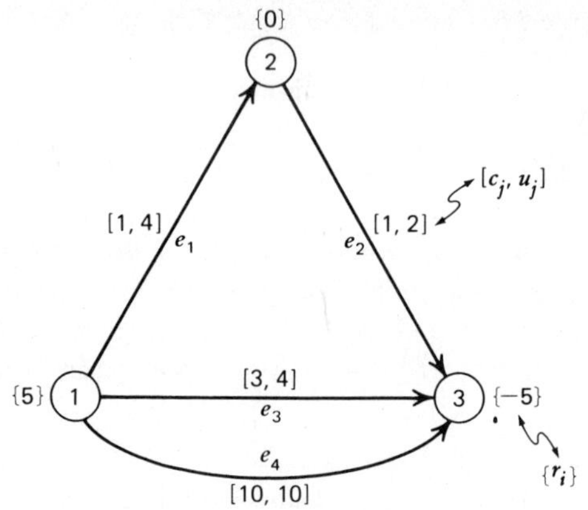

FIGURE 4.2 Sample network for out-of-kilter algorithm.

3 *Dual Phase.* Execute ALG 4.2 with the tree developed in Step 2. If e_s is out-of-kilter, return to Step 2; otherwise return to Step 1.

Consider again Example 2.1. This problem is illustrated in Figure 4.2. The variable x_5 has been omitted from the network since the sum of the three equality constraints implies that $x_5=0$. In our previous discussions concerning the problem in this example the column associated with x_5 was required to ensure full row rank of the constraint matrix. The out-of-kilter algorithm has no such requirement, and hence x_5 may be dropped from the problem.

ALG 4.3: **0** (Initialization). Let $[x_1 \ \ x_2 \ \ x_3 \ \ x_4]=[0 \ \ 0 \ \ 4 \ \ 1]$, and let $[\pi_1 \ \ \pi_2 \ \ \pi_3]=[5 \ \ 0 \ \ -1]$.

1 (Find Out-of-Kilter Arc). $\bar{c}_4=-4, x_4=1, \Rightarrow e_4$ out-of-kilter.

ALG 4.1: **0** (Initialization). $\hat{\mathfrak{N}}=\{3\}, \Delta_3=1, \hat{\mathcal{Q}}=\phi$.

1 (Candidate Arcs). $\psi_1=\{e_2\}, \psi_2=\phi$.

2 (Append New Arc). $e_k=e_2, \Delta_2 \leftarrow \min[1,2]=1$, $\hat{\mathfrak{N}}=\{3,2\}$, and $\hat{\mathcal{Q}}=\{e_2\}$. (The current tree is illustrated in Figure 4.3*a*.)

1 (Candidate Arcs). $\psi_1=\{e_1\}, \psi_2=\phi$.

2 (Append New Arc). $e_k=e_1, \Delta_1=\min[1,4]=1$, $\hat{\mathfrak{N}}=\{3,2,1\}$, and $\hat{\mathcal{Q}}=\{e_2,e_1\}$.

3 (Breakthrough). Cycle is $\{1,e_1,2,e_2,3,e_4,1\}$, and new flows are $[x_1 \ \ x_2 \ \ x_3 \ \ x_4]=[1 \ \ 1 \ \ 4 \ \ 0]$.

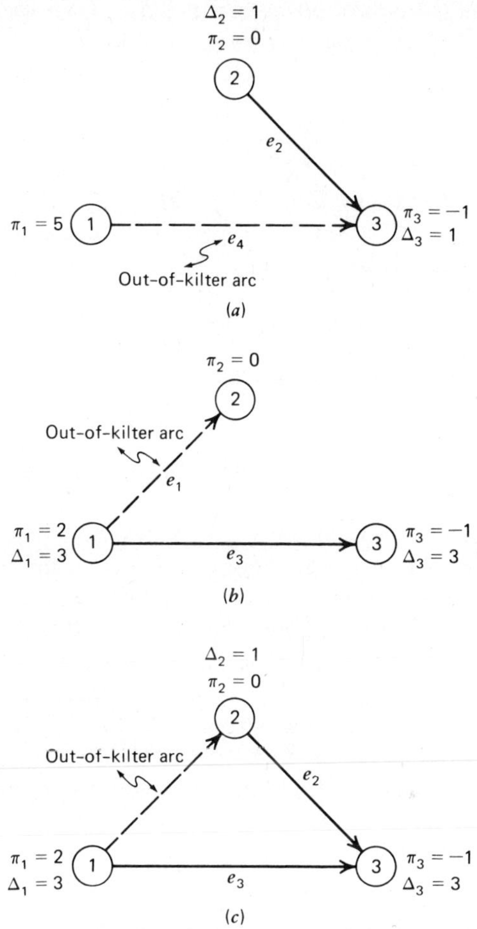

FIGURE 4.3 Trees associated with various phases of sample problem.

ALG 4.3: **1** (Find Out-of-Kilter Arc). $\bar{c}_1=4, x_1=1, \Rightarrow e_1$ out of-kilter.

ALG 4.1: **0** (Initialization). $\hat{\mathcal{N}}=\{1\}, \Delta_1=3, \hat{\mathcal{A}}=\phi$.

1 (Candidate Arcs). $\psi_1=\psi_2=\phi$.

ALG 4.2: **1** (Determine Arc Incident on Tree). $\psi_1=\phi, \psi_2=\{e_1,e_3\}$.

2 (Determine Maximum Change). $\theta=\min[|\bar{c}_1|, |\bar{c}_3|]=\min[4,3]=3$.

3 (Reduce Duals). $\pi_1=5-3=2$.

ALG 4.1: **0** (Initialization). $\hat{\mathfrak{N}}=\{1\}, \Delta_1=3, \hat{\mathcal{Q}}=\phi$.
1 (Candidate Arcs). $\psi_1=\phi, \psi_2=\{e_3\}$.
2 (Append New Arc). $e_k=e_3$, $\Delta_3=\min[3,4]=3$, $\hat{\mathfrak{N}}=\{1,3\}$, and $\hat{\mathcal{Q}}=\{e_3\}$. (The current tree is illustrated in Figure 4.3*b*.)
1 (Candidate Arcs). $\psi_1=\{e_2\}, \psi_2=\phi$.
2 (Append New Arc). $e_k=e_2$, $\Delta_2=\min[3,1]=1$, $\hat{\mathfrak{N}}=\{1,3,2\}$, and $\hat{\mathcal{Q}}=\{e_3,e_2\}$. (The current tree is illustrated in Figure 4.3*c*.)
3 (Breakthrough). Cycle is $\{2,e_2,3,e_3,1,e_1,2\}$, and new flows are $[x_1 \quad x_2 \quad x_3 \quad x_4]=[2 \quad 2 \quad 3 \quad 0]$.

ALG 4.3: **1** (Find Out-of-Kilter Arc). All arcs in-kilter!

4.6 FINITENESS OF OUT-OF-KILTER ALGORITHM

To prove finiteness we assume that **r**, **u**, **c**, **x**, and $\boldsymbol{\pi}$ are integer vectors. Because of this integrality assumption, each time we achieve breakthrough and change flows in the primal phase, the kilter number for e_s is reduced by at least 1. Each time we change the dual variables in the dual phase, either the tree developed in the next primal phase has at least one more node (and arc) or the system kilter number is reduced by at least 1. In applying the primal and dual phases repeatedly to a single out-of-kilter arc, we will be unable to enlarge the tree indefinitely, for eventually a spanning tree for the graph will be produced in the primal phase. The spanning tree together with e_s must contain a cycle. Thus, in a finite number of steps, $\Sigma_{e_j \in \mathcal{Q}} \mathcal{K}_j$ can be reduced to 0.

EXERCISES

4.1 Solve the problem described in the Table E4.1 using the out-of-kilter algorithm. Begin with the dual variables and initial flows shown in the table. In Step 1 of ALG 4.3, always select the out-of-kilter arc with the smallest arc number, (i.e., if arc 1 is out-of-kilter, select it; if not, try 2, 3, etc.).

4.2 Solve Exercise 2.8 using the out-of-kilter algorithm (ALG 4.3). Begin with all the dual variables equal to 0.

4.3 Solve Exercise 2.9 using the out-of-kilter algorithm (ALG 4.3). Begin with all the dual variables equal to 10.

Table E4.1 Problem Description

Node Number	Requirements	Dual Variables
1	10	10
2	0	0
3	0	0
4	−10	10

Arc Number	"From" Node	"To" Node	Unit Cost	Arc Capacity	Initial Flows
1	1	2	1	10	5
2	1	3	4	5	5
3	2	3	2	9	0
4	2	4	6	10	5
5	3	4	3	8	5

4.4 Solve the transportation problem of Exercise 2.10 using the out-of-kilter algorithm (ALG 4.3). Begin with all the dual variables equal to −5.

4.5 Solve the assignment problem of Exercise 2.11 using the out-of-kilter algorithm (ALG 4.3). Begin with all the dual variables equal to 2.

4.6 Develop a data structure for computer implementation of ALG 4.3.

4.7 Extend the ideas of the out-of-kilter algorithm to solve **LP**. What are the advantages and disadvantages of this algorithm as compared to the primal simplex method?

4.8 Tables 4.4 and 4.5 could be changed to read as shown in Tables E4.8*a* and E4.8*b*.

Table E4.8*a*

	$\bar{c}_j < 0$	$\bar{c}_j = 0$	$\bar{c}_j > 0$
$x_j = u_j$	∞	∞	∞
$0 < x_j < u_j$	$-\bar{c}_j$		
$x_j = 0$	$-\bar{c}_j$		

, and

Table E4.8*b*

	$\bar{c}_j < 0$	$\bar{c}_j = 0$	$\bar{c}_j > 0$
$x_j = u_j$			$\bar{c}_j$
$0 < x_j < u_j$			$\bar{c}_j$
$x_j = 0$	∞	∞	∞

Would these changes cause any problems with the out-of-kilter algorithm? If so, what difficulties would arise? If not, could one expect convergence to be speeded by these changes?

4.9 Devise an algorithm to obtain a good set of dual variables with which to begin ALG 3.3.

NOTES AND REFERENCES

Section 4.1

The necessary and sufficient conditions used to obtain the optimality conditions were first developed by Kuhn and Tucker [128] in 1950.

Sections 4.2, 4.3, 4.4, 4.5, and 4.6

The out-of-kilter algorithm was developed by Fulkerson [70]. A presentation similar to that in Ref. [70] can be found in Ford and Fulkerson [65]. Our own presentation is a combination of the ideas of Bazaraa and Jarvis [20] and Aashtiani and Magnanti [1]. A different point of view for this algorithm is given by Barr, Glover, and Klingman [11]. Clasen [34] developed the first out-of-kilter code. Computational experience comparing out-of-kilter codes with primal simplex codes can also be found in Barr, Glover, and Klingman [11], Glover and Klingman [86], and Hatch [95].

CHAPTER 5

The Simplex Method for the Generalized Network Problem

The *generalized flow problem* is a special case of LP in which each column of the constraint matrix has at most two nonzero entries. Formally, the problem may be stated as follows:

$$\left.\begin{array}{ll} \min & \mathbf{cx} \\ \text{s.t.} & \mathbf{Gx}=\mathbf{r} \\ & \mathbf{0}\leqslant \mathbf{x}\leqslant \mathbf{u} \end{array}\right\} \quad \text{GP},$$

where $\mathbf{G}$ is an $\bar{I}\times\bar{J}$ matrix such that, for each j, $G_{ij}\neq 0$ for at most two $i\varepsilon\{1,\ldots,\bar{I}\}$. In many practical applications the two nonzero entries in each column are of opposite sign. This enables one to model flows that increase or decrease, such as financial models involving interest rates or transmission problems that incur losses. Note that every two-constraint LP is a GP. For our discussion we assume that the rank of $\mathbf{G}$ is $\bar{I}$. If for some problem the constraint matrix does not have full row rank, artificials may be added to obtain GP in the prescribed form.

We call GP a flow problem because one may associate a graph with any GP. This graph consists of undirected arcs, in contrast to the network graph used in Chapters 3 and 4. Let the arcs be defined by

$$e_j = \begin{cases} \{i,k\}, & \text{if } G_{ij} \neq 0,\ G_{kj} \neq 0, \text{ and } i<k, \\ \{i,0\}, & \text{if } G_{ij} \neq 0 \text{ and } G_{kj}=0 \text{ for all } k \neq i. \end{cases}$$

Letting $\mathcal{A}=\{e_1,\ldots,e_{\bar{J}}\}$ and $\mathcal{N}=\{1,\ldots,\bar{I}\}$, we can write the associated graph as $\mathcal{G}=[\mathcal{N},\mathcal{A}]$. Arcs of the form $\{i,0\}$ are called *root arcs* with corresponding *root node i*.

Consider the following example:

$$\begin{array}{llllllll}
\text{min} & x_1 & + & 2x_2 & + & 5x_3 - x_4 & - 2x_5 + x_6 & \\
\text{s.t.} & 2x_1 & + & x_2 & + & 9x_3 & & = 31 \\
 & 3x_1 & & & & -2x_4 & + \ x_5 & = -1 \\
 & & - & x_2 & & & + x_6 & = -2 \\
 & & & & & x_3 - 5x_4 & + x_6 & = -7
\end{array}$$

$$0 \leqslant x_1 \leqslant 1, \qquad 0 \leqslant x_2 \leqslant 2, \qquad 0 \leqslant x_3 \leqslant 3,$$
$$0 \leqslant x_4 \leqslant 4, \qquad 0 \leqslant x_5 \leqslant 5, \qquad 0 \leqslant x_6 \leqslant 6.$$

The corresponding graph is illustrated in Figure 5.1*a*. Arc e_5 is called a root arc, and node 2 will be referred to as a root node. We will also illustrate the network with the arc multipliers, as shown in Figure 5.1*b*.

In this chapter we present a specialization of the primal simplex algorithm that exploits the underlying graphical structure of GP. As with the primal simplex method on a graph of Chapter 3, the simplex operations can be carried out on the graph $\mathcal{G}$, thereby eliminating the need for matrix operations.

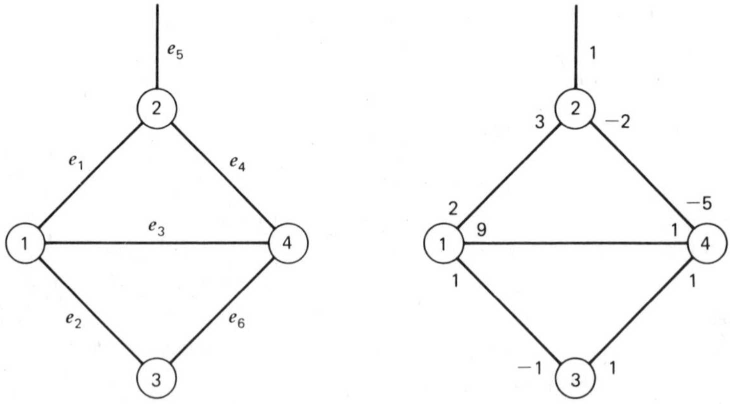

(*a*) (*b*)

FIGURE 5.1 Sample graphs. (*a*) Sample graph with arc numbers. (*b*) Sample graph with arc multipliers.

5.1 FUNDAMENTAL ALGEBRAIC RESULTS

In this section we present the fundamental algebraic results used in the specialization of the primal simplex algorithm for GP. Consider an $n \times n$ matrix $\mathbf{H}$ of the form

$$\mathbf{H}=\left[\begin{array}{c|c|c|c|c|c} a_1 & & & & & b_n \\ b_1 & a_2 & & & & \\ & b_2 & a_3 & & & \\ & & & \ddots & & \\ & & & & a_{n-1} & \\ & & & & b_{n-1} & a_n \end{array}\right].$$

Throughout this section we assume that the rank of $\mathbf{H}$ is n, and we develop formulae for the analytical solution to the two systems, $\mathbf{Hx}=\alpha\mathbf{e}^1$ and $\boldsymbol{\pi}\mathbf{H}=\mathbf{c}$, for any scalar α and any n-component vector $\mathbf{c}$. These formulae will be developed by the application of Cramer's rule.

We begin by finding the determinant of $\mathbf{H}$.

Proposition 5.1

$$\mathrm{Det}(\mathbf{H})=a_1a_2\cdots a_n+(-1)^{n-1}b_1b_2\cdots b_n.$$

Proof. Using the Laplace expansion of the determinant of $\mathbf{H}$ along the first row, we obtain

$$\mathrm{Det}(\mathbf{H})=a_1\,\mathrm{Det}\left\{\left[\begin{array}{c|c|c|c|c} a_2 & & & & \\ b_2 & a_3 & & & \\ & b_3 & a_4 & & \\ & & & \ddots & \\ & & & & a_n \end{array}\right]\right\}$$

$$+(-1)^{n-1}b_n\,\mathrm{Det}\left\{\left[\begin{array}{c|c|c|c} b_1 & a_2 & & \\ & b_2 & & \\ & & \ddots & \\ & & b_{n-2} & a_{n-1} \\ & & & b_{n-1} \end{array}\right]\right\}.$$

Using the result that the determinant of a triangular matrix is the product of the diagonal elements, we obtain

$$=a_1a_2\cdots a_n+(-1)^{n-1}b_1b_2\cdots b_n. \qquad \blacksquare$$

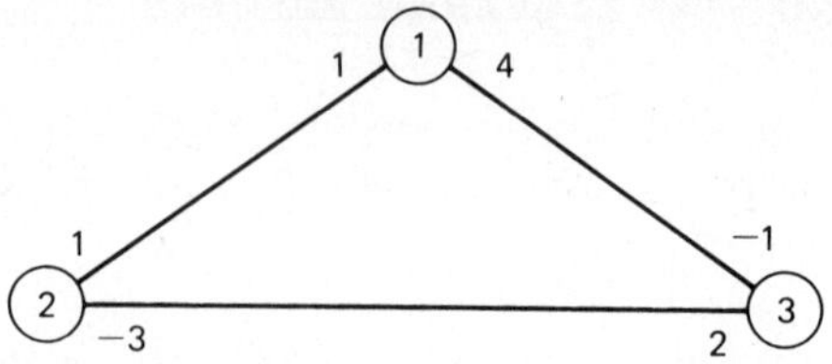

FIGURE 5.2 Example of a generalized network.

Proposition 5.1 can be used to show that the columns associated with the generalized graph illustrated in Figure 5.2 are linearly independent, that is,

$$\operatorname{Det}\left\{\begin{bmatrix} 1 & 4 & \\ & -1 & 2 \\ 1 & & -3 \end{bmatrix}\right\} = (1)(-3)(-1) + (-1)^2(1)(2)(4) = 11.$$

Hence the columns of **G** corresponding to a cycle in an undirected graph may be linearly independent, whereas the columns of a node-arc incidence matrix corresponding to a cycle in a directed graph are necessarily linearly dependent (see Corollary 3.5).

Corollary 5.2

If any $b_j = 0$ for $1 \leqslant j \leqslant n$, then $\operatorname{Det}(\mathbf{H}) = a_1 a_2 \cdots a_n$.

Consider the matrix formed from **H** which has the first row replaced by the vector **c**, that is, let

$$\mathbf{H}_1 = \left[\begin{array}{c|c|c|c|c|c} c_1 & c_2 & c_3 & \cdots & c_{n-1} & c_n \\ b_1 & a_2 & & & & \\ & b_2 & a_3 & & & \\ & & & \ddots & & \\ & & & & a_{n-1} & \\ & & & & b_{n-1} & a_n \end{array}\right].$$

Given $\mathbf{H}_1$, we know by Cramer's rule that, for the system $\mathbf{xH} = \mathbf{c}$, $x_1 = \operatorname{Det}(\mathbf{H}_1)/\operatorname{Det}(\mathbf{H})$. The determinant of $\mathbf{H}_1$ is given by the following proposition.

Proposition 5.3

$$\operatorname{Det}(\mathbf{H}_1) = c_1 a_2 a_3 \cdots a_n - b_1 c_2 a_3 a_4 \cdots a_n + b_1 b_2 c_3 a_4 \cdots a_n$$
$$+ \cdots + (-1)^{n-1} b_1 b_2 \cdots b_{n-1} c_n$$
$$= \sum_{i=1}^{n} (-1)^{i-1} \left[\prod_{j=1}^{i-1} b_j \right] c_i \left[\prod_{j=i+1}^{n} a_j \right].$$

(We use the convention that $\Pi_{j=j_i}^{j_k} p_j = 1$ when $j_i > j_k$.)

Proof. The Laplace expansion of the determinant of $\mathbf{H}_1$ along the first row is given as follows:

$$\operatorname{Det}(\mathbf{H}_1) = c_1 \operatorname{Det}\left\{ \begin{bmatrix} a_2 & & & & \\ b_2 & a_3 & & & \\ & b_3 & a_4 & & \\ & & & \ddots & \\ & & & & a_n \end{bmatrix} \right\}$$

$$+ \sum_{i=2}^{i=n-1} (-1)^{i-1} c_i \operatorname{Det}\left\{ \begin{bmatrix} b_1 & a_2 & & & & & & & & \\ & b_2 & a_3 & & & & & & & \\ & & b_3 & & & & & & & \\ & & & \ddots & & & & & & \\ & & & & b_{i-1} & & & & & \\ & & & & & a_{i+1} & & & & \\ & & & & & b_{i+1} & a_{i+2} & & & \\ & & & & & & b_{i+2} & a_{i+3} & & \\ & & & & & & & & \ddots & \\ & & & & & & & & & a_n \end{bmatrix} \right\}$$

$$+ (-1)^{n-1} c_n \operatorname{Det}\left\{ \begin{bmatrix} b_1 & a_2 & & & \\ & b_2 & & & \\ & & \ddots & & \\ & & & b_{n-2} & a_{n-1} \\ & & & & b_{n-1} \end{bmatrix} \right\}$$

$$= c_1 a_2 a_3 \cdots a_n - b_1 c_2 a_3 a_4 \cdots a_n + b_1 b_2 c_3 a_4 a_5 \cdots a_n - \cdots +$$
$$(-1)^{n-1} b_1 b_2 \cdots b_{n-1} c_n.$$

■

Using Propositions 5.2 and 5.3, we may now solve $\pi\mathbf{H}=\mathbf{c}$. Consider the following.

Proposition 5.4

If $a_j \neq 0$ and $b_j \neq 0$ for $1 \leqslant j \leqslant n$ and $\text{Det}(\mathbf{H}) \neq 0$, the solution to the system

$$\pi\mathbf{H}=\mathbf{c}$$

is given by

$$\left\{\begin{aligned} \pi_1 &= \frac{c_1/a_1 + w_1c_2/a_2 + w_1w_2c_3/a_3 + \cdots + w_1w_2\cdots w_{n-1}c_n/a_n}{1 - w_1w_2\cdots w_n} \\ &= \frac{\sum_{i=1}^{n}\left\{\left[\prod_{j=1}^{i-1} w_j\right]c_i/a_i\right\}}{1 - w_1w_2\cdots w_n} \\ \pi_{i+1} &= \frac{c_i - a_i\pi_i}{b_i} \qquad \text{for} \quad i=1,\ldots,n-1 \end{aligned}\right\}, \tag{5.1}$$

where $w_i = -b_i/a_i$ for $i=1,\ldots,n$.

Proof. By Cramer's rule $\pi_1 = \text{Det}(\mathbf{H}_1)/\text{Det}(\mathbf{H})$. By Propositions 5.1 and 5.3

$$\pi_1 = \frac{\sum_{i=1}^{n}(-1)^{i-1}\left[\prod_{j=1}^{i-1} b_j\right]c_i\left[\prod_{j=i+1}^{n} a_j\right]}{a_1a_2\cdots a_n + (-1)^{n-1}b_1b_2\cdots b_n}$$

$$= \frac{\sum_{i=1}^{n}(-1)^{i-1}\left[\prod_{j=1}^{i-1} b_j\right]c_i\left[\prod_{j=i+1}^{n} a_j\right]}{a_1a_2\cdots a_n - (-b_1)(-b_2)\cdots(-b_n)}.$$

Multiplying both the numerator and the denominator by $[a_1a_2\cdots a_n]^{-1}$, we obtain π_1 in the form given by (5.1). When backward substitution is used, the other equations given in (5.1) are immediate. ∎

Consider the example presented in Figure 5.3. By Proposition 5.4

$$\pi_1 = \frac{c_1/a_1 + w_1 c_2/a_2 + w_1 w_2 c_3/a_3}{1 - w_1 w_2 w_3}$$

$$= \frac{2/1 + (-2)(0)/(-3) + (-2)(\frac{1}{3})(-2)/1}{1-(-2)(\frac{1}{3})(-1)}$$

$$= 10,$$

$$\pi_2 = \frac{c_1 - a_1 \pi_1}{b_1} = \frac{2-(1)(10)}{2} = -4,$$

$$\pi_3 = \frac{c_2 - a_2 \pi_2}{b_2} = \frac{0-(-3)(-4)}{1} = -12.$$

Therefore, if the graph illustrated in Figure 5.3 corresponds to a basis, the π_i's correspond to the dual variables.

Consider now the matrix formed from $\mathbf{H}$ by replacing the first column of $\mathbf{H}$ by $\alpha \mathbf{e}^1$, that is, let

$$\mathbf{H}_2 = \left[\begin{array}{c|c|c|c|c|c|c} \alpha & & & & & & b_n \\ & a_2 & & & & & \\ & b_2 & a_3 & & & & \\ & & b_3 & a_4 & & & \\ & & & & \ddots & & \\ & & & & & a_{n-1} & \\ & & & & & b_{n-1} & a_n \end{array}\right].$$

We know that, for the system $\mathbf{Hx} = \alpha \mathbf{e}^1$, $x_1 = \mathrm{Det}(\mathbf{H}_2)/\mathrm{Det}(\mathbf{H})$.

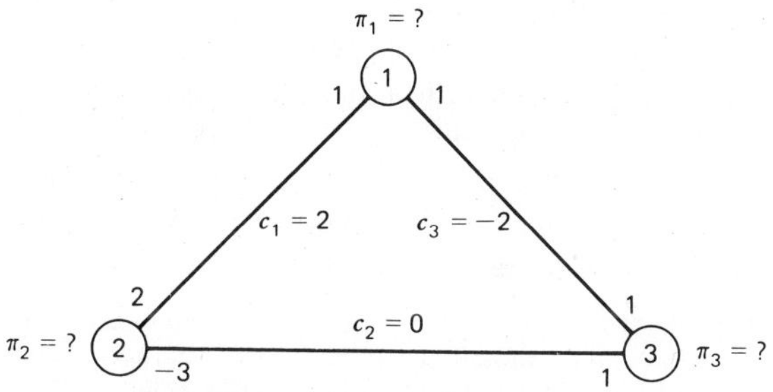

FIGURE 5.3 Sample generalized network.

Proposition 5.5

If $a_j \neq 0$ and $b_j \neq 0$ for all $1 \leqslant j \leqslant n$ and $\mathrm{Det}(\mathbf{H}) \neq 0$, the solution to the system $\mathbf{Hx} = \alpha \mathbf{e}^1$ for any α is given by

$$\left\{ \begin{array}{lll} x_1 & = \dfrac{\alpha / a_1}{1 - w_1 w_2 \cdots w_n} & \\ x_{i+1} & = \dfrac{-b_i x_i}{a_{i+1}} & \text{for} \quad i = 1, \ldots, n-1 \end{array} \right\}, \tag{5.2}$$

where $w_i = -b_i / a_i$ for $i = 1, \ldots, n$.

Proof. By Cramer's rule $x_1 = \mathrm{Det}(\mathbf{H}_2)/\mathrm{Det}(\mathbf{H})$. By Corollary 5.2 $\mathrm{Det}(\mathbf{H}_2) = \alpha a_2 \cdots a_n$. Then from Proposition 5.1

$$x_1 = \frac{\alpha a_2 \cdots a_n}{a_1 a_2 \cdots a_n - (-b_1)(-b_2) \cdots (-b_n)}.$$

Multiplying both the numerator and the denominator by $[a_1 a_2 \cdots a_n]^{-1}$, we obtain x_1 in the form given by (5.2). When backward substitution is used, the other equations given in (5.2) are immediate. ■

5.2 BASIS CHARACTERIZATION

In this section we show that a basis for GP has a graph-theoretic structure which can be exploited when applying the simplex steps.

The idea of connectedness introduced in Section 3.1 carries over in a natural manner from directed to undirected graphs. We leave the details as an exercise to the reader (e.g., both path and cycle must be appropriately redefined). As a result Propositions 3.1, 3.2, 3.6–3.9, and 3.11 are also immediately available for undirected graphs.

We now extend the definition of a connected graph to include a graph whose node set is a singleton. A *component* of a graph $\mathcal{G}$, is a maximal connected subgraph of $\mathcal{G}$. Any graph may be partitioned into a set of components. Furthermore, this partitioning is unique, and some components may be one-node subgraphs.

Let the matrix $\mathbf{B}$ be a basis for GP, and let $\mathcal{A}_B = \{e_j : \mathbf{G}(j) \text{ is a column of } \mathbf{B}\}$. Then $\mathcal{G}_B = [\mathcal{N}, \mathcal{A}_B]$ denotes the graph associated with the basis $\mathbf{B}$. Denote the components of $\mathcal{G}_B$ by $\mathcal{G}^1 = [\mathcal{N}^1, \mathcal{A}^1], \ldots, \mathcal{G}^p = [\mathcal{N}^p, \mathcal{A}^p]$, where it is understood that, if $\mathcal{G}_B$ is connected, $p = 1$. In this section we show that

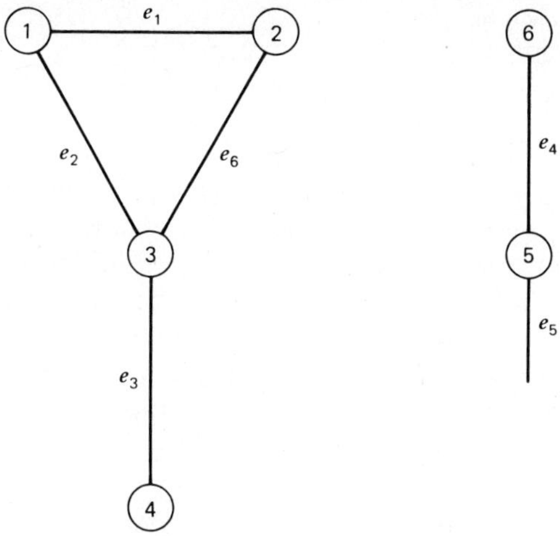

FIGURE 5.4 Graphical structure.

each of the components $\mathcal{G}^1,\ldots,\mathcal{G}^p$ either is a rooted tree or contains exactly one cycle.

Consider the following 6×6 nonsingular matrix, which corresponds to a basis for some generalized flow problem:

$$\mathbf{B}=\overbrace{\begin{bmatrix} 1 & 1 & & & & \\ -2 & & & & & 2 \\ & -3 & 2 & & & 4 \\ & & 5 & & & \\ & & & -1 & 1 & \\ & & & -1 & & \end{bmatrix}}^{\text{arcs: } e_1\; e_2\; e_3\; e_4\; e_5\; e_6} \begin{matrix} \text{nodes} \\ 1 \\ 2 \\ 3 \\ 4 \\ 5 \\ 6 \end{matrix}$$

The graphical structure corresponding to $\mathbf{B}$ is illustrated in Figure 5.4, where $\mathcal{G}^1=[\{1,2,3,4\},\{e_1,e_2,e_3,e_6\}]$ and $\mathcal{G}^2=[\{5,6\},\{e_4,e_5\}]$.

Proposition 5.6

The graph $\mathcal{G}^t$ for each $t=1,\ldots,p$ can have at most one root arc.

Proof. Let t be any element of $\{1,\ldots,p\}$. Suppose $\mathcal{G}^t$ has two root arcs with root nodes l_1 and l_2. We will show that the columns of $\mathbf{B}$ are linearly dependent. If $l_1=l_2$, the dependency is obvious. If $l_1\neq l_2$, there exists a path

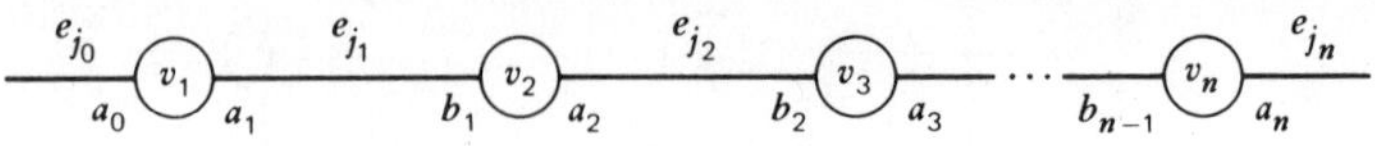

FIGURE 5.5 Path connecting two root nodes.

connecting l_1 to l_2 since $\mathcal{G}^t$ is connected. This path is illustrated in Figure 5.5, where $v_1 = l_1$, $v_n = l_2$, and the nonzero coefficients for $\mathbf{G}(j_k)$ are given by a_k and b_k for rows (nodes) v_k and v_{k+1}, respectively. Consider the columns of $\mathbf{B}$ corresponding to the arcs in the path connecting l_1 to l_2 together with columns corresponding to the root arcs. Removing all rows that have only zeros, we equate a column for one root arc to a linear combination of the remaining columns. This system can be placed in the form

$$\begin{bmatrix} a_1 & & & & & \\ b_1 & a_2 & & & & \\ & b_2 & a_3 & & & \\ & & & \ddots & & \\ & & & & a_{n-1} & \\ & & & & b_{n-1} & a_n \end{bmatrix} \begin{bmatrix} \lambda_1 \\ \lambda_2 \\ \lambda_3 \\ \vdots \\ \lambda_{n-1} \\ \lambda_n \end{bmatrix} = \begin{bmatrix} a_0 \\ 0 \\ 0 \\ \vdots \\ 0 \\ 0 \end{bmatrix}. \tag{5.3}$$

The solution to 5.3 is given by

$$\left\{ \begin{aligned} \lambda_1 &= \frac{a_0}{a_1} \\ \lambda_i &= \frac{-b_{i-1}\lambda_{i-1}}{a_i} \qquad \text{for} \quad i = 2,\ldots,n \end{aligned} \right\}.$$

Therefore one column of $\mathbf{B}$, that is, $\mathbf{G}(j_0)$, may be expressed as a linear combination of other columns of $\mathbf{B}$, namely, $\mathbf{G}(j_1),\ldots,\mathbf{G}(j_n)$, contradicting the linear independence of the basis. Therefore $\mathcal{G}^t$ cannot contain two root arcs. ■

Proposition 5.7

The graph $\mathcal{G}^t$ for $t = 1,\ldots,p$ cannot contain both a root arc and a cycle.

Proof. Let t be any element of $\{1,\ldots,p\}$. Suppose $\mathcal{G}^t$ contains a root arc with corresponding root node l and a cycle. We will show that the columns of $\mathbf{B}$ are linearly dependent, contradicting $\mathbf{B}$ a basis.

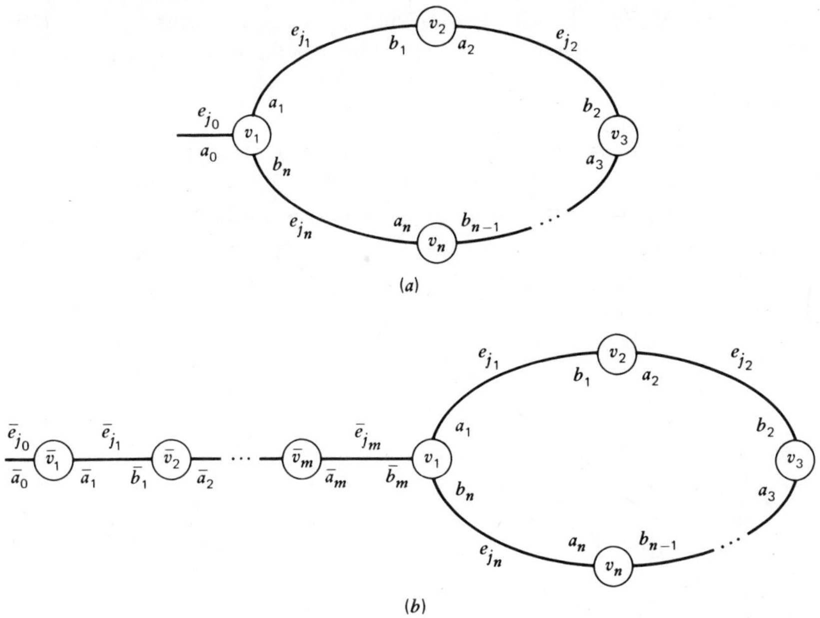

FIGURE 5.6 Connected graphs having both a root arc and a cycle. (*a*) Cycle with root node in cycle. (*b*) Cycle and path from root to cycle.

Case 1 Suppose l is a node in the cycle. This cycle is illustrated in Figure 5.6*a*, where $v_1 = v_{n+1} = l$ and the nonzero coefficients for $\mathbf{G}(j_k)$ are given by a_k and b_k for rows v_k and v_{k+1}, respectively. Consider the columns of $\mathbf{B}$ corresponding to the arcs in the cycle together with the column corresponding to the root arc. Removing all rows that have only zeros, we equate the column for the root arc to a linear combination of the remaining columns. The system can be placed in the form

$$\left[\begin{array}{c|c|c|c|c|c}
a_1 & & & & & b_n \\
b_1 & a_2 & & & & \\
 & b_2 & a_3 & & & \\
 & & & \ddots & & \\
 & & & & a_{n-1} & \\
 & & & & b_{n-1} & a_n
\end{array}\right]
\begin{bmatrix} \lambda_1 \\ \lambda_2 \\ \lambda_3 \\ \vdots \\ \lambda_{n-1} \\ \lambda_n \end{bmatrix}
=
\begin{bmatrix} a_0 \\ 0 \\ 0 \\ \vdots \\ 0 \\ 0 \end{bmatrix}. \tag{5.4}$$

Let λ^* denote the solution to (5.4) as given by Proposition 5.5. Since $a_0 \neq 0$, $\lambda_1 \neq 0$ and we have expressed one column of

B, that is, $\mathbf{G}(j_0)$, in terms of other columns of **B**, namely, $\mathbf{G}(j_1),\ldots,\mathbf{G}(j_n)$, contradicting the linear independence of the basis. Therefore $\mathcal{G}^t$ cannot have both a root arc and a cycle.

Case 2 Suppose l is a node not in the cycle. Since $\mathcal{G}^t$ is connected, there exists a path connecting l to each node in the cycle. This path along with the cycle is illustrated in Figure 5.6*b*. Consider the columns of **B** corresponding to the cycle, root arc, and connecting path. Removing all rows that have only zeros, we equate the column for the root arc to a linear combination of the remaining columns. This system can be placed in the form

$$\left[\begin{array}{cccccc|cccccc}
\bar{a}_1 & & & & & & & & & & & \\
\bar{b}_1 & \bar{a}_2 & & & & & & & & & & \\
 & \bar{b}_2 & \bar{a}_3 & & & & & & & & & \\
 & & & \ddots & & & & & & & & \\
 & & & & \bar{a}_{m-1} & & & & & & & \\
 & & & & \bar{b}_{m-1} & \bar{a}_m & & & & & & \\
\hline
 & & & & & \bar{b}_m & a_1 & & & & & b_n \\
 & & & & & & b_1 & a_2 & & & & \\
 & & & & & & & b_2 & a_3 & & & \\
 & & & & & & & & & \ddots & & \\
 & & & & & & & & & & a_{n-1} & \\
 & & & & & & & & & & b_{n-1} & a_n
\end{array}\right]
\left[\begin{array}{c}
\bar{\lambda}_1 \\ \bar{\lambda}_2 \\ \bar{\lambda}_3 \\ \vdots \\ \bar{\lambda}_{m-1} \\ \bar{\lambda}_m \\ \hline \lambda_1 \\ \lambda_2 \\ \lambda_3 \\ \vdots \\ \lambda_{n-1} \\ \lambda_n
\end{array}\right]
=
\left[\begin{array}{c}
\bar{a}_0 \\ 0 \\ 0 \\ \vdots \\ 0 \\ 0 \\ \hline 0 \\ 0 \\ 0 \\ \vdots \\ 0 \\ 0
\end{array}\right]. \tag{5.5}$$

Since the first m equations of (5.5) are triangular,

$$\left\{\begin{array}{ll}
\bar{\lambda}_1 = \dfrac{\bar{a}_0}{\bar{a}_1} & \\[2ex]
\bar{\lambda}_i = \dfrac{-\bar{b}_{i-1}\bar{\lambda}_{i-1}}{\bar{a}_i} & \text{for } i=2,\ldots,m
\end{array}\right\}.$$

By Proposition 5.5

$$\left\{ \begin{aligned} \lambda_1 &= \frac{-\bar{b}_m \bar{\lambda}_m / a_1}{1 - w_1 w_2 \cdots w_n} \\ \lambda_{i+1} &= \frac{-b_i \lambda_i}{a_{i+1}} \qquad \text{for} \quad i = 1,\ldots,n-1 \end{aligned} \right\}.$$

We have expressed one column of **B** as a linear combination of other columns of **B**, contradicting the linear independence of the basis. Therefore $\mathcal{G}^t$ cannot contain both a root and a cycle. ∎

Proposition 5.8

The graph $\mathcal{G}^t$ for $t = 1,\ldots,p$ can have at most one cycle.

Proof. Let t be any element of $\{1,\ldots,p\}$. Suppose $\mathcal{G}^t$ contains two cycles. We will show that the columns of **B** are linearly dependent, contradicting **B** a basis.

Case 1 Suppose there is a node, say l, that appears in both cycles. These cycles are illustrated in Figure 5.7, where $v_1 = \bar{v}_1 = l$. Let

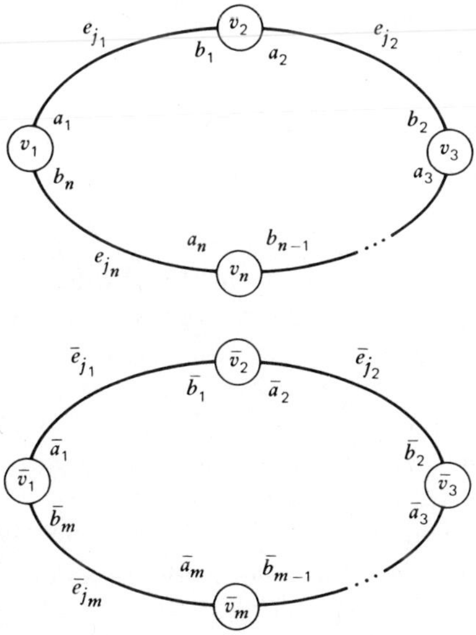

FIGURE 5.7 Two cycles in $\mathcal{G}^t$.

$\mathbf{C}^1$ and $\mathbf{C}^2$ be defined as follows:

$$\mathbf{C}^1=\left[\begin{array}{c:c:c:c:c:c}
a_1 & & & & & b_n \\
b_1 & a_2 & & & & \\
 & b_2 & a_3 & & & \\
 & & & \ddots & & \\
 & & & & a_{n-1} & \\
 & & & & b_{n-1} & a_n
\end{array}\right]$$

and

$$\mathbf{C}^2=\left[\begin{array}{c:c:c:c:c:c}
\bar{a}_1 & & & & & \bar{b}_m \\
\bar{b}_1 & \bar{a}_2 & & & & \\
 & \bar{b}_2 & \bar{a}_3 & & & \\
 & & & \ddots & & \\
 & & & & \bar{a}_{m-1} & \\
 & & & & \bar{b}_{m-1} & \bar{a}_m
\end{array}\right].$$

Choose any $\alpha \neq 0$. Then by Proposition 5.5 there exists $\boldsymbol{\lambda}^1$ and $\boldsymbol{\lambda}^2$ with each component not zero such that $\mathbf{C}^1\boldsymbol{\lambda}^1=\alpha\mathbf{e}^1$ and $\mathbf{C}^2\boldsymbol{\lambda}^2=-\alpha\mathbf{e}^1$. Since each cycle must contain at least one arc not in the other cycle, the foregoing implies that we can form a nontrivial linear combination of columns from **B**, corresponding to the arcs from one or both cycles, which equals **0**. This contradicts the linear independence of the basis. Therefore $\mathcal{G}'$ cannot contain two cycles.

Case 2 Suppose the cycles have no node in common. Then there is a path from any node in one cycle to any node in the other. These cycles and a path connecting a node in one to a node in the other are illustrated in Figure 5.8. Consider the columns of **B** corresponding to the arcs in the cycles and connecting path. Removing all rows that have only zeros, we equate a linear combination of those columns to **0**. The system can be placed

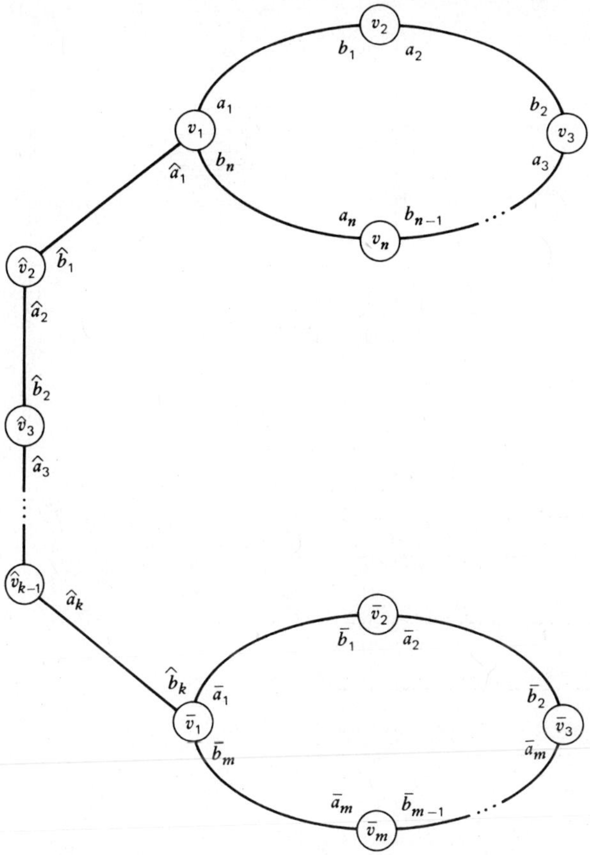

FIGURE 5.8 Two cycles in $\mathcal{G}^t$ connected by a path.

in the form

$$\begin{bmatrix} \mathbf{H}^1 & \hat{a}_1 & & & \\ & \hat{b}_1 & \mathbf{H}^2 & \hat{a}_k & \\ & & & \hat{b}_k & \mathbf{H}^3 \end{bmatrix} \begin{bmatrix} \boldsymbol{\lambda}^1 \\ \boldsymbol{\lambda}^2 \\ \boldsymbol{\lambda}^3 \end{bmatrix} = \mathbf{0}, \qquad (5.6)$$

where

$$
\mathbf{H}^1 = \left[\begin{array}{c|c|c|c|c|c}
a_1 & & & & & \\
b_1 & a_2 & & & & \\
 & b_2 & a_3 & & & \\
 & & & \ddots & & \\
 & & & & a_{n-1} & \\
 & & & & b_{n-1} & a_n
\end{array}\right],
$$

$$
\mathbf{H}^2 = \left[\begin{array}{c|c|c|c|c|c}
\hat{a}_2 & & & & & \\
\hat{b}_2 & \hat{a}_3 & & & & \\
 & \hat{b}_3 & \hat{a}_4 & & & \\
 & & & \ddots & & \\
 & & & & \hat{a}_{k-1} & \\
 & & & & \hat{b}_{k-1} & \hat{a}_k
\end{array}\right],
$$

and

$$
\mathbf{H}^3 = \left[\begin{array}{c|c|c|c|c|c}
\bar{a}_1 & & & & & \bar{b}_m \\
\bar{b}_1 & \bar{a}_2 & & & & \\
 & \bar{b}_2 & \bar{a}_3 & & & \\
 & & & \ddots & & \\
 & & & & \bar{a}_{m-1} & \\
 & & & & \bar{b}_{m-1} & \bar{a}_m
\end{array}\right].
$$

By Proposition 5.5 there exists $\boldsymbol{\lambda}^1 \neq \mathbf{0}$ and $\boldsymbol{\lambda}^3 \neq \mathbf{0}$ so that $\mathbf{H}^1\boldsymbol{\lambda}^1 = \mathbf{e}^1$ and $\mathbf{H}^3\boldsymbol{\lambda}^3 = \mathbf{e}^1$. This implies that (5.6) has a nontrivial solution, contradicting the linear independence of the basis. Hence $\mathcal{G}^t$ cannot contain two cycles. ■

We will now use the results obtained above to show that each connected subgraph of $\mathcal{G}_B$ is either a rooted tree or a graph with one cycle. We call a connected graph having exactly one cycle a *one-tree*. Note that for both a rooted tree and a one-tree the number of arcs equals the number of nodes. Furthermore, since $\mathbf{B}$ is $\bar{I} \times \bar{I}$, $\mathcal{G}_B$ has $\bar{I}$ arcs and $\bar{I}$ nodes.

Proposition 5.9

The graph $\mathcal{G}^t$ for $t = 1, \ldots, p$ is either a rooted tree or a one-tree.

Proof. Let n^t and a^t be the number of nodes and arcs, respectively, associated with $\mathcal{G}^t$. By Propositions 5.6–5.8, $a^t \leqslant n^t$. Since each $\mathcal{G}^t$ contains a spanning tree, $a^t \geqslant n^t - 1$ by Proposition 3.6. Since $\mathbf{B}$ is a basis, we have $\Sigma_{t=1}^{p} n^t = \Sigma_{t=1}^{p} a^t = \bar{I}$. The only solution to the integral system

$$\left\{\begin{array}{l} \sum\limits_{t=1}^{p} n^t = \bar{I} \\ \sum\limits_{t=1}^{p} a^t = \bar{I} \\ n^t - 1 \leqslant a^t \leqslant n^t, \quad t = 1,\ldots,p \end{array}\right\}$$

is $a^t = n^t$, $t = 1,\ldots,p$. This implies that each $\mathcal{G}^t$ is either a rooted tree or a one-tree.

Using the connected components of $\mathcal{G}_B$ (i.e., $\mathcal{G}^1 = [\mathfrak{N}^1, \mathcal{Q}^1], \ldots, \mathcal{G}^p = [\mathfrak{N}^p, \mathcal{Q}^p]$), we can display the basis $\mathbf{B}$ in block diagonal form as follows:

$$\mathbf{B} = \underbrace{\left[\begin{array}{c|c|c|c} \mathbf{B}^1 & & & \\ \hline & \mathbf{B}^2 & & \\ \hline & & \ddots & \\ \hline & & & \mathbf{B}^p \end{array}\right]}_{\text{arcs in } \mathcal{Q}^1 \quad \text{arcs in } \mathcal{Q}^2 \quad \cdots \quad \text{arcs in } \mathcal{Q}^p} \begin{array}{l} \}\text{nodes in } \mathfrak{N}^1 \\ \}\text{nodes in } \mathfrak{N}^2 \\ \quad \vdots \\ \}\text{nodes in } \mathfrak{N}^p \end{array}. \tag{5.7}$$

Since each $\mathcal{G}^t$ has the same number of arcs and nodes, each $\mathbf{B}^t$ is square. Also, $\text{Det}(\mathbf{B}^t) \neq 0$ for each t since otherwise $\mathbf{B}$ would not be a basis. ■

Consider the following basis from the example presented in Figure 5.1;

$$\begin{array}{c} \begin{array}{cccc} e_2 & e_3 & e_5 & e_6 \end{array} \\ \left[\begin{array}{r|r|r|r} 1 & 9 & & \\ & & 1 & \\ -1 & & & 1 \\ & 1 & & 1 \end{array}\right]. \end{array}$$

The corresponding graph $\mathcal{G}_B$ is illustrated in Figure 5.9, with connected

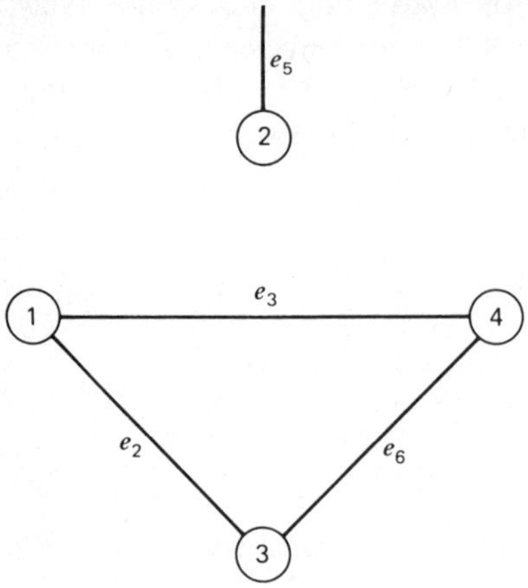

FIGURE 5.9 Graphical basis.

components $\mathcal{G}^1=[\{2\},\{e_5\}]$ and $\mathcal{G}^2=[\{1,3,4\},\ \{e_2,e_3,e_6\}]$. Using these components, we may display $\mathbf{B}$ in block diagonal form as follows:

$$\mathbf{B}=\begin{array}{c} \\ \end{array}\left[\begin{array}{c|c:c:c} e_5 & e_2 & e_3 & e_6 \\ 1 & & & \\ \hline & 1 & 9 & \\ & -1 & & 1 \\ & & 1 & 1 \end{array}\right]\begin{array}{c} \text{nodes} \\ 2 \\ 1 \\ 3 \\ 4 \end{array}.$$

5.3 DUAL VARIABLE CALCULATION

In this section we indicate how the structure of the basis $\mathbf{B}$ may be used to efficiently calculate the dual variables $\boldsymbol{\pi}=\mathbf{c}^B\mathbf{B}^{-1}$, that is, we wish to solve the system $\boldsymbol{\pi}\mathbf{B}=\mathbf{c}^B$. Partitioning $\boldsymbol{\pi}$ and $\mathbf{c}^B$ to be compatible with $\mathbf{B}$ as shown in (5.7), we obtain

$$[\boldsymbol{\pi}^1 \vdots \boldsymbol{\pi}^2 \vdots \cdots \vdots \boldsymbol{\pi}^p]\left[\begin{array}{c:c:c:c} \mathbf{B}^1 & & & \\ & \mathbf{B}^2 & & \\ & & \ddots & \\ & & & \mathbf{B}^p \end{array}\right]=[\mathbf{c}^1 \vdots \mathbf{c}^2 \vdots \cdots \vdots \mathbf{c}^p].$$

Hence we must solve p systems, each of the form $\boldsymbol{\pi}^t\mathbf{B}^t=\mathbf{c}^t$. Let $\mathcal{G}^t$ denote the graph associated with $\mathbf{B}^t$. Then $\mathcal{G}^t$ is connected and is either a rooted tree or a one-tree.

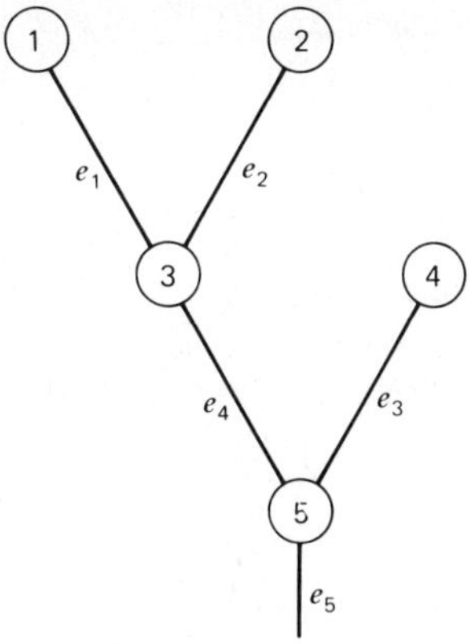

FIGURE 5.10 Rooted tree.

Suppose $\mathcal{G}^t$ is a rooted tree. Then, by arguments identical to those used in the proof of Proposition 3.17, $\mathbf{B}^t$ can be shown to be triangular. Hence $\boldsymbol{\pi}^t\mathbf{B}^t = \mathbf{c}^t$ is a triangular system.

Consider the rooted tree $\mathcal{G}^t$ illustrated in Figure 5.10 with the corresponding basis component

$$\mathbf{B}^t = \begin{array}{c} \\ \\ \\ \\ \\ \end{array} \begin{array}{ccccc} e_1 & e_3 & e_2 & e_4 & e_5 \end{array} \quad \text{nodes}$$

$$\mathbf{B}^t = \left[\begin{array}{c|c|c|c|c} G_{11} & & & & \\ & G_{43} & & & \\ & & G_{22} & & \\ G_{31} & & G_{32} & G_{34} & \\ & G_{53} & & G_{54} & G_{55} \end{array}\right] \begin{array}{c} 1 \\ 4 \\ 2 \\ 3 \\ 5 \end{array} . \tag{5.8}$$

Then the solution to $\boldsymbol{\pi}^t\mathbf{B}^t = \mathbf{c}^t$ reduces to

$$\begin{array}{lllll} G_{11}\pi_1^t & & & + G_{31}\pi_3^t & & = c_1^t \\ & G_{43}\pi_4^t & & & + G_{53}\pi_5^t & = c_2^t \\ & & G_{22}\pi_2^t & + G_{32}\pi_3^t & & = c_3^t, \\ & & & G_{34}\pi_3^t & + G_{54}\pi_5^t & = c_4^t \\ & & & & G_{55}\pi_5^t & = c_5^t \end{array}$$

where the order of the components of $\boldsymbol{\pi}^t$ corresponds to the row order of (5.8). In general, for a basis component $\mathbf{B}^t$ with corresponding rooted tree

$\mathcal{G}^t$, having root node l and root arc e_k, $\boldsymbol{\pi}^t\mathbf{B}^t = \mathbf{c}^t$ reduces to

$$G_{lk}\pi_l = c_k,$$

$$G_{F(j),j}\pi_{F(j)} + G_{T(j),j}\pi_{T(j)} = c_j \qquad \text{for all } e_j \varepsilon \mathcal{G}^t.$$

Therefore the following algorithm may be used to iteratively determine $\boldsymbol{\pi}$.

ALG 5.1 DETERMINATION OF DUAL VARIABLES FOR A ROOTED TREE

0 *Initialization.* Let $\mathcal{G}^t = [\mathcal{N}^t, \mathcal{A}^t]$ denote the basis tree with root node l and root arc e_k. Set $\pi_l \leftarrow c_k / G_{lk}$, $\mathcal{N}^L \leftarrow \{l\}$, and $\mathcal{N}^U \leftarrow \mathcal{N}^t - \{l\}$.

1 *Find an Arc with One Node Labeled.* Let $e_j = \{i,k\} \varepsilon \mathcal{A}^t$ be such that $i \varepsilon \mathcal{N}^U$ and $k \varepsilon \mathcal{N}^L$. If no such arc exists, go to Step 3.

2 *Label New Node.* Set $\pi_i \leftarrow (c_j - G_{kj}\pi_k)/G_{ij}$, $\mathcal{N}^L \leftarrow \mathcal{N}^L \cup \{i\}$, $\mathcal{N}^U \leftarrow \mathcal{N}^U - \{i\}$, and go to Step 1.

3 *Find an Arc with One Node Labeled.* Let $e_j = \{k,i\} \varepsilon \mathcal{A}^t$ be such that $i \varepsilon \mathcal{N}^U$ and $k \varepsilon \mathcal{N}^L$. If no such arc exists, terminate; otherwise go to Step 2.

We now consider the case in which $\mathcal{G}^t$ is a one-tree. Note that a one-tree may be decomposed into a tree and one additional arc. Hence, except for the one nonzero entry corresponding to the extra arc, $\mathbf{B}^t$ would be triangular. Consider the one-tree illustrated in Figure 5.11 with the corresponding basis component

$$\mathbf{B}^t = \begin{array}{c} \\ \\ \\ \\ \\ \\ \end{array}\left[\begin{array}{c|c|c|c|c} e_1 & e_2 & e_5 & e_3 & e_4 \\ G_{11} & & & & \\ & G_{52} & & & \\ G_{21} & & G_{25} & & G_{24} \\ & & G_{35} & G_{33} & \\ & G_{42} & & G_{43} & G_{44} \end{array}\right] \begin{array}{c} \text{nodes} \\ 1 \\ 5 \\ 2 \\ 3 \\ 4 \end{array}.$$

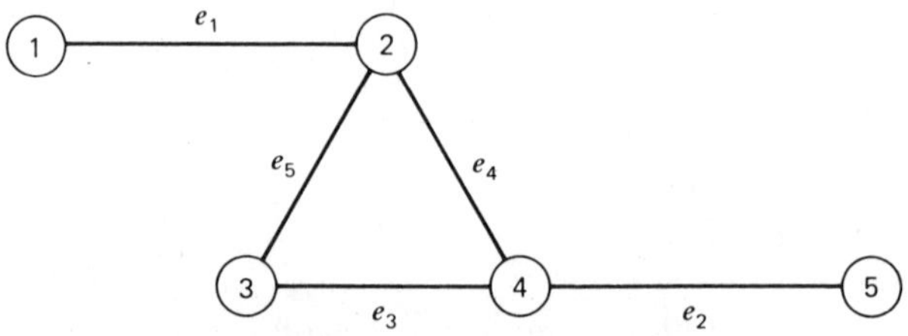

FIGURE 5.11 Sample One-Tree.

Then the solution to the system $\boldsymbol{\pi}^t\mathbf{B}^t = \mathbf{c}^t$ reduces to

$$\begin{array}{llllll}
G_{11}\pi_1^t & + G_{21}\pi_2^t & & & & = c_1^t \\
G_{52}\pi_5^t & & & & + G_{42}\pi_4^t & = c_2^t \\
 & G_{25}\pi_2^t & + G_{35}\pi_3^t & & & = c_3^t \\
 & & G_{33}\pi_3^t & & + G_{43}\pi_4^t & = c_4^t \\
 & G_{24}\pi_2^t & & & + G_{44}\pi_4^t & = c_5^t
\end{array} \tag{5.9}$$

This system may be solved by first solving the $3x3$ system given by the last three equations and then using backward substitution on the first two equations. Note that the solution to the last three equations can be obtained by application of Proposition 5.4.

In general, the strategy is to solve for all components of $\boldsymbol{\pi}$ in $\mathcal{G}^t$ corresponding to nodes in the cycle and then obtain all other components of $\boldsymbol{\pi}$ in $\mathcal{G}^t$ by backward substitution. Consider the cycle of $\mathcal{G}^t$ as illustrated in Figure 5.12. Then we wish to solve the system

$$\begin{bmatrix} a_1 & & & & & b_n \\ b_1 & a_2 & & & & \\ & b_2 & a_3 & & & \\ & & & \ddots & & \\ & & & & a_{n-1} & \\ & & & & b_{n-1} & a_n \end{bmatrix} \begin{bmatrix} \pi_{v_1} \\ \pi_{v_2} \\ \pi_{v_3} \\ \vdots \\ \pi_{v_{n-1}} \\ \pi_{v_n} \end{bmatrix} = \begin{bmatrix} c_{j_1} \\ c_{j_2} \\ c_{j_3} \\ \vdots \\ c_{j_{n-1}} \\ c_{j_n} \end{bmatrix}. \tag{5.10}$$

But the solution to (5.10) is given by Proposition 5.4. Therefore the following algorithm may be used to iteratively determine $\boldsymbol{\pi}$.

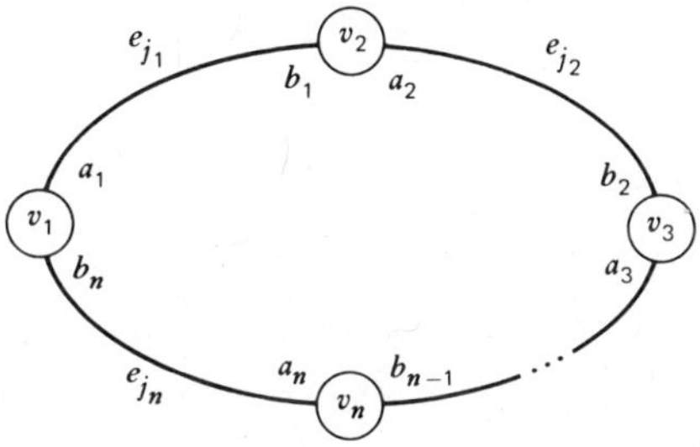

FIGURE 5.12 Sample cycle.

ALG 5.2 DETERMINATION OF DUAL VARIABLES FOR A ONE-TREE

0 *Initialization.* Let $\mathcal{G}^t = [\mathcal{N}^t, \mathcal{Q}^t]$ denote the one-tree with cycle $C = \{s_1, e_{j_1}, s_2, e_{j_2}, \ldots, s_n, e_{j_n}, s_{n+1}\}$. Set $\mathcal{N}^L \leftarrow \{s_1, s_2, \ldots, s_{n+1}\}$ and $\mathcal{N}^U \leftarrow \mathcal{N}^t - \mathcal{N}^L$.

1 *Find Duals on Cycle.* Apply (5.1) to obtain $\boldsymbol{\pi}$ associated with the nodes of the cycle.

2 *Backward Substitution.* Apply ALG 5.1, beginning at Step 1.

5.4 COLUMN UPDATE

In this section we indicate how the structure of the basis $\mathbf{B}$ may be used to efficiently calculate the updated column $\mathbf{y} = \mathbf{B}^{-1}\mathbf{G}(j)$, for a given j. Suppose $e_j = \{i, k\}$; then $\mathbf{G}(j) = \mathrm{G}_{ij}\mathbf{e}^i + \mathrm{G}_{kj}\mathbf{e}^k$, and we must solve the system

$$\mathbf{B}\mathbf{y} = \mathbf{G}(j) = \mathrm{G}_{ij}\mathbf{e}^i + \mathrm{G}_{kj}\mathbf{e}^k, \tag{5.11}$$

where $\mathbf{e}^0$ is taken to be a vector of zeros. The solution to (5.11) may be obtained by solving the two systems $\mathbf{B}\mathbf{w} = \mathrm{G}_{ij}\mathbf{e}^i$ and $\mathbf{B}\mathbf{z} = \mathrm{G}_{kj}\mathbf{e}^k$ and letting $\mathbf{y} = \mathbf{w} + \mathbf{z}$. Therefore let us examine the system $\mathbf{B}\mathbf{x} = \alpha\mathbf{e}^i$ for some i. Partitioning $\mathbf{x}$ and $\mathbf{B}$ as shown in (5.7), we obtain

$$\left[\begin{array}{c|c|c|c|c|c|c} \mathbf{B}^1 & & & & & & \\ & \ddots & & & & & \\ & & \mathbf{B}^{t-1} & & & & \\ & & & \mathbf{B}^t & & & \\ & & & & \mathbf{B}^{t+1} & & \\ & & & & & \ddots & \\ & & & & & & \mathbf{B}^p \end{array}\right] \begin{bmatrix} \mathbf{x}^1 \\ \vdots \\ \mathbf{x}^{t-1} \\ \mathbf{x}^t \\ \mathbf{x}^{t+1} \\ \vdots \\ \mathbf{x}^p \end{bmatrix} = \alpha \begin{bmatrix} \mathbf{0} \\ \vdots \\ \mathbf{0} \\ \mathbf{e}^r \\ \mathbf{0} \\ \vdots \\ \mathbf{0} \end{bmatrix} \tag{5.12}$$

for an appropriate selection of r. The solution of (5.12) is given by

$$x^i = \begin{cases} 0 & \text{for } i \neq t \\ (\mathbf{B}^t)^{-1}\alpha\mathbf{e}^r & \text{for } i = t. \end{cases}$$

Thus we need only solve the system $\mathbf{B}^t\mathbf{x}^t = \alpha\mathbf{e}^r$. Let $\mathcal{G}^t$ denote the connected graph corresponding to $\mathbf{B}^t$. Then $\mathcal{G}^t$ is either a rooted tree or a one-tree.

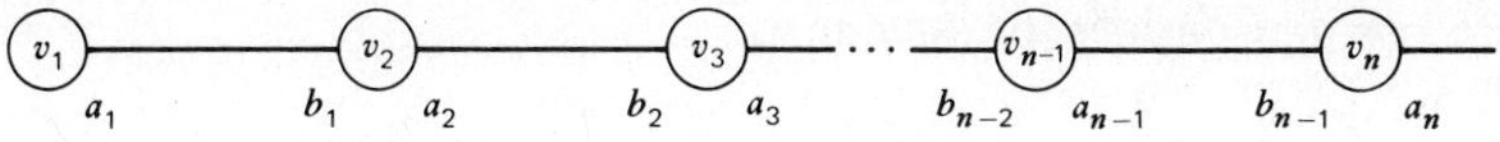

FIGURE 5.13 Path from v_1 to the root node.

Suppose $\mathcal{G}^t$ is a rooted tree. Let the nonzero component in $\mathbf{e}^r$ be associated with row (node) v_1. Then either v_1 is the root node, or there exists a unique path in $\mathcal{G}^t$ from v_1 to the root node, as illustrated by Figure 5.13. For either case we may define the corresponding triangular system

$$\left[\begin{array}{c:c:c:c:c:c}
a_1 & & & & & \\
b_1 & a_2 & & & & \\
 & b_2 & a_3 & & & \\
 & & & \ddots & & \\
 & & & & a_{n-1} & \\
 & & & & b_{n-1} & a_n
\end{array}\right]
\begin{bmatrix} \lambda_1 \\ \lambda_2 \\ \lambda_3 \\ \vdots \\ \lambda_{n-1} \\ \lambda_n \end{bmatrix}
=
\begin{bmatrix} \alpha \\ 0 \\ 0 \\ \vdots \\ 0 \\ 0 \end{bmatrix}. \tag{5.13}$$

Then

$$\left\{\begin{array}{ll}
\lambda_1 = \dfrac{\alpha}{a_1} & \\[2ex]
\lambda_i = \dfrac{-b_{i-1}\lambda_{i-1}}{a_i} & \text{for } i=2,\ldots,n
\end{array}\right\} \tag{5.14}$$

solves (5.13). Then the nonzero components of $\mathbf{x}^t$ correspond to the columns of $\mathbf{B}^t$ that appear in (5.13). Let the first n columns of $\mathbf{B}^t$ correspond to those in (5.13). Then $\mathbf{x}^t = [\boldsymbol{\lambda}|\mathbf{0}]$.

Suppose $\mathcal{G}^t$ is a one-tree. Let node $\bar{v}_1$ be associated with the nonzero component in $\mathbf{e}^r$. Suppose $\bar{v}_1$ is in the cycle as illustrated in Figure 5.14,

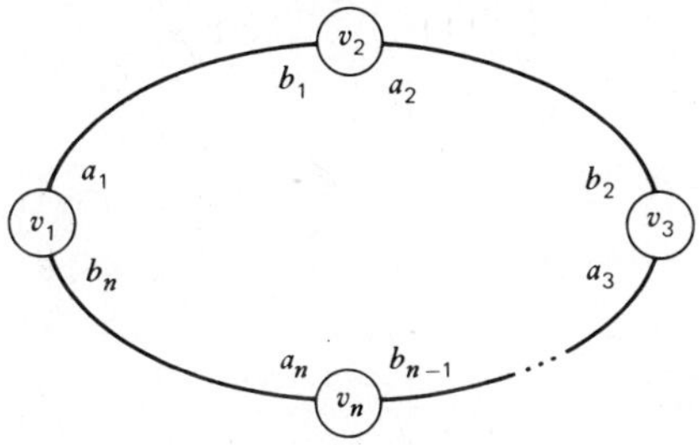

FIGURE 5.14 A cycle.

where $\bar{v}_1 = v_1$. Consider the system

$$\begin{bmatrix} a_1 & & & & & b_n \\ b_1 & a_2 & & & & \\ & b_2 & a_3 & & & \\ & & & \ddots & & \\ & & & & a_{n-1} & \\ & & & & b_{n-1} & a_n \end{bmatrix} \begin{bmatrix} \lambda_1 \\ \lambda_2 \\ \lambda_3 \\ \vdots \\ \lambda_{n-1} \\ \lambda_n \end{bmatrix} = \begin{bmatrix} \alpha \\ 0 \\ 0 \\ \vdots \\ 0 \\ 0 \end{bmatrix}. \tag{5.15}$$

Proposition 5.5 gives the solution to (5.15), that is,

$$\left\{ \begin{aligned} \lambda_1 &= \frac{\alpha / a_1}{1 - w_1 w_2 \cdots w_n} \\ \lambda_{i+1} &= \frac{-b_i \lambda_i}{a_{i+1}} \qquad \text{for } i = 1, \ldots, n-1 \end{aligned} \right\}. \tag{5.16}$$

Letting the first n columns of $\mathbf{B}^t$ correspond to those of (5.15), we have $\mathbf{x}^t = [\boldsymbol{\lambda} ; \mathbf{0}]$.

Suppose $\bar{v}_1$ is not in the cycle. Then there exists some path from $\bar{v}_1$ to a node in the cycle, as illustrated in Figure 5.15. Consider the system

$$\begin{bmatrix} \bar{a}_1 & & & & & & & & & & \\ \bar{b}_1 & \bar{a}_2 & & & & & & & & & \\ & \bar{b}_2 & \bar{a}_3 & & & & & & & & \\ & & & \ddots & & & & & & & \\ & & & & \bar{a}_m & & & & & & \\ & & & & \bar{b}_m & a_1 & & & & & b_n \\ & & & & & b_1 & a_2 & & & & \\ & & & & & & b_2 & a_3 & & & \\ & & & & & & & & \ddots & & \\ & & & & & & & & & a_{n-1} & \\ & & & & & & & & & b_{n-1} & a_n \end{bmatrix} \begin{bmatrix} \bar{\lambda}_1 \\ \bar{\lambda}_2 \\ \bar{\lambda}_3 \\ \vdots \\ \bar{\lambda}_m \\ \hline \lambda_1 \\ \lambda_2 \\ \lambda_3 \\ \vdots \\ \lambda_{n-1} \\ \lambda_n \end{bmatrix} = \begin{bmatrix} \alpha \\ 0 \\ 0 \\ \vdots \\ 0 \\ 0 \\ 0 \\ 0 \\ \vdots \\ 0 \\ 0 \end{bmatrix}. \tag{5.17}$$

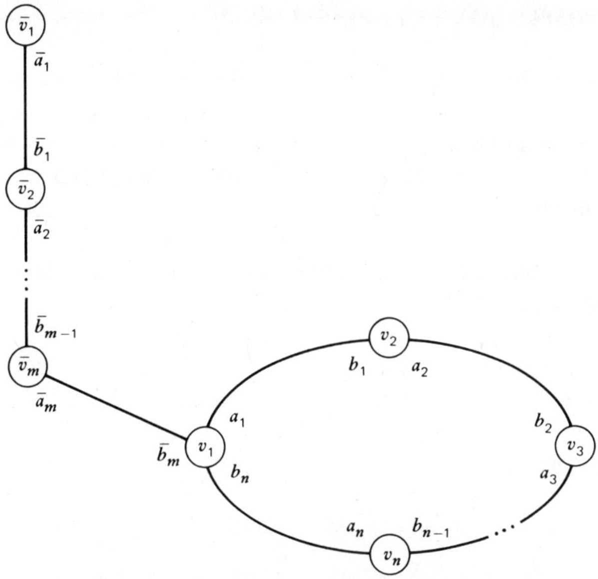

FIGURE 5.15 Path from $\bar{v}_1$ to a node in the cycle.

Since the first m equations of (5.17) are triangular,

$$\left.\begin{aligned} \bar{\lambda}_1 &= \frac{\alpha}{\bar{a}_1}, \\ \bar{\lambda}_i &= \frac{-\bar{b}_{i-1}\bar{\lambda}_{i-1}}{\bar{a}_i} \qquad \text{for } i=2,\ldots,m. \end{aligned}\right\} \tag{5.18}$$

By Proposition 5.5

$$\left\{\begin{aligned} \lambda_1 &= \frac{-\bar{b}_m\bar{\lambda}_m/a_1}{1-w_1w_2\cdots w_n} \\ \lambda_{i+1} &= \frac{b_i\lambda_i}{a_{i+1}} \qquad \text{for } i=1,\ldots,n-1. \end{aligned}\right\}. \tag{5.19}$$

Letting the first $m+n$ columns of $\mathbf{B}^t$ correspond to those of (5.17), we have $\mathbf{x}^t=[\bar{\boldsymbol{\lambda}}\,|\,\boldsymbol{\lambda}\,|\,\mathbf{0}]$.

5.5 PRIMAL SIMPLEX SPECIALIZATION

Given the algorithms for calculating the dual variables and the formulae for updating a column, we may now present a specialization of the primal simplex algorithm. This specialization allows one to perform the simplex operations directly on the basis graph $\mathcal{G}_B$, thereby eliminating the need for matrix operations.

ALG 5.3 PRIMAL SIMPLEX SPECIALIZATION FOR THE GENERALIZED FLOW PROBLEM

0 *Initialization.* Let $[\mathbf{x}^B \mid \mathbf{x}^N]$ be a basic feasible solution with basis graph $\mathcal{G}_B$ having connected components $\mathcal{G}^1, \ldots, \mathcal{G}^p$. Calculate $\boldsymbol{\pi}$ using ALG 5.1 and ALG 5.2.

1 *Pricing.* Let

$$\begin{aligned}\psi_1 = \{e_j = \{m,n\} : x_j^N = 0 \quad &\text{and} \quad G_{mj}\pi_m + G_{nj}\pi_n - c_j > 0\} \cup \\ \{e_j = \{m,0\} : x_j^N = 0 \quad &\text{and} \quad G_{mj}\pi_m - c_j > 0\}\end{aligned}$$

and

$$\begin{aligned}\psi_2 = \{e_j = \{m,n\} : x_j^N = u_j \quad &\text{and} \quad G_{mj}\pi_m + G_{nj}\pi_n - c_j < 0\} \cup \\ \{e_j = \{m,0\} : x_j^N = u_j \quad &\text{and} \quad G_{mj}\pi_m - c_j < 0\}.\end{aligned}$$

If $\psi_1 \cup \psi_2 = \phi$, terminate with $[\mathbf{x}^B \mid \mathbf{x}^N]$ an optimum; otherwise select $e_k \varepsilon \psi_1 \cup \psi_2$ and set

$$\delta \leftarrow \begin{cases} +1 & \text{if } e_k \varepsilon \psi_1, \\ -1 & \text{if } e_k \varepsilon \psi_2. \end{cases}$$

2 *Column Update* [i.e., $\mathbf{y} \leftarrow \mathbf{B}^{-1}\mathbf{G}(k)$].

Case 1 $e_k = \{m,0\}$, node m in $\mathcal{G}^t$, and $\mathcal{G}^t$ is a rooted tree. The nonzero components of $\mathbf{y}$ are obtained by applying (5.14).

Case 2 $e_k = \{m,0\}$, node m in $\mathcal{G}^t$, and $\mathcal{G}^t$ is a one-tree. The nonzero components of $\mathbf{y}$ are obtained by applying either (5.16) or (5.18) and (5.19), whichever is appropriate.

Case 3 $e_k = \{m,n\}$. Arc e_k is viewed as two arcs of the form $\bar{e}_k = \{m,0\}$ and $\hat{e}_k = \{n,0\}$, and case 1 or 2 is applied to each of the arcs $\bar{e}_k$ and $\hat{e}_k$.

3 *Ratio Test.* Set

$$\Delta_1 \leftarrow \min_{\substack{1<j<\bar{I} \\ \sigma(y_j)=\delta}} \left\{ \frac{x_j^B}{|y_j|}, \infty \right\},$$

$$\Delta_2 \leftarrow \min_{\substack{1<j<\bar{I} \\ -\sigma(y_j)=\delta}} \left\{ \frac{u_j^B - x_j^B}{|y_j|}, \infty \right\},$$

$$\Delta \leftarrow \min\{\Delta_1, \Delta_2, u_k^N\}.$$

4 *Update Flows.* Set $x_k \leftarrow x_k + \Delta\delta$ and $\mathbf{x}^B \leftarrow \mathbf{x}^B - \Delta\delta\mathbf{y}$. If $\Delta = u_k^N$, return to Step 1.

5 *Pivot.* Let

$$\psi_3 = \{j : \sigma(y_j) = \delta \quad \text{and} \quad x_j^B = 0\},$$

and

$$\psi_4 = \{j : -\sigma(y_j) = \delta \quad \text{and} \quad x_j^B = u_j^B\}.$$

Select any $l \in \psi_3 \cup \psi_4$. Replace $\mathbf{G}(l)$ in $\mathbf{B}$ with $\mathbf{G}(k)$, update the dual variables, and return to Step 1.

Consider the following sample problem:

$$\begin{array}{llcl} \min & 5x_1 + 10x_2 \qquad + 100x_4 + 100x_5 & & \\ \text{s.t.} & 4x_1 \qquad - x_3 & = & 0 \\ & 2x_1 + 3x_2 \qquad + x_4 & = & 4 \\ & \qquad - 4x_2 - 2x_3 \qquad - x_5 & = & -8 \end{array}$$

$$0 \leqslant x_1 \leqslant 1, 0 \leqslant x_2 \leqslant 2, 0 \leqslant x_3 \leqslant 3, 0 \leqslant x_4 \leqslant 8, 0 \leqslant x_5 \leqslant 8.$$

The graph corresponding to this example is shown in Figure 5.16.

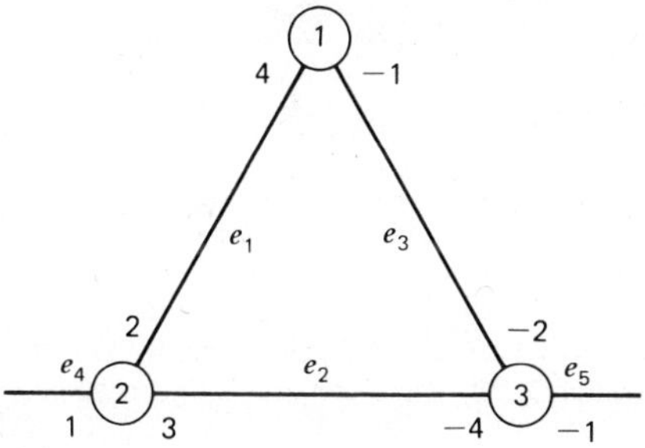

FIGURE 5.16 Sample graph.

We now solve this example using ALG 5.3. The basis graphs and ratio test graphs are given in Figure 5.17.

0 (Initialization) Let $[x_1 \quad x_2 \quad x_3 \quad x_4 \quad x_5]=[0.75 \quad 0 \quad 3 \quad 2.5 \quad 2]$, where $\mathbf{x}^B=[x_1 \quad x_4 \quad x_5]$ and $\mathbf{x}^N=[x_2 \quad x_3]$. Then the basis $\mathbf{B}$ is as follows:

$$\mathbf{B}=\begin{array}{c} \\ \\ \\ \\ \end{array}\left[\begin{array}{cc|c} e_1 & e_4 & e_5 \\ 4 & & \\ 2 & 1 & \\ \hline & & -1 \end{array}\right]\begin{array}{c} \text{nodes} \\ 1 \\ 2 \\ 3 \end{array}.$$

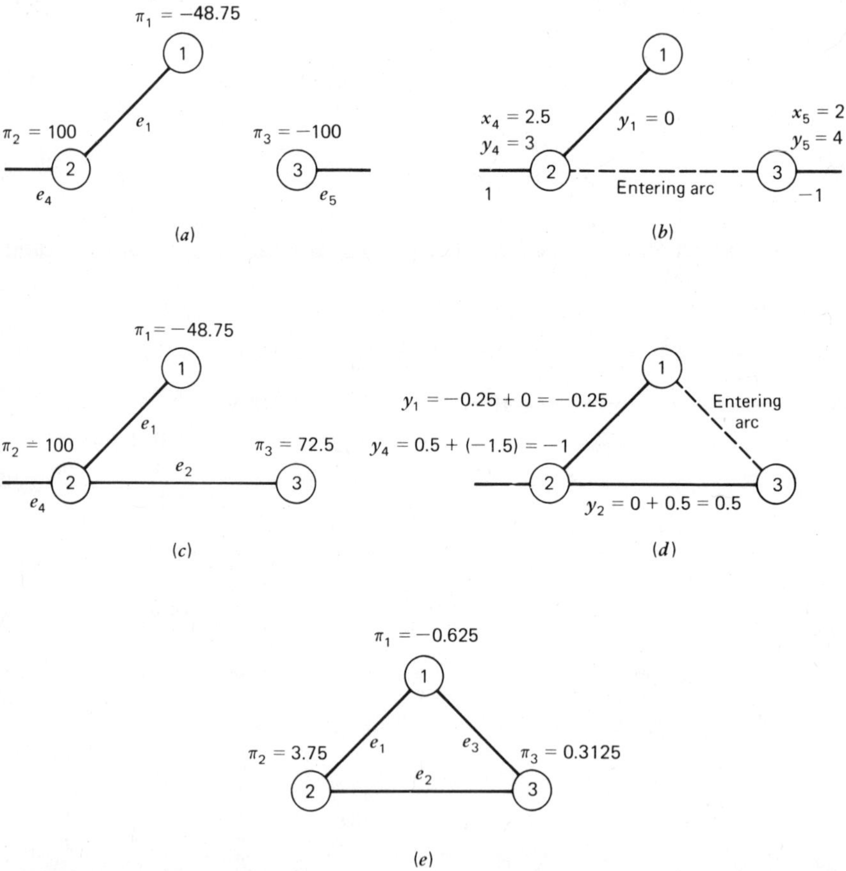

FIGURE 5.17 Basis graphs and ratio test graphs. (*a*) Basis graph for iteration 1. (*b*) Ratio test graph for iteration 1. (*c*) Basis graph for iteration 2. (*d*) Ratio test graph for iteration 2. (*e*) Basis graph for iteration 3.

The dual variables are obtained by applying ALG 5.1 to solve the systems:

$$[\pi_1 \quad \pi_2]\left[\begin{array}{c|c} 4 & \\ \hline 2 & 1 \end{array}\right]=[5 \quad 100] \qquad \text{and} \qquad \pi_3[-1]=100,$$

yielding

$$\pi_1=-48.75, \qquad \pi_2=100, \qquad \pi_3=-100.$$

1 (Pricing at Iteration 1)

For e_2: $3\pi_2-4\pi_3-c_2=300+400-10>0 \Rightarrow e_2 \varepsilon \psi_1$.
For e_3: $-\pi_1-2\pi_3-c_3=48.75+200-0>0 \Rightarrow e_3 \not\varepsilon \psi_2$.
Then $\psi_1=\{e_2\}, \psi_2=\phi, e_k=e_2$, and $\delta=1$.

2 (Column Update). We apply (5.14) to solve the systems

$$[1]y_4=3 \qquad \text{and} \qquad [-1]y_5=-4$$

corresponding to paths from nodes 2 and 3 to root nodes.

3 (Ratio Test)

$$\Delta_1=\min\left\{\frac{x_4}{y_4},\frac{x_5}{y_5}\right\}=\min\left\{\frac{2.5}{3},\frac{2}{4}\right\}=0.5,$$

$$\Delta_2=\infty,$$

$$\Delta=\min\{0.5,\infty,2\}=0.5.$$

4 (Update Flows)

$$x_2=0+(0.5)(1)=0.5,$$

$$x_4=2.5-(0.5)(1)(3)=1,$$

$$x_5=2-(0.5)(1)(4)=0.$$

5 (Pivot) $\psi_3=\{5\}$ and $\psi_4=\phi$. Therefore the new basis is as follows:

$$\begin{array}{c} \\ \\ \\ \end{array}\begin{array}{ccc} e_1 & e_2 & e_4 \end{array} \quad \text{nodes}$$

$$\left[\begin{array}{c|c|c} 4 & & \\ & -4 & \\ 2 & 3 & 1 \end{array}\right] \quad \begin{array}{c} 1 \\ 3 \\ 2 \end{array},$$

with $\mathbf{x}^B=[x_1 \quad x_2 \quad x_4]$ and $\mathbf{x}^N=[x_3 \quad x_5]$.

1 (Pricing at Iteration 2). $[x_1 \;\; x_2 \;\; x_3 \;\; x_4 \;\; x_5]=[0.75 \;\; 0.5 \;\; 3 \;\; 1 \;\; 0]$.

$$\text{For } e_3\text{: } -\pi_1-2\pi_3-c_3=48.75-145-0<0\Rightarrow e_3\varepsilon\psi_2.$$
$$\text{For } e_5\text{: } -\pi_3-c_5=-72.5-100<0\Rightarrow e_5\not\varepsilon\psi_1.$$

Therefore $e_k=e_3$ and $\delta=-1$.

2 (Column Update). We solve

$$\left[\begin{array}{c:c:c} 4 & & \\ & -4 & \\ 2 & 3 & 1 \end{array}\right]\begin{bmatrix} y_1 \\ y_2 \\ y_4 \end{bmatrix}=\begin{bmatrix} -1 \\ -2 \\ \\ \end{bmatrix}.$$

3 (Ratio Test)

$$\Delta_1=\min\left\{\frac{x_1}{|y_1|},\frac{x_4}{|y_4|}\right\}=\min\left\{\frac{0.75}{0.25},\frac{1}{1}\right\}=1,$$
$$\Delta_2=\min\left\{\frac{u_2-x_2}{y_2}\right\}=3,$$
$$\Delta=\min\{1,3,3\}=1.$$

4 (Update Flows)

$$x_3=3+1(-1)=2,$$
$$x_1=0.75-(1)(-1)(-0.25)=0.5,$$
$$x_2=0.5-(1)(-1)(0.5)=1,$$
$$x_4=1-(1)(-1)(-1)=0.$$

5 (Pivot). $\psi_3=\{4\}$ and $\psi_4=\phi$. Now we solve

$$[\pi_1 \;\; \pi_2 \;\; \pi_3]\left[\begin{array}{c:c:c} 4 & & -1 \\ 2 & 3 & \\ & -4 & -2 \end{array}\right]=[5 \;\; 10 \;\; 0].$$

From (5.1) we get $w_1=-\frac{1}{2}, w_2=\frac{4}{3}, w_3=-\frac{1}{2}$, and

$$\pi_1=\frac{\frac{5}{4}+(-\frac{1}{2})(10)/3+(-\frac{1}{2})(4/3)(0)/(-2)}{1-(-\frac{1}{2})(4/3)(-\frac{1}{2})}=-.625,$$
$$\pi_2=\frac{5-4(-.625)}{2}=3.75,$$
$$\pi_3=\frac{10-3(3.75)}{(-4)}=0.3125.$$

1 (Pricing at Iteration 2). $[x_1\ x_2\ x_3\ x_4\ x_5]=[0.5\ 1\ 2\ 0\ 0]$.

$$\text{For } e_4\text{: } \pi_2 - c_4 = 3.75 - 100 < 0 \Rightarrow e_4 \not\in \psi_1$$
$$\text{For } e_5\text{: } -\pi_3 - c_5 = -0.3125 - 100 < 0 \Rightarrow e_5 \not\in \psi_1.$$

Optimality attained!

The discussions in Chapter 3 regarding an initial feasible solution and pricing strategies either apply directly or can be easily modified for ALG 5.3. Furthermore, the data structures and algorithms presented in Appendix B can be extended for ALG 5.3.

EXERCISES

5.1 Solve the following generalized network flow problem using ALG 5.3:

$$\begin{array}{lrl} \min & 2x_1 + x_2 + x_3 + 100x_4 + 8x_5 & \\ \text{s.t.} & x_1 \quad + 3x_3 + x_4 - 2x_5 = & 2 \\ & -2x_1 + x_2 \quad = & 0 \\ & 3x_2 - 4x_3 - 5x_4 - x_5 = & -10 \\ & x_i \geq 0. & \end{array}$$

Begin with the solution $[x_1\ x_2\ x_3\ x_4\ x_5]=[0\ 0\ 0\ 2\ 0]$. Give the optimal dual variables for this problem.

5.2 Solve the following generalized network flow problem using ALG 5.3:

$$\begin{array}{lrl} \min & 2x_1 + 3x_2 + 8x_3 + 8x_4 & \\ \text{s.t.} & x_1 + x_2 \quad = & 9 \\ & -7x_1 \quad + 4x_3 \quad + x_5 = & 0 \\ & - x_2 - x_3 + x_4 \quad = & 2 \\ & -3x_4 - x_5 = & -24 \end{array}$$

$$0 \leq x_1 \leq 3, \quad 0 \leq x_2 \leq 4, \quad 0 \leq x_3 \leq 2, \quad 0 \leq x_4 \leq 9, \quad 0 \leq x_5 \leq 3.$$

Add four artificial variables, and begin with an all-artificial start.

5.3 Solve the following generalized network flow problem using ALG 5.3:

$$\begin{array}{lrl} \min & x_1 - 4x_2 & \\ \text{s.t.} & 3x_1 \quad + x_3 = & -7 \\ & x_1 - x_2 \quad = & -7 \\ & 2x_2 + x_3 = & 9 \\ & x_1 \leq 0, \quad x_2 \leq 3, \quad x_3 \leq 7. & \end{array}$$

Append three artificial variables, and begin with an all-artificial start. Display your answer graphically.

5.4 Solve the following generalized network flow problem using ALG 5.3:

$$\begin{array}{llcl} \max & 3x_1+8x_2+15x_3 & & \\ \text{s.t.} & 3x_1 \quad\quad + \quad x_3 & = & 28 \\ & x_1 - x_2 & = & 1 \\ & \quad 2x_2 + x_3 & = & 19 \end{array}$$

$$1 \leqslant x_1 \leqslant 7, \quad 4 \leqslant x_2 \leqslant 10, \quad 7 \leqslant x_3 \leqslant 8.$$

Append three artificials and begin with an all-artificial start. (Optimal cost equals 174.)

5.5 Solve the following generalized flow problem:

$$\begin{array}{llcl} \min & 10x_1+20x_2+30x_3+ \quad\quad\quad\quad 40x_6 & & \\ \text{s.t.} & x_1+x_2 & = & 2 \\ & x_1 \quad - 3x_3-2x_4 & = & -2 \\ & \quad 2x_2+x_3 \quad +0.5x_5 & = & 3 \\ & \quad\quad -x_4-x_5-x_6 & = & -1 \end{array}$$

$$0 \leqslant x_1 \leqslant 2, \quad 0 \leqslant x_2 \leqslant 2, \quad 0 \leqslant x_3 \leqslant 2, \quad 0 \leqslant x_4 \leqslant 1, \quad 0 \leqslant x_5 \leqslant 1, \quad 0 \leqslant x_6 \leqslant 3.$$

Begin with the solution $x_1 = x_2 = x_3 = x_6 = 1, x_4 = x_5 = 0$.

5.6 Show that any **GP** can be transformed into a problem in which every column has at least one $+1$. Discuss the specializations that could be made in ALG 5.3 if this were done.

5.7 Devise an algorithm to develop a row of the inverse of a basis for **GP**.

5.8 Consider the allocation problem in which production time in three machine centers must be allocated to four products. The cost and production time data are given in Table E5.8.

Table E5.8 Machine Center

	Red		White		Blue	
Product	Unit Cost ($)	Processing Time (min/unit)	Unit Cost ($)	Processing Time (min/unit)	Unit Cost ($)	Processing Time (min/unit)
W	7	2	10	1	4	3
X	8	2	2	4	6	1
Y	1	5	11	8	9	3
Z	12	9	3	6	5	7

Suppose the demands for products W, X, Y, and Z are 10, 20, 30, and 40, respectively; and the processing times (in minutes) for machine centers red, white, and blue are 60, 240, and 280, respectively. Formulate this allocation problem as a generalized network problem.

5.9 Obtain a proof of Proposition 5.3, expanding Det($\mathbf{H}_1$) by the last column and proceeding recursively.

5.10 Develop an algorithm for the identification of components of a graph.

5.11 The network problem $\mathcal{NP}$ is just a specialization of **GP**. Show the correspondence of (5.14), when applied to $\mathcal{NP}$, to (3.5). Show the correspondence of ALG 5.1, when applied to $\mathcal{NP}$, to ALG 3.2.

5.12 Define both path and cycle for an undirected graph.

5.13 Define both $F(j)$ and $T(j)$ for undirected arcs so that the proof in Appendix A is valid also for trees in an undirected graph.

NOTES AND REFERENCES

The results of this chapter may be traced to Dantzig's [47] linear programming textbook. This algorithm was also discussed by Johnson [112]. Data structures for implementation have been studied by Glover, Klingman, and Stutz [80] and by Elam, Glover, and Klingman [53]. Computational studies have been performed by Maurras [141], Glover, Hultz, Klingman, and Stutz [84], and Langley [130]. The code of Glover, Hultz, and Klingman [81], which is based on the ideas presented in this chapter, is approximately 50 times faster than a general linear programming code. An extension of the out-of-kilter algorithm to the generalized network problem can be found in Jewell [110]. Held and Karp [96] were the first to use the term "one-tree" to describe a connected graph with exactly one cycle. An application involving a large financial model is given in Rutenberg [163]. These ideas have also been used by Ross and Soland [159] to solve the generalized assignment problem.

CHAPTER 6

The Multicommodity Network Flow Problem

In this chapter we present primal algorithms for solving the *multicommodity network flow problem*. Problems of this type arise when several items (commodities) share arcs in a capacitated network. Mathematically this problem takes the form

$$\left.\begin{array}{ll} \min & \sum_k \mathbf{c}^k\mathbf{x}^k \\ \text{s.t.} & \mathbf{A}\mathbf{x}^k=\mathbf{r}^k, k=1,\dots,K \\ & \sum_k \mathbf{D}^k\mathbf{x}^k \leqslant \mathbf{b} \\ & \mathbf{0}\leqslant \mathbf{x}^k \leqslant \mathbf{u}^k, k=1,\dots,K \end{array}\right\}\text{MP},$$

where $\mathbf{A}$ is a node-arc incidence matrix for the graph $\mathcal{G}=[\mathcal{N},\mathcal{A}]$ and $\mathbf{D}^k$ for $k=1,\dots,K$ are diagonal matrices. The jth component of $\mathbf{b}$ is called the *mutual arc capacity* of arc e_j. This capacity may be shared among the K commodities. $\mathbf{D}^k_{jj}$ serves as a weighting factor for x_j^k with respect to this limitation and may be used to change units if the mutual capacity b_j is expressed in units different from those used for x_j^k. If $\mathbf{D}^k_{jj}=0$ for some k, then x_j^k is omitted from the mutual capacity constraint b_j. In many cases $\mathbf{D}^k=\mathbf{I}$ for $k=1,\dots,K$. We assume that, for each commodity, total supply equals total demand, that is, $\mathbf{1}\mathbf{r}^k=0$ for $k=1,\dots,K$.

Table 6.1 Farm Production (bushels)

Crop	Farm				Total
	A	*B*	*C*	*D*	Production
Corn	100	50	210	150	510
Wheat	150	180	75	150	555

It is possible to generalize MP to allow for commodity-dependent subgraphs (instead of using $\mathcal{G}$ for each commodity) or to impose limits on flow through a node. To do so involves no mathematical difficulties but greatly complicates the notation. In the interest of clarity, we have omitted these generalizations in the description of both the model and algorithms.

Before proceeding with the development of algorithms for MP, we first present an example of a multicommodity network flow problem. Consider the case of a farmer who owns four different farms (A,B,C,D) in the state of Iowa. Each of the farms raises both corn and wheat, and the anticipated yield for the current year is as shown in Table 6.1. The produce from these farms may be sold at auctions in Chicago, St. Louis, and/or Peoria. The anticipated selling prices and the demands at the three auctions are given in Tables 6.2 and 6.3. Each farm has a truck that will be

Table 6.2 Selling Prices ($/bushel)

Crop	Market		
	Chicago	St. Louis	Peoria
Corn	17	16	14
Wheat	15	16	13

Table 6.3 Market Demands (bushels)

Crop	Market			
	Chicago	St. Louis	Peoria	Total Demand
Corn	200	200	100	500
Wheat	200	150	200	550

Table 6.4 Transportation Costs ($/bushel)

Farm	Market			
	Chicago	St. Louis	Peoria	Truck Capacity (bushels)
A	3	2	1	200
B	4	1	1	250
C	2	2	1	300
D	4	3	1	250

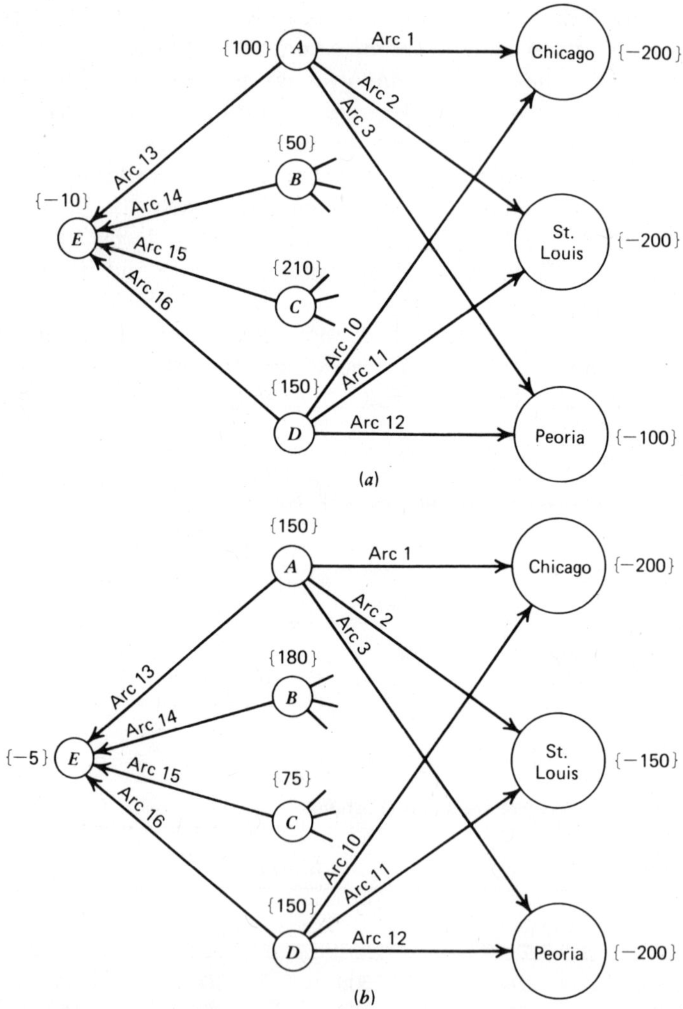

FIGURE 6.1 The distribution network. (*a*) Network associated with corn. (*b*) Network associated with wheat.

used to transport the corn and wheat to the auctions. The transportation cost and the truck capacities are as shown in Table 6.4. We assume that each truck has two storage compartments with a movable divider between the two. Therefore both corn and wheat can be shipped together on the same truck, and the capacity of the truck can be shared by the two commodities. However, farm A's truck can carry only 125 bushels of corn even if no wheat is carried. We assume that the auctions in the three cities are held on a single day, thereby precluding multiple trips to the same auction. The problem of maximizing the sum of the profits from the four farms can be modeled as a multicommodity network flow problem.

The underlying network structure associated with this problem is shown in Figure 6.1. The node requirements, arc specifications, and mutual capacity restrictions are given in Tables 6.5, 6.6, and 6.7.

In this chapter we present primal algorithms that exploit the network structure of MP. Two basic approaches have been employed to develop specialized techniques for MP, *partitioning* and *decomposition*. Partitioning approaches seek to exploit the special structure of MP by means of basis partitioning, so that portions of each basis have the characteristics of minimal cost network flow problem bases. Decomposition methods place MP in a form where a master optimization problem coordinates the solution of subproblems, each of which is a minimal cost network flow problem. Decomposition approaches may be further divided into *price-directive* and *resource-directive* methods. In price-directive methods a pricing mechanism is used by the master problem to generate subproblems; whereas in resource-directive methods the master problem makes allocations of the mutual arc capacities to individual commodities in order to generate subproblems.

Table 6.5 Requirements (bushels)

Node	Requirements	
Number	Commodity 1	Commodity 2
1	100	150
2	50	180
3	210	75
4	150	150
5	−200	−200
6	−200	−150
7	−100	−200
8	−10	−5

Table 6.6 Arc Specifications

Arc Number	"From" Node	"To" Node	Commodity	Unit Cost	Arc Capacity
1	1	5	1	−14	125
1	1	5	2	−12	∞
2	1	6	1	−14	125
2	1	6	2	−14	∞
3	1	7	1	−13	125
3	1	7	2	−12	∞
4	2	5	1	−13	∞
4	2	5	2	−11	∞
⋮					
12	4	7	1	−13	∞
12	4	7	2	−12	∞
13	1	8	1	0	∞
13	1	8	2	0	∞
⋮					
16	4	8	1	0	∞
16	4	8	2	0	∞

Table 6.7 Mutual Capacities (bushels)

Arc Number	Mutual Capacity
1	200
2	200
3	200
4	250
5	250
6	250
7	300
8	300
9	300
10	250
11	250
12	250
13	∞
14	∞
15	∞
16	∞

6.1 PRIMAL PARTITIONING

In this section we present a specialization of the primal simplex algorithm for MP. This specialization involves partitioning the basis in an attempt to exploit the underlying network structure.

6.1.1 Algebraic Results

Consider the matrix $\mathbf{E}$ of the form

$$\mathbf{E}=\left[\begin{array}{c:c:c} \mathbf{L}_1 & \mathbf{R}_1 & \mathbf{0} \\ \hdashline \mathbf{L}_2 & \mathbf{R}_2 & \mathbf{0} \\ \hdashline \mathbf{L}_3 & \mathbf{R}_3 & \mathbf{I} \end{array}\right]\begin{array}{l} \}n_1 \\ \}n_2. \\ \}n_3 \end{array}$$
$$\qquad\underbrace{\ }_{n_1}\ \underbrace{\ }_{n_2}\ \underbrace{\ }_{n_3}$$

Note that $\mathbf{L}_1$ and $\mathbf{R}_2$ are both square. It will be shown in a subsequent section that bases for multicommodity network problems assume this general form. Therefore we will examine the formulae required for solving the systems $\pi\mathbf{E}=\mathbf{c}$ and $\mathbf{E}\mathbf{y}=\mathbf{a}$, which are the operations required for determining the dual variables and updating the columns in the simplex algorithm.

Proposition 6.1

If $\text{Det}(\mathbf{L}_1)\neq 0$ and $\text{Det}(\mathbf{E})\neq 0$, then $\text{Det}(\mathbf{R}_2-\mathbf{L}_2\mathbf{L}_1^{-1}\mathbf{R}_1)\neq 0$.

Proof. Let

$$T_1=\left[\begin{array}{c:c:c} \mathbf{I} & -\mathbf{L}_1^{-1}\mathbf{R}_1 & \\ \hdashline & \mathbf{I} & \\ \hdashline & & \mathbf{I} \end{array}\right] \quad \text{and} \quad \mathbf{T}_2=\left[\begin{array}{c:c:c} \mathbf{I} & & \\ \hdashline & \mathbf{I} & \\ \hdashline & \mathbf{L}_3\mathbf{L}_1^{-1}\mathbf{R}_1-\mathbf{R}_3 & \mathbf{I} \end{array}\right].$$

Then $\text{Det}(\mathbf{T}_1)=\text{Det}(\mathbf{T}_2)=1$. Consider $\mathbf{E}\mathbf{T}_1\mathbf{T}_2$, that is,

$$\left[\begin{array}{c:c:c} \mathbf{L}_1 & \mathbf{0} & \mathbf{0} \\ \hdashline \mathbf{L}_2 & \mathbf{R}_2-\mathbf{L}_2\mathbf{L}_1^{-1}\mathbf{R}_1 & \mathbf{0} \\ \hdashline \mathbf{L}_3 & \mathbf{0} & \mathbf{I} \end{array}\right]\}n_2.$$
$$\qquad\underbrace{\qquad\qquad}_{n_2}$$

Then $\text{Det}(\mathbf{ET}_1\mathbf{T}_2)=\text{Det}(\mathbf{E})\ \text{Det}(\mathbf{T}_1)\ \text{Det}(\mathbf{T}_2)=\text{Det}(\mathbf{E})\neq 0$. But $\text{Det}(\mathbf{ET}_1\mathbf{T}_2)\neq 0$ implies that $\mathbf{ET}_1\mathbf{T}_2$ has full column rank. Then $\mathbf{R}_2-\mathbf{L}_2\mathbf{L}_1^{-1}\mathbf{R}_1$ must have full column rank since otherwise $\mathbf{ET}_1\mathbf{T}_2$ would not have full column rank. Therefore $\text{Det}(\mathbf{R}_2-\mathbf{L}_2\mathbf{L}_1^{-1}\mathbf{R}_1)\neq 0$. ■

Proposition 6.2

If $\text{Det}(\mathbf{L}_1)\neq 0$ and $\text{Det}(\mathbf{E})\neq 0$, the solution to

$$\begin{bmatrix} \mathbf{L}_1 & \mathbf{R}_1 & \\ \mathbf{L}_2 & \mathbf{R}_2 & \\ \mathbf{L}_3 & \mathbf{R}_3 & \mathbf{I} \end{bmatrix} \begin{pmatrix} \mathbf{x} \\ \mathbf{y} \\ \mathbf{z} \end{pmatrix} = \begin{pmatrix} \mathbf{u} \\ \mathbf{r} \\ \mathbf{w} \end{pmatrix}$$

is given by

$$\begin{aligned} \mathbf{y} &= \left(\mathbf{R}_2-\mathbf{L}_2\mathbf{L}_1^{-1}\mathbf{R}_1\right)^{-1}\left(\mathbf{r}-\mathbf{L}_2\mathbf{L}_1^{-1}\mathbf{u}\right), \\ \mathbf{x} &= \mathbf{L}_1^{-1}\mathbf{u}-\mathbf{L}_1^{-1}\mathbf{R}_1\mathbf{y}, \\ \mathbf{z} &= \mathbf{w}-\mathbf{L}_3\mathbf{L}_1^{-1}\mathbf{u}-\left(\mathbf{R}_3-\mathbf{L}_3\mathbf{L}_1^{-1}\mathbf{R}_1\right)\mathbf{y}. \end{aligned}$$

Proof. By block multiplication

$$\mathbf{L}_1\mathbf{x}+\mathbf{R}_1\mathbf{y} \qquad =\mathbf{u}, \tag{6.1}$$

$$\mathbf{L}_2\mathbf{x}+\mathbf{R}_2\mathbf{y} \qquad =\mathbf{r}, \tag{6.2}$$

$$\mathbf{L}_3\mathbf{x}+\mathbf{R}_3\mathbf{y}+\mathbf{z}=\mathbf{w}. \tag{6.3}$$

From (6.1) $\mathbf{L}_1\mathbf{x}=\mathbf{u}-\mathbf{R}_1\mathbf{y}$. Since $\mathbf{L}_1$ is invertible,

$$\mathbf{x}=\mathbf{L}_1^{-1}\mathbf{u}-\mathbf{L}_1^{-1}\mathbf{R}_1\mathbf{y}. \tag{6.4}$$

Substituting (6.4) into (6.2) yields $\mathbf{L}_2(\mathbf{L}_1^{-1}\mathbf{u}-\mathbf{L}_1^{-1}\mathbf{R}_1\mathbf{y})+\mathbf{R}_2\mathbf{y}=\mathbf{r}$ or simply $(\mathbf{R}_2-\mathbf{L}_2\mathbf{L}_1^{-1}\mathbf{R}_1)\mathbf{y}=\mathbf{r}-\mathbf{L}_2\mathbf{L}_1^{-1}\mathbf{u}$. By Proposition 6.1 $(\mathbf{R}_2-\mathbf{L}_2\mathbf{L}_1^{-1}\mathbf{R}_1)$ is invertible, so

$$\mathbf{y}=\left(\mathbf{R}_2-\mathbf{L}_2\mathbf{L}_1^{-1}\mathbf{R}_1\right)^{-1}\left(\mathbf{r}-\mathbf{L}_2\mathbf{L}_1^{-1}\mathbf{u}\right). \tag{6.5}$$

From (6.3) we have $\mathbf{z}=\mathbf{w}-\mathbf{L}_3\mathbf{x}-\mathbf{R}_3\mathbf{y}$. Substituting for $\mathbf{x}$ from (6.4), we obtain

$$\mathbf{z}=\mathbf{w}-\mathbf{L}_3\left(\mathbf{L}_1^{-1}\mathbf{u}-\mathbf{L}_1^{-1}\mathbf{R}_1\mathbf{y}\right)-\mathbf{R}_3\mathbf{y}$$

or simply

$$\mathbf{z}=\mathbf{w}-\mathbf{L}_3\mathbf{L}_1^{-1}\mathbf{u}-(\mathbf{R}_3-\mathbf{L}_3\mathbf{L}_1^{-1}\mathbf{R}_1)\mathbf{y}, \tag{6.6}$$

and (6.4), (6.5), and (6.6) constitute the given solution. ■

Proposition 6.3

If $\mathrm{Det}(\mathbf{L}_1)\neq 0$ and $\mathrm{Det}(\mathbf{E})\neq 0$, the solution to

$$(\mathbf{x}\,\vdots\,\mathbf{y}\,\vdots\,\mathbf{z})\left[\begin{array}{c:c:c} \mathbf{L}_1 & \mathbf{R}_1 & \\ \hdashline \mathbf{L}_2 & \mathbf{R}_2 & \\ \hdashline \mathbf{L}_3 & \mathbf{R}_3 & \mathbf{I} \end{array}\right]=(\mathbf{u}\,\vdots\,\mathbf{r}\,\vdots\,\mathbf{0})$$

is given by

$$\mathbf{z}=\mathbf{0},$$
$$\mathbf{y}=(\mathbf{r}-\mathbf{u}\mathbf{L}_1^{-1}\mathbf{R}_1)(\mathbf{R}_2-\mathbf{L}_2\mathbf{L}_1^{-1}\mathbf{R}_1)^{-1},$$
$$\mathbf{x}=(\mathbf{u}-\mathbf{y}\mathbf{L}_2)\mathbf{L}_1^{-1}.$$

Proof. By block multiplication

$$\mathbf{x}\mathbf{L}_1+\mathbf{y}\mathbf{L}_2+\mathbf{z}\mathbf{L}_3=\mathbf{u}, \tag{6.7}$$
$$\mathbf{x}\mathbf{R}_1+\mathbf{y}\mathbf{R}_2+\mathbf{z}\mathbf{R}_3=\mathbf{r}, \tag{6.8}$$
$$\mathbf{z}=\mathbf{0}. \tag{6.9}$$

From (6.7) and (6.9) we have $\mathbf{x}\mathbf{L}_1+\mathbf{y}\mathbf{L}_2=\mathbf{u}$. Since $\mathbf{L}_1$ is invertible,

$$\mathbf{x}=(\mathbf{u}-\mathbf{y}\mathbf{L}_2)\mathbf{L}_1^{-1}. \tag{6.10}$$

Substituting (6.9) and (6.10) into (6.8), we obtain

$$(\mathbf{u}-\mathbf{y}\mathbf{L}_2)\mathbf{L}_1^{-1}\mathbf{R}_1+\mathbf{y}\mathbf{R}_2=\mathbf{r} \quad \text{or} \quad \mathbf{y}(\mathbf{R}_2-\mathbf{L}_2\mathbf{L}_1^{-1}\mathbf{R}_1)=\mathbf{r}-\mathbf{u}\mathbf{L}_1^{-1}\mathbf{R}_1.$$

By Proposition 6.1, $(\mathbf{R}_2-\mathbf{L}_2\mathbf{L}_1^{-1}\mathbf{R}_1)$ is invertible. Therefore

$$\mathbf{y}=(\mathbf{r}-\mathbf{u}\mathbf{L}_1^{-1}\mathbf{R}_1)(\mathbf{R}_2-\mathbf{L}_2\mathbf{L}_1^{-1}\mathbf{R}_1)^{-1}, \tag{6.11}$$

and (6.9), (6.10), and (6.11) constitute the given solution. ■

6.1.2 Basis Characterization

To obtain a problem having equality constraints and full row rank, we append slacks and artificials to MP to obtain

$$
\begin{aligned}
\min \quad & \sum_k \mathbf{c}^k \mathbf{x}^k \\
\text{s.t.} \quad & \mathbf{A}\mathbf{x}^k + \mathbf{e}^l a_k = \mathbf{r}^k, \qquad \text{for all } k && (6.12) \\
& \sum_k \mathbf{D}^k \mathbf{x}^k + \mathbf{s} = \mathbf{b} && (6.13) \\
& \mathbf{0} \leqslant \mathbf{x}^k \leqslant \mathbf{u}^k \text{ and } 0 \leqslant a_k \leqslant 0 \text{ for all } k, \mathbf{s} \geqslant \mathbf{0}.
\end{aligned}
$$

The constraint matrix associated with (6.12) and (6.13) may be displayed as follows:

$$
\hat{\mathbf{A}} = \left(\begin{array}{cc|cc|c|cc|c}
\mathbf{A} & \mathbf{e}^l & & & & & & \\ \hline
 & & \mathbf{A} & \mathbf{e}^l & & & & \\ \hline
 & & & & \ddots & & & \\ \hline
 & & & & & \mathbf{A} & \mathbf{e}^l & \\ \hline
\mathbf{D}^1 & \mathbf{0} & \mathbf{D}^2 & \mathbf{0} & \cdots & \mathbf{D}^K & \mathbf{0} & \mathbf{I}
\end{array}\right)
=
\left[\begin{array}{c|c|c|c|c}
\bar{\mathbf{A}} & & & & \\ \hline
 & \bar{\mathbf{A}} & & & \\ \hline
 & & \ddots & & \\ \hline
 & & & \bar{\mathbf{A}} & \\ \hline
\bar{\mathbf{D}}^1 & \bar{\mathbf{D}}^2 & \cdots & \bar{\mathbf{D}}^K & \mathbf{I}
\end{array}\right]
\begin{array}{l}
\}\bar{I} \\ \}\bar{I} \\ \vdots \\ \}\bar{I} \\ \}\bar{J}
\end{array},
$$

$$
\qquad\qquad\qquad\qquad\qquad\qquad \underbrace{\quad}_{\bar{J}+1} \; \underbrace{\quad}_{\bar{J}+1} \; \cdots \; \underbrace{\quad}_{\bar{J}+1} \; \underbrace{\quad}_{\bar{J}}
$$

where we have assumed that $\mathcal{G}$ has $\bar{I}$ nodes and $\bar{J}$ arcs.

Proposition 6.4

Every basis for $\hat{\mathbf{A}}$ may be placed in the form

$$
\left[\begin{array}{c|c|c|c|c|c|c}
\mathbf{B}^1 & & & \mathbf{R}^1 & & & \\ \hline
 & \ddots & & & \ddots & & \\ \hline
 & & \mathbf{B}^K & & & \mathbf{R}^K & \\ \hline
\mathbf{P}^1 & \cdots & \mathbf{P}^K & \mathbf{T}^1 & \cdots & \mathbf{T}^K & \\ \hline
\mathbf{S}^1 & \cdots & \mathbf{S}^K & \mathbf{U}^1 & \cdots & \mathbf{U}^K & \mathbf{I}
\end{array}\right]
\begin{array}{l}
\}\bar{I} \\ \vdots \\ \}\bar{I}, \\ \}q \\ \}p
\end{array}
\qquad (6.14)
$$

$$
\underbrace{\quad}_{\bar{I}} \; \cdots \; \underbrace{\quad}_{\bar{I}} \; \underbrace{\qquad\qquad\qquad}_{\bar{J}}
$$

where $\mathbf{B}^1, \ldots, \mathbf{B}^K$ are bases for $\bar{\mathbf{A}}$. (Note that any of the partitions $\mathbf{R}^1, \ldots, \mathbf{R}^K$ may be empty.)

Proof. Let $\hat{\mathbf{B}}$ be a basis for $\hat{\mathbf{A}}$. Then $\hat{\mathbf{B}}$ may be partitioned as follows:

$$\hat{\mathbf{B}} = \left[\begin{array}{c|c|c|c} \mathbf{V}^1 & & & \\ \hline & \ddots & & \\ \hline & & \mathbf{V}^K & \\ \hline \mathbf{X}^1 & \cdots & \mathbf{X}^K & \mathbf{W} \end{array}\right] \begin{array}{l} \}\bar{I} \\ \vdots \\ \\ \}\bar{I} \end{array}.$$

Let $k \varepsilon \{1,\ldots,K\}$. Suppose the rank of $\mathbf{V}^k < \bar{I}$. Then some rows of $\hat{\mathbf{B}}$ are dependent, a contradiction. Therefore the rank of $\mathbf{V}^k$ is $\bar{I}$. Then $\mathbf{V}^k$ may be displayed as $\mathbf{V}^k = [\mathbf{B}^k \,\vdots\, \mathbf{R}^k]$, where the rank of $\mathbf{B}^k$ is $\bar{\mathbf{I}}$, and it is understood that the partition corresponding to $\mathbf{R}^k$ may be empty. ■

Note that (6.14) takes the same form as $\mathbf{E}$ in Section 6.1.1, where

$$\mathbf{L}_1 = \left(\begin{array}{c|c|c} \mathbf{B}^1 & & \\ \hline & \ddots & \\ \hline & & \mathbf{B}^K \end{array}\right), \qquad \mathbf{R}_1 = \left(\begin{array}{c|c|c} \mathbf{R}^1 & & \\ \hline & \ddots & \\ \hline & & \mathbf{R}^K \end{array}\right),$$

$$\mathbf{L}_2 = (\mathbf{P}^1 \,\vdots\, \ldots \,\vdots\, \mathbf{P}^K), \qquad \mathbf{R}_2 = (\mathbf{T}^1 \,\vdots\, \ldots \,\vdots\, \mathbf{T}^K),$$

$$\mathbf{L}_3 = (\mathbf{S}^1 \,\vdots\, \ldots \,\vdots\, \mathbf{S}^K), \qquad \mathbf{R}_3 = (\mathbf{U}^1 \,\vdots\, \ldots \,\vdots\, \mathbf{U}^K).$$

Furthermore, $\mathbf{L}_1$ is invertible, and

$$\mathbf{L}_1^{-1} = \left(\begin{array}{c|c|c} (\mathbf{B}^1)^{-1} & & \\ \hline & \ddots & \\ \hline & & (\mathbf{B}^K)^{-1} \end{array}\right).$$

The matrix corresponding to $(\mathbf{R}_2 - \mathbf{L}_2\mathbf{L}_1^{-1}\mathbf{R}_1)$ from Proposition 6.1 is called the *working basis* or *cycle matrix* and will be denoted by $\mathbf{Q}$, that is,

$$\mathbf{Q} = (\mathbf{T}^1 \,\vdots\, \cdots \,\vdots\, \mathbf{T}^K) - (\mathbf{P}^1 \,\vdots\, \cdots \,\vdots\, \mathbf{P}^K) \left(\begin{array}{c|c|c} (\mathbf{B}^1)^{-1} & & \\ \hline & \ddots & \\ \hline & & (\mathbf{B}^K)^{-1} \end{array}\right) \left(\begin{array}{c|c|c} \mathbf{R}^1 & & \\ \hline & \ddots & \\ \hline & & \mathbf{R}^K \end{array}\right)$$

$$= \left(\mathbf{T}^1 - \mathbf{P}^1(\mathbf{B}^1)^{-1}\mathbf{R}^1 \,\vdots\, \cdots \,\vdots\, \mathbf{T}^K - \mathbf{P}^K(\mathbf{B}^K)^{-1}\mathbf{R}^K\right).$$

Let $\Omega = \{e_{m_1},\ldots,e_{m_q}\} \subset \mathscr{E}$ denote the index set corresponding to the rows of $\mathbf{R}_2 = [\mathbf{T}^1,\ldots,\mathbf{T}^K]$. The index set Ω is ordered so that e_{m_r} denotes the arc corresponding to the rth row of R_2. Let $\Delta^k = \{e_{j_1},\ldots,e_{j_k}\} \subset \mathscr{Q}$ denote the index set corresponding to the columns of R^k. The index set Δ^k is ordered so that e_{j_s} denotes the arc corresponding to the sth column of R^k. Using these index sets we obtain the substitution formula $T^k_{rs} = D^k_{m_r j_r}$. (A similar substitution formula can be obtained for the elements of P^k by defining an index set, analogous to Δ^k, corresponding to the columns of B^k.) Using these index sets, we will show that $\mathbf{Q}$ can be generated from the basis trees.

Recall that the kth partition of $\mathbf{Q}$ is given by $\mathbf{T}^k - \mathbf{P}^k(\mathbf{B}^k)^{-1}\mathbf{R}^k$. Let $\mathscr{T}^k$ denote the rooted spanning tree corresponding to $\mathbf{B}^k$. Then, for any arc $e_{j_s} \varepsilon \Delta^k$, there is a unique path in $\mathscr{T}^k$, say P_s, linking $T(j_s)$ to $F(j_s)$. Let $O(s)$ denote the orientation sequence corresponding to P_s. We shall say that an arc, $e_j \varepsilon P_s$, has *normal orientation* if the component of $O(s)$ corresponding to e_j is $+1$ and has *reverse orientation* if this condition is not met. We now indicate a means for generating $\mathbf{Q}$ from $\mathscr{T}^k$, $\mathbf{R}^k$, and the index sets.

Proposition 6.5

Choose the commodity index k. Let r (the row index) be any element of $\{1,\ldots,q\}$, and let s (the column index) be any element of $\{1,\ldots,v_k\}$. Let P_s be the path in $\mathscr{T}^k$ linking $T(j_s)$ and $F(j_s)$. Then $[\mathbf{T}^k - \mathbf{P}^k(\mathbf{B}^k)^{-1}\mathbf{R}^k]_{rs}$ is given by

$D^k_{m_r m_r}$	if $m_r = j_s$, where $e_{j_s} \varepsilon \Delta^k$ and $e_{m_r} \varepsilon \Omega$;
$D^k_{m_r m_r}$	if $e_{m_r} \varepsilon P_s$ with normal orientation and $e_{m_r} \varepsilon \Omega$;
$-D^k_{m_r m_r}$	if $e_{m_r} \varepsilon P_s$ with reverse orientation and $e_{m_r} \varepsilon \Omega$;
0	otherwise.

Proof. Let $\mathbf{P}^k_r$ denote the rth row of $\mathbf{P}^k$, and let $\boldsymbol{\beta}^s$ denote the sth column of $(\mathbf{B}^k)^{-1}\mathbf{R}^k$.

Case 1 Suppose $m_r = j_s$, where $e_{m_r} \varepsilon \Omega$ and $e_{j_s} \varepsilon \Delta^k$. Then

$$T^k_{rs} = D^k_{m_r m_r} \quad \text{and} \quad P^k_r = 0.$$

Hence

$$Q^k_{rs} = D^k_{m_r m_r}.$$

Case 2 Suppose $e_{m_r} \varepsilon P_s$ with normal orientation, where $e_{m_r} \varepsilon \Omega$, and suppose e_{m_r} is in the tth column of $\mathbf{B}^k$. Then

$$\mathrm{Q}_{rs}^k = \mathrm{T}_{rs}^k - \mathrm{P}_r^k \boldsymbol{\beta}^s$$

$$= 0 - \mathrm{D}_{m_r m_r}^k \mathbf{e}^t \begin{bmatrix} \\ -1 \\ \\ \end{bmatrix} \leftarrow t\text{th}$$

$$= \mathrm{D}_{m_r m_r}^k.$$

Case 3 Suppose $e_{m_r} \varepsilon P_s$ with reverse orientation, where $e_{m_r} \varepsilon \Omega$ and suppose e_{m_r} is in the tth column of $\mathbf{B}^k$. Then

$$\mathrm{Q}_{rs}^k = \mathrm{T}_{rs}^k - \mathrm{P}_r^k \boldsymbol{\beta}^s$$

$$= 0 - \mathrm{D}_{m_r m_r}^k \mathbf{e}^t \begin{bmatrix} \\ 1 \\ \\ \end{bmatrix} \leftarrow t\text{th}$$

$$= -\mathrm{D}_{m_r m_r}^k.$$

Case 4 Suppose $e_{m_r} \not\varepsilon P_s$ and $e_{m_r} \neq e_{j_s}$, where $e_{m_r} \varepsilon \Omega$ and $e_{j_s} \varepsilon \Delta^k$. Suppose $P_r^k = 0$. Then, clearly,

$$\mathrm{Q}_{rs}^k = 0 \qquad \text{since } \mathbf{T}_{rs}^k = 0.$$

Suppose $P_r^k = \mathrm{D}_{m_r m_r}^k \mathbf{e}^t$. Then

$$\mathrm{Q}_{rs}^k = \mathrm{T}_{rs}^k - \mathbf{P}_r^k \boldsymbol{\beta}^s$$

$$= 0 - \mathrm{D}_{m_r m_r}^k \mathbf{e}^t \begin{bmatrix} \\ 0 \\ \\ \end{bmatrix} \leftarrow t\text{th}$$

$$= 0.$$

■

The matrix $\mathbf{Q}$ is called the cycle matrix because $\mathbf{Q}$ can be generated as shown in Proposition 6.5 by generating the cycles corresponding to columns of each $\mathbf{R}_k$.

Consider the two-commodity problem illustrated in Figure 6.2*a*, where the supply for nodes 1, 2, and 3 for both commodities is 2 and the demand at nodes 4, 5, and 6 for both commodities is 2. The individual bounds u_j^k are infinite for all arcs and both commodities, and the mutual capacity for arc e_1 is 2 and for all others is 3. Also, $\mathbf{D}^1 = \mathbf{D}^2 = \mathbf{I}$. Consider the basis shown as Display 1. Then $\Omega = \{1,2,4\}$, $\Delta^1 = \{4\}$, $\Delta^2 = \{3,8\}$, and $\mathfrak{T}^1, \mathfrak{T}^2$ are as shown in Figure 6.2*b*. Thus

1	1																				
		1											1								2
			1	1																	3
−1													−1								4
	−1	−1	−1																		5
				−1	1																6
						1	1							1							1
								1	1												2
										1					1						3
						−1				−1											4
							−1	−1							−1						5
									−1		1			−1							6
1						1															e_1
	1						1														e_2
													1								e_4
														1		1					e_3
		1						1									1				e_5
									1									1			e_6
										1									1		e_7
			1												1					1	e_8
				1																	1	e_9

Display 1

$$\mathbf{P}^1 = \left(\begin{array}{c|c|c|c|c|c} 1 & & & & & \\ \hline & 1 & & & & \\ \hline & & & & & \end{array}\right), \qquad \mathbf{T}^1 = \left(\begin{array}{c} \\ \hline \\ \hline 1 \end{array}\right)\begin{array}{l} e_1 \\ e_2, \\ e_4 \end{array} \qquad \mathbf{P}^2 = \left(\begin{array}{c|c|c|c|c|c} 1 & & & & & \\ \hline & 1 & & & & \\ \hline & & & & & \end{array}\right), \qquad \mathbf{T}^2 = \mathbf{0}.$$

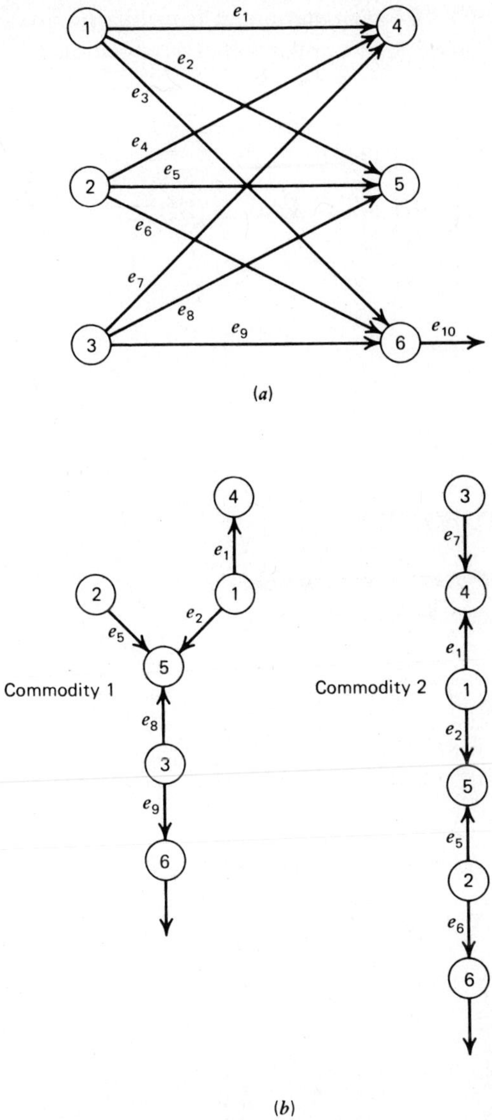

FIGURE 6.2 Two-commodity sample problem. (*a*) Graph for example. (*b*) Basis tree for each commodity.

$(\mathbf{B}^1)^{-1}\mathbf{R}^1$ and $(\mathbf{B}^2)^{-1}\mathbf{R}^2$ are determined from the orientation sequences as illustrated in Figure 6.3, by application of Proposition 3.3. Therefore

$$(\mathbf{B}^1)^{-1}\mathbf{R}^1=\left[\begin{array}{r} 1 \\ \hline -1 \\ \hline 1 \\ \hline 0 \\ \hline 0 \\ \hline 0 \end{array}\right]\begin{array}{l} e_1 \\ e_2 \\ e_5 \\ e_8 \\ e_9 \\ e_{10} \end{array} \quad \text{and} \quad (\mathbf{B}^2)^{-1}\mathbf{R}^2=\left[\begin{array}{r|r} 0 & -1 \\ \hline 1 & 1 \\ \hline -1 & 0 \\ \hline 1 & 0 \\ \hline 0 & 1 \\ \hline 0 & 0 \end{array}\right]\begin{array}{l} e_1 \\ e_2 \\ e_5 \\ e_6 \\ e_7 \\ e_{10} \end{array}.$$

Then

$$\mathbf{T}^1-\mathbf{P}^1(\mathbf{B}^1)^{-1}\mathbf{R}^1=\left[\begin{array}{c} \\ \hline \\ \hline 1 \end{array}\right]-\left[\begin{array}{c|c|c|c|c|c} 1 & & & & & \\ \hline & 1 & & & & \\ \hline 0 & & & & & \end{array}\right]\left[\begin{array}{r} 1 \\ \hline -1 \\ \hline 1 \\ \hline \\ \hline \\ \hline \\ \end{array}\right]$$

$$=\left[\begin{array}{r} -1 \\ 1 \\ 1 \end{array}\right],$$

and

$$\mathbf{T}^2-\mathbf{P}^2(\mathbf{B}^2)^{-1}\mathbf{R}^2=\mathbf{0}-\left[\begin{array}{c|c|c|c|c|c} 1 & & & & & \\ \hline & 1 & & & & \\ \hline & & & & & \end{array}\right]\left[\begin{array}{r|r} & -1 \\ \hline 1 & 1 \\ \hline -1 & \\ \hline 1 & \\ \hline & 1 \\ \hline & \end{array}\right]$$

$$=\left[\begin{array}{r|r} & 1 \\ \hline -1 & -1 \\ \hline & \end{array}\right].$$

Hence

$$\mathbf{Q}=\left[\begin{array}{r|r|r} -1 & & 1 \\ \hline 1 & -1 & -1 \\ \hline 1 & & \end{array}\right].$$

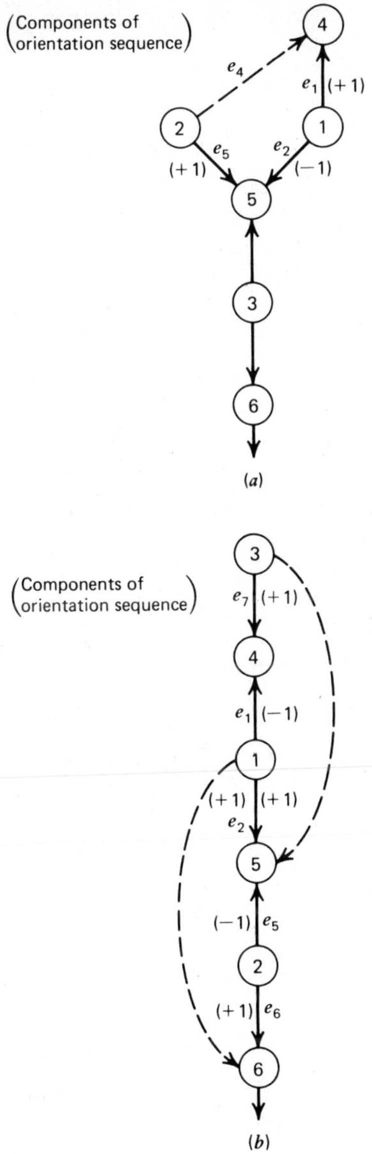

FIGURE 6.3 Orientation sequences. (*a*) Orientation sequence for $\mathcal{T}^1$. (*b*) Orientation sequences for $\mathcal{T}^2$.

6.1.3 Dual Variable Calculation

In this section we indicate how the structure of the basis $\hat{\mathbf{B}}$ may be used to efficiently calculate the dual variables $\boldsymbol{\pi}=\mathbf{c}_B\hat{\mathbf{B}}^{-1}$, that is, we wish to solve the system $\boldsymbol{\pi}\hat{\mathbf{B}}=\mathbf{c}_B$. This is equivalent to

$$(\boldsymbol{\pi}^1 \mid \boldsymbol{\pi}^2 \mid \boldsymbol{\pi}^3)\left[\begin{array}{c|c|c} \mathbf{L}_1 & \mathbf{R}_1 & \\ \hline \mathbf{L}_2 & \mathbf{R}_2 & \\ \hline \mathbf{L}_3 & \mathbf{R}_3 & \mathbf{I} \end{array}\right]=(\mathbf{c}_B^1 \mid \mathbf{c}_B^2 \mid \mathbf{0}),$$

where

$$\mathbf{L}_1=\left[\begin{array}{c|c|c} \mathbf{B}^1 & & \\ \hline & \ddots & \\ \hline & & \mathbf{B}^K \end{array}\right], \qquad \mathbf{R}_1=\left[\begin{array}{c|c|c} \mathbf{R}^1 & & \\ \hline & \ddots & \\ \hline & & \mathbf{R}^K \end{array}\right],$$

$$\mathbf{L}_2=(\mathbf{P}^1 \mid \cdots \mid \mathbf{P}^K), \qquad \mathbf{R}_2=(\mathbf{T}^1 \mid \cdots \mid \mathbf{T}^K),$$

$$\mathbf{L}_3=(\mathbf{S}^1 \mid \cdots \mid \mathbf{S}^K), \qquad \mathbf{R}_3=(\mathbf{U}^1 \mid \cdots \mid \mathbf{U}^K).$$

By partitioning $\mathbf{c}_B^1$ and $\boldsymbol{\pi}^1$ to be compatible with $\mathbf{L}_1$, partitioning $\mathbf{c}_B^2$ and $\boldsymbol{\pi}^2$ to be compatible with $\mathbf{L}_2$, and applying Proposition 6.3, we obtain

$$\boldsymbol{\pi}^3=\mathbf{0},$$

$$\boldsymbol{\pi}^2=\left\{(\mathbf{c}_B^{21} \mid \cdots \mid \mathbf{c}_B^{2K}) - (\mathbf{c}_B^{11} \mid \cdots \mid \mathbf{c}_B^{1K})\left[\begin{array}{c|c|c} (\mathbf{B}^1)^{-1} & & \\ \hline & \ddots & \\ \hline & & (\mathbf{B}^K)^{-1} \end{array}\right]\left[\begin{array}{c|c|c} \mathbf{R}^1 & & \\ \hline & \ddots & \\ \hline & & \mathbf{R}^K \end{array}\right]\right\}\mathbf{Q}^{-1} \tag{6.15}$$

and

$$\boldsymbol{\pi}^1 = (\boldsymbol{\pi}^{11} \vdots \ldots \vdots \boldsymbol{\pi}^{1K}) = \left\{ (\mathbf{c}_B^{11} \vdots \ldots \vdots \mathbf{c}_B^{1K}) - \boldsymbol{\pi}^2 (\mathbf{P}^1 \vdots \ldots \vdots \mathbf{P}^K) \right\} \left[\begin{array}{c|c|c} (\mathbf{B}^1)^{-1} & & \\ \hline & \ddots & \\ \hline & & (\mathbf{B}^K)^{-1} \end{array} \right]. \tag{6.16}$$

Simplifying (6.15), we obtain

$$\boldsymbol{\pi}^2 = \left(\mathbf{c}_B^{21} - \mathbf{c}_B^{11} (\mathbf{B}^1)^{-1} \mathbf{R}^1 \vdots \ldots \vdots \mathbf{c}_B^{2K} - \mathbf{c}_B^{1K} (\mathbf{B}^K)^{-1} \mathbf{R}^K \right) \mathbf{Q}^{-1}. \tag{6.17}$$

By using (3.4), we can generate the components $\mathbf{c}_B^{2k} - \mathbf{c}_B^{1k}(\mathbf{B}^k)^{-1}\mathbf{R}^k$ from the basis tree $\mathcal{T}^k$. Once these have been determined, $\boldsymbol{\pi}^2$ can be obtained by postmultiplying by $\mathbf{Q}^{-1}$.

Let us now examine (6.16). Postmultiplying by $\mathbf{L}_1$ yields the system

$$\left\{ \begin{array}{ll} \boldsymbol{\pi}^{11}\mathbf{B}^1 & = \mathbf{c}_B^{11} - \boldsymbol{\pi}^2 \mathbf{P}^1 \\ \vdots & \\ \boldsymbol{\pi}^{1K}\mathbf{B}^K & = \mathbf{c}_B^{1K} - \boldsymbol{\pi}^2 \mathbf{P}^K \end{array} \right\}. \tag{6.18}$$

Letting $\gamma^k = \mathbf{c}_B^{1k} - \boldsymbol{\pi}^2 \mathbf{P}^k$, we have

$$\boldsymbol{\pi}^{1k}\mathbf{B}^k = \gamma^k. \tag{6.19}$$

Hence ALG 3.2 can be used to solve (6.19) for $k = 1, \ldots, K$, where γ^k plays the role of the basic costs. Since $\mathbf{P}^k$ has at most one nonzero entry in each column, $\boldsymbol{\pi}^{2k}\mathbf{P}^k$ can be generated from the original data. Therefore (6.16) can be computed without matrix operations.

6.1.4 Column Update

In this section we indicate how the structure of the basis $\hat{\mathbf{B}}$ can be used to efficiently calculate an updated column, that is, $\mathbf{y} = \hat{\mathbf{B}}^{-1}\bar{\mathbf{a}}_j$, where $\bar{\mathbf{a}}_j$ is a nonbasic column of (6.12)–(6.13). When $\bar{\mathbf{a}}_j$ is partitioned appropriately, this is equivalent to

$$\left[\begin{array}{c|c|c} \mathbf{L}_1 & \mathbf{R}_1 & \\ \hline \mathbf{L}_2 & \mathbf{R}_2 & \\ \hline \mathbf{L}_3 & \mathbf{R}_3 & \mathbf{I} \end{array} \right] \left[\begin{array}{c} \mathbf{y}^1 \\ \hline \mathbf{y}^2 \\ \hline \mathbf{y}^3 \end{array} \right] = \left[\begin{array}{c} \boldsymbol{\alpha} \\ \hline \boldsymbol{\beta} \\ \hline \boldsymbol{\gamma} \end{array} \right],$$

where

$$\mathbf{L}_1=\begin{bmatrix}\mathbf{B}^1 & & \\ & \ddots & \\ & & \mathbf{B}^K\end{bmatrix},\qquad \mathbf{R}_1=\begin{bmatrix}\mathbf{R}^1 & & \\ & \ddots & \\ & & \mathbf{R}^K\end{bmatrix},$$

$$\mathbf{L}_2=(\mathbf{P}^1 \mid \cdots \mid \mathbf{P}^K),\qquad \mathbf{R}_2=(\mathbf{T}^1 \mid \cdots \mid \mathbf{T}^K),$$

$$\mathbf{L}_3=(\mathbf{S}^1 \mid \cdots \mid \mathbf{S}^K),\qquad \mathbf{R}_3=(\mathbf{U}^1 \mid \cdots \mid \mathbf{U}^K).$$

By an appropriate partitioning of $\mathbf{y}^1$, $\mathbf{y}^2$, $\boldsymbol{\alpha}$, and $\boldsymbol{\beta}$, and application of Proposition 6.2, we have

$$\mathbf{y}^2=\begin{bmatrix}\mathbf{y}^{21}\\ \vdots \\ \mathbf{y}^{2K}\end{bmatrix}=\mathbf{Q}^{-1}\left\{\begin{bmatrix}\boldsymbol{\beta}^1\\ \vdots \\ \boldsymbol{\beta}^K\end{bmatrix}-(\mathbf{P}^1 \mid \cdots \mid \mathbf{P}^K)\begin{bmatrix}(\mathbf{B}^1)^{-1} & & \\ & \ddots & \\ & & (\mathbf{B}^K)^{-1}\end{bmatrix}\begin{bmatrix}\boldsymbol{\alpha}^1\\ \vdots \\ \boldsymbol{\alpha}^K\end{bmatrix}\right\}$$

$$=\mathbf{Q}^{-1}\begin{bmatrix}\boldsymbol{\beta}^1-\mathbf{P}^1(\mathbf{B}^1)^{-1}\boldsymbol{\alpha}^1\\ \vdots \\ \boldsymbol{\beta}^K-\mathbf{P}^K(\mathbf{B}^K)^{-1}\boldsymbol{\alpha}^K\end{bmatrix} \tag{6.20}$$

and

$$\mathbf{y}^1=\begin{bmatrix}\mathbf{y}^{11}\\ \vdots \\ \mathbf{y}^{1K}\end{bmatrix}=\begin{bmatrix}(\mathbf{B}^1)^{-1} & & \\ & \ddots & \\ & & (\mathbf{B}^K)^{-1}\end{bmatrix}\begin{bmatrix}\boldsymbol{\alpha}^1\\ \vdots \\ \boldsymbol{\alpha}^K\end{bmatrix}$$

$$-\begin{bmatrix}(\mathbf{B}^1)^{-1} & & \\ & \ddots & \\ & & (\mathbf{B}^K)^{-1}\end{bmatrix}\begin{bmatrix}\mathbf{R}^1 & & \\ & \ddots & \\ & & \mathbf{R}^K\end{bmatrix}\begin{bmatrix}\mathbf{y}^{21}\\ \vdots \\ \mathbf{y}^{2K}\end{bmatrix}$$

$$=\begin{Bmatrix}(\mathbf{B}^1)^{-1}\boldsymbol{\alpha}^1-(\mathbf{B}^1)^{-1}\mathbf{R}^1\mathbf{y}^{21}\\ \vdots \\ (\mathbf{B}^K)^{-1}\boldsymbol{\alpha}^K-(\mathbf{B}^K)^{-1}\mathbf{R}^K\mathbf{y}^{2K}\end{Bmatrix}. \tag{6.21}$$

Likewise, $\mathbf{y}^3$ is given by

$$\mathbf{y}_3=\gamma-(\mathbf{S}^1 \vdots \cdots \vdots \mathbf{S}^K)\begin{bmatrix}(\mathbf{B}^1)^{-1}\boldsymbol{\alpha}^1\\ \vdots \\ (\mathbf{B}^K)^{-1}\boldsymbol{\alpha}^K\end{bmatrix}$$

$$-\left[(\mathbf{U}^1 \vdots \cdots \vdots \mathbf{U}^K)-(\mathbf{S}^1 \vdots \cdots \vdots \mathbf{S}^K)\begin{bmatrix}(\mathbf{B}^1)^{-1} & & \\ & \ddots & \\ & & (\mathbf{B}^K)^{-1}\end{bmatrix}\cdot\begin{bmatrix}\mathbf{R}^1 & & \\ & \ddots & \\ & & \mathbf{R}^K\end{bmatrix}\right]\begin{bmatrix}\mathbf{y}^{21}\\ \vdots \\ \mathbf{y}^{2K}\end{bmatrix}$$

$$\mathbf{y}^3=\gamma-\mathbf{S}^1\mathbf{y}^{11}-\cdots-\mathbf{S}^K\mathbf{y}^{1K}-\mathbf{U}^1\mathbf{y}^{21}-\cdots-\mathbf{U}^K\mathbf{y}^{2K}. \qquad (6.22)$$

But (3.5) can be used to generate $(\mathbf{B}^k)^{-1}\boldsymbol{\alpha}^k$ and $(\mathbf{B}^k)^{-1}\mathbf{R}^k$ for all k. Furthermore, $\mathbf{S}^k$ and $\mathbf{U}^k$ have at most one nonzero in each column, and therefore $\mathbf{y}^3$ may be generated from $\mathcal{T}^1,\ldots,\mathcal{T}^k$.

6.1.5 Primal Partitioning Algorithm

In this section we restate the primal simplex algorithm (ALG 2.2) and indicate which formulae can be used to expedite the calculations when ALG 2.2 is applied to a multicommodity network flow problem. Consider the following.

ALG 6.1 PRIMAL PARTITIONING ALGORITHM (ALG 2.2 SPECIALIZED FOR MP)

0 *Initialization.* Let $[\mathbf{x}^B \vdots \mathbf{x}^N]$ be a basic feasible solution with basis **B** and nonbasic columns **N**. Root arcs, slack arcs, and artificial arcs may be appended to $\mathbf{A}\mathbf{x}^k=\mathbf{r}^k$, and slack variables may be appended to $\Sigma_k\mathbf{D}^k\mathbf{x}^k \leqslant \mathbf{b}$ to obtain an initial feasible solution.

This initial basis will assume the form

$$\mathbf{B} = \left[\begin{array}{c|c|c|c} \mathbf{B}^1 & & & \\ \hline & \ddots & & \\ \hline & & \mathbf{B}^K & \\ \hline & & & \mathbf{I} \end{array}\right].$$

1 *Pricing.* Let

$$\psi_1 = \{ i : x_i^N = 0 \text{ and } c_i^N - \mathbf{c}^B \mathbf{B}^{-1} \mathbf{N}(i) < 0 \}$$

and

$$\psi_2 = \{ i : x_i^N = u_i^N \text{ and } c_i^N = \mathbf{c}^B \mathbf{B}^{-1} \mathbf{N}(i) > 0 \}.$$

If $\psi_1 \cup \psi_2 = \phi$, terminate with $[\mathbf{x}^B \,|\, \mathbf{x}^N]$ optimal; otherwise select $k \,\varepsilon\, \psi_1 \cup \psi_2$ and set

$$\delta \leftarrow \begin{cases} 1 & \text{if } k \,\varepsilon\, \psi_1, \\ -1 & \text{if } k \,\varepsilon\, \psi_2. \end{cases}$$

The calculation $c_i^N - \mathbf{c}^B \mathbf{B}^{-1} \mathbf{N}(i)$ is made by finding $\boldsymbol{\pi} = \mathbf{c}^B \mathbf{B}^{-1}$ and using $\boldsymbol{\pi}$ to obtain $c_i^N - \boldsymbol{\pi} \mathbf{N}(i)$. For MP $\boldsymbol{\pi}$ is partitioned into $[\boldsymbol{\pi}^{11}, \ldots, \boldsymbol{\pi}^{1K} \,|\, \boldsymbol{\pi}^2 \,|\, \boldsymbol{\pi}^3]$ corresponding to (6.14), where $\boldsymbol{\pi}^3 = \mathbf{0}$. Use (6.17) to calculate $\boldsymbol{\pi}^2$ and (6.19) to calculate $\boldsymbol{\pi}^1 = [\boldsymbol{\pi}^{11}, \ldots, \boldsymbol{\pi}^{1K}]$.

2 *Ratio Test.* Set $\mathbf{y} \leftarrow \mathbf{B}^{-1} \mathbf{N}(k)$. Set

$$\Delta_1 \leftarrow \min_{\sigma(y_j) = \delta} \left\{ \frac{x_j^B}{|y_j|}, \infty \right\},$$

$$\Delta_2 \leftarrow \min_{-\sigma(y_j) = \delta} \left\{ \frac{u_j^B - x_j^B}{|y_j|}, \infty \right\},$$

$$\Delta \leftarrow \min \{ \Delta_1, \Delta_2, u_k^N \}.$$

For MP y is partitioned into

$$\mathbf{y}=\begin{bmatrix}\mathbf{y}^1\\ \hdashline \mathbf{y}^2\\ \hdashline \mathbf{y}^3\end{bmatrix}=\begin{bmatrix}\mathbf{y}^{11}\\ \vdots\\ \mathbf{y}^{1K}\\ \hdashline \mathbf{y}^{21}\\ \vdots\\ \mathbf{y}^{2K}\\ \hdashline \mathbf{y}^3\end{bmatrix}\qquad \text{corresponding to (6.14).}$$

Use (6.20) to find $\mathbf{y}^2=[\mathbf{y}^{21},\ldots,\mathbf{y}^{2K}]$, use (6.21) to find $\mathbf{y}^1=[\mathbf{y}^{11},\ldots,\mathbf{y}^{1K}]$, and use (6.22) to obtain $\mathbf{y}^3$.

3 *Update.* Set $x_k^N \leftarrow x_k^N+\Delta\delta$, and $x^B \leftarrow x^B-\Delta\delta y$. If $\Delta=u_k^N$, return to Step 1.

4 *Pivot.* Let

$$\psi_3=\{j:\sigma(y_j)=\delta \text{ and } x_j^B=0\}$$

and

$$\psi_4=\{j:-\sigma(y_j)=\delta \text{ and } x_j^B=u_j^B\}.$$

Select any $l \varepsilon \psi_3 \cup \psi_4$. In the basis, replace $\mathbf{B}(l)$ with $\mathbf{N}(k)$ and return to Step 1.

Note that the basic formulae used to execute Alg 6.1 involve the cycle matrix inverse $\mathbf{Q}^{-1}$. Efficient implementation of this algorithm requires that one have available a good procedure for maintaining and updating the inverse of this matrix. The updating formulae for $\mathbf{Q}^{-1}$ are given in Appendix E.

We now present an example illustrating these ideas. Consider the multicommodity network flow problem described in Table 6.8 with supplies and demands as given in Table 6.9. Suppose we begin with the initial solution given in Table 6.10 (also shown in Figure 6.4). After appending roots to both commodities, arcs 3 and 6 to commodity 1, and arc 1 to commodity 2, we can show the initial basis as in Display 2,

nodes	e_9	e_1	e_5	e_3	e_6	e_{10}	e_9	e_2	e_1	e_4	e_6	e_{10}	e_7	s_1	s_2	s_3	s_4	s_5	s_6	s_7	s_8	
3	1																					
4		−1																				
5			−1																			
1		1		1																		
2			1		1																	
6	−1			−1	−1	1																
3							1						1									
5								−1														
1								1	1													
4									−1	−1			−1									
2										1	1											
6							−1				−1	1										
	1						1															e_9
		1							1					1								e_1
								1							1							e_2
				1												1						e_3
										1							1					e_4
			1															1				e_5
					1						1								1			e_6
													1							1		e_7
																					1	e_8

Display 2

Table 6.8 Example of Multicommodity Problem

			Commodity 1			Commodity 2			
Arc Number, j	Directed Away from Node	Directed toward Node	Unit Cost, c_j^1	Arc Capacity, u_j^1	Flow at Optimality	Unit Cost, c_j^2	Arc Capacity, u_j^2	Flow at Optimality	Mutual Arc Capacity
1	1	4	1	2	1.5	4	2	0.5	2
2	1	5	8	2	0.0	2	2	1.5	3
3	1	6	9	2	0.5	8	3	0.0	3
4	2	4	10	2	0.0	3	2	1.5	3
5	2	5	1	3	2.0	3	3	0.0	3
6	2	6	4	2	0.0	2	3	0.5	3
7	3	4	4	2	0.5	18	2	0.0	3
8	3	5	10	2	0.0	4	2	0.5	3
9	3	6	4	3	1.5	3	2	1.5	3

Table 6.9 Requirements for Sample Problem

Node	Commodity 1		Commodity 2	
Number	Supply	Demand	Supply	Demand
1	2	0	2	0
2	2	0	2	0
3	2	0	2	0
4	0	2	0	2
5	0	2	0	2
6	0	2	0	2

and $\mathbf{x}_B = [2\ \ 2\ \ 2\ \ 0\ \ 0\ \ 0 \,\vdots\, 1\ \ 2\ \ 0\ \ 1\ \ 1\ \ 0 \,\vdots\, 1 \,\vdots\, 0\ \ 1\ \ 3\ \ 2\ \ 1\ \ 2\ \ 2\ \ 3]$. Therefore we see that R^1 is empty. From Figure 6.5 and Proposition 6.5 we see that $\mathbf{Q} = [-\mathbf{1}]$.

Following ALG 6.1, we are now ready to determine the dual variables for use in the pricing operation. By (6.17), we obtain $\boldsymbol{\pi}^2 = [\mathbf{c}_B^{22} - \mathbf{c}_B^{12}(\mathbf{B}^2)^{-1}\mathbf{R}^2]\mathbf{Q}^{-1}$, where $\mathbf{c}_B^{22} = 18$, $\mathbf{c}_B^{12} = [3\ \ 2\ \ 4\ \ 3\ \ 2\ \ 0]$, and $(\mathbf{B}^2)^{-1}\mathbf{R}^2 = [1\ \ 0\ \ 0\ \ 1\ \ -1\ \ 0]$. Hence $\boldsymbol{\pi}^2 = -14$. By (6.18), $\boldsymbol{\pi}^{11}\mathbf{B}^1 = \mathbf{c}_B^{11} - \boldsymbol{\pi}^2\mathbf{P}^1$ and $\boldsymbol{\pi}^{12}\mathbf{B}^2 = \mathbf{c}_B^{12} - \boldsymbol{\pi}^2\mathbf{P}^2$, where $\mathbf{c}_B^{11} = [4\ \ 1\ \ 1\ \ 9\ \ 4\ \ 0]$ and $\mathbf{P}^1 = \mathbf{P}^2 = [1\ \ 0\ \ 0\ \ 0\ \ 0\ \ 0]$. By Figure 6.6, we have $\boldsymbol{\pi}^{11} = [9\ \ 4\ \ 18\ \ 8\ \ 3\ \ 0]$ and $\boldsymbol{\pi}^{12} = [3\ \ 2\ \ 17\ \ -1\ \ 1\ \ 0]$. The reduced cost for arc 8, commodity 2, is $\pi_3^2 - \pi_5^2 - c_8^2 = 17 - 1 - 4 = 12$. Hence arc 8, commodity 2, prices favorably, and $\delta \leftarrow 1$.

Table 6.10 Initial Flows for Sample Problem

	Flow	
Arc	Commodity 1	Commodity 2
1	2	0
2	0	2
3	0	0
4	0	1
5	2	0
6	0	1
7	0	1
8	0	0
9	2	1

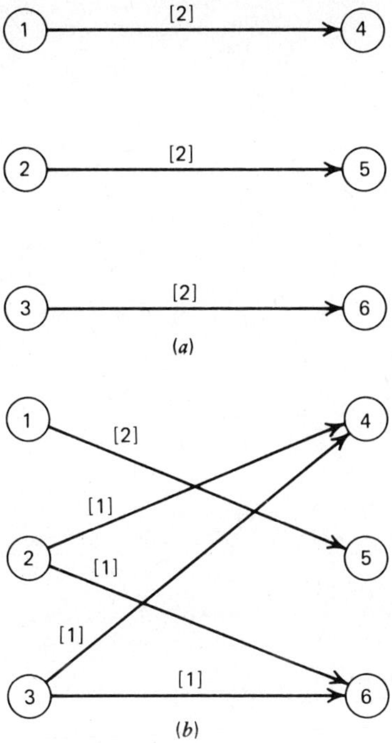

FIGURE 6.4 Initial flows for example. (*a*) Initial flows for commodity 1. (*b*) Initial flows for commodity 2.

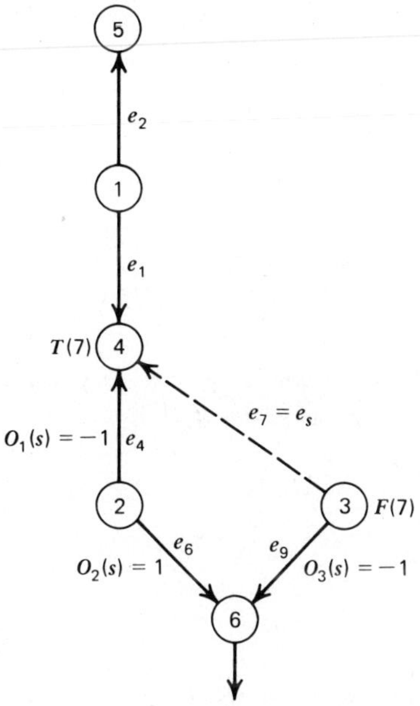

FIGURE 6.5 Cycle formed with e_7 and $\mathcal{T}^2$.

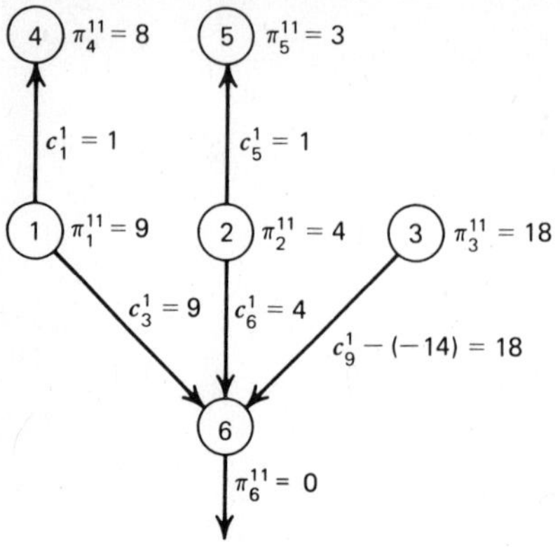

(*a*)

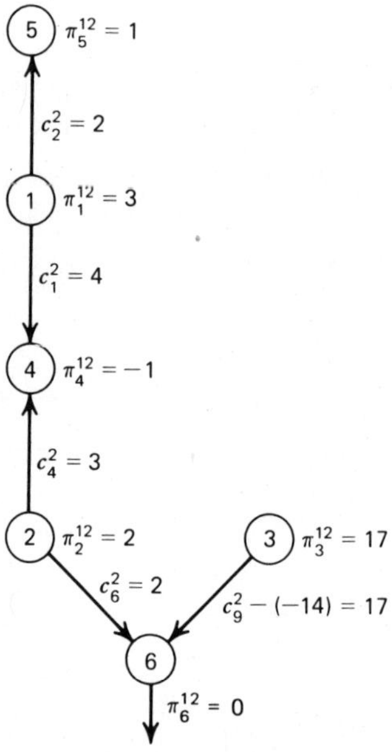

(*b*)

FIGURE 6.6 Determination of π^1. (*a*) Calculation of π^{11}. (*b*) Calculation of π^{12}.

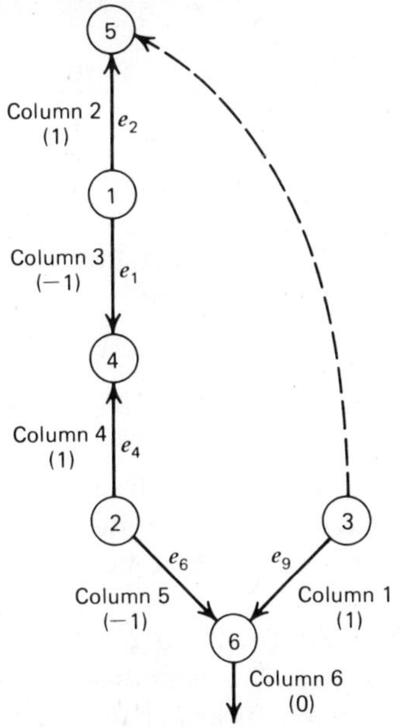

FIGURE 6.7 Calculation of orientation sequence for arc 8, commodity 2.

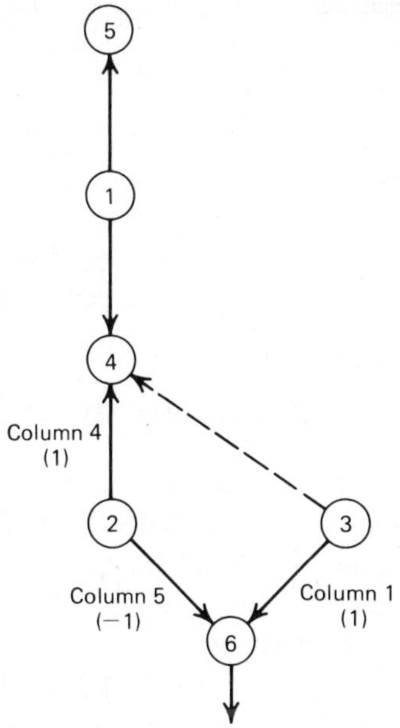

FIGURE 6.8 Calculation of $(\mathbf{B})^{-1}\mathbf{R}^2$.

Let us now examine the ratio test. By (6.20), $\mathbf{y}^{22}=\mathbf{Q}^{-1}[\boldsymbol{\beta}^2-\mathbf{P}^2(\mathbf{B}^2)^{-1}\boldsymbol{\alpha}^2]$, where $\boldsymbol{\beta}^2=[0]$, $\mathbf{P}^2=[1\quad 0\quad 0\quad 0\quad 0\quad 0]$, and $(\mathbf{B}^2)^{-1}\boldsymbol{\alpha}^2=[1\quad 1\quad -1\quad 1\quad -1\quad 0]$ by Figure 6.7. Therefore $\mathbf{y}^{22}=1$. By (6.21), $\mathbf{y}^{11}=(\mathbf{B}^1)^{-1}\boldsymbol{\alpha}^1=0$ and $\mathbf{y}^{12}=(\mathbf{B}^2)^{-1}\boldsymbol{\alpha}^2-(\mathbf{B}^2)^{-1}\mathbf{R}^2\mathbf{y}^{22}$. By Figure 6.8, $(\mathbf{B}^2)^{-1}\mathbf{R}^2\mathbf{y}^{22}=[1\quad 0\quad 0\quad 1\quad -1\quad 0](1)$. Therefore $\mathbf{y}^{12}=[1\quad 1\quad -1\quad 1\quad -1\quad 0]-[1\quad 0\quad 0\quad 1\quad -1\quad 0]=[0\quad 1\quad -1\quad 0\quad 0\quad 0]$. By (6.22),

$$\mathbf{y}^3=\boldsymbol{\gamma}-\mathbf{S}^2\mathbf{y}^{12}-\mathbf{U}^2\mathbf{y}^{22}$$

$$=\left[\begin{array}{c} \\ \hline \\ \hline \\ \hline \\ \hline \\ \hline \\ \hline \\ \hline 1\end{array}\right]-\left[\begin{array}{c|c|c|c|c|c} & & 1 & & & \\ \hline & 1 & & & & \\ \hline & & & & & \\ \hline & & & 1 & & \\ \hline & & & & & \\ \hline & & & & 1 & \\ \hline & & & & & \\ \hline & & & & & \end{array}\right]\left[\begin{array}{c} \\ \hline 1 \\ \hline -1 \\ \hline \\ \hline \\ \hline \end{array}\right]-\left[\begin{array}{c} \\ \hline \\ \hline \\ \hline \\ \hline \\ \hline \\ \hline 1 \\ \hline \end{array}\right][1]=\left[\begin{array}{c}1 \\ \hline -1 \\ \hline \\ \hline \\ \hline \\ \hline \\ \hline -1 \\ \hline 1\end{array}\right].$$

Therefore

$$y=[0\ \ 0\ \ 0\ \ 0\ \ 0\ \ 0 \vdots 0\ \ 1\ \ -1\ \ 0\ \ 0\ \ 0 \vdots 1 \vdots 1\ \ -1\ \ 0\ \ 0\ \ 0\ \ 0\ \ -1\ \ 1],$$

$$x_B=[2\ \ 2\ \ 2\ \ 0\ \ 0\ \ 0 \vdots 1\ \ 2\ \ 0\ \ 1\ \ 1\ \ 0 \vdots 1 \vdots 0\ \ 1\ \ 3\ \ 2\ \ 1\ \ 2\ \ 2\ \ 3],$$

$$\Delta_1=\min\{2,1,0,3\}=0, \qquad \Delta_2=\min\{(2-0),(\infty-1),(\infty-2)\}=2.$$

Therefore $\Delta=0$, and s_1 leaves the basic set. This concludes the example, which demonstrates one pricing operation and one ratio test.

6.2 PRICE-DIRECTIVE DECOMPOSITION

In this section we describe what is called price-directive or Dantzig-Wolfe decomposition for MP. This technique decomposes MP into a master program (a general linear program) and K minimal cost network flow problems.

For each $k \varepsilon \{1,\ldots,K\}$ let $\Lambda^k=\{\mathbf{x}^k:\mathbf{A}\mathbf{x}^k=\mathbf{r}^k, \mathbf{0}\leqslant \mathbf{x}^k\leqslant \mathbf{u}^k\}$, where each component of $\mathbf{u}^k$ is assumed to be finite so that Λ^k is bounded. It is well known that Λ^k has a finite number of extreme points and that each $\mathbf{x}^k \varepsilon \Lambda^k$ can be represented as a convex combination of these extreme points. Let n^k denote the number of extreme points of Λ^k. We assume $n^k \geqslant 1$ since otherwise MP has no solution. Let the extreme points of Λ^k be ordered $1,\ldots,n^k$, and let $\mathbf{x}^k(j)$ denote the jth extreme point of Λ^k. Thus for any $\mathbf{x}^k \varepsilon \Lambda^k$ there is an n^k component vector $\boldsymbol{\lambda}^k$ such that

$$\mathbf{x}^k=\sum_j \mathbf{x}^k(j)\lambda_j^k,$$

where $\mathbf{1}\boldsymbol{\lambda}^k=1$ and $\boldsymbol{\lambda}^k \geqslant \mathbf{0}$.

Adding slacks to the mutual capacity constraints and using the representation for $\mathbf{x}^k$ given above, we can write MP as follows:

$$\left.\begin{array}{lll} \min & \sum_{j,k} \mathbf{c}^k\mathbf{x}^k(j)\lambda_j^k & \\ \text{s.t.} & \sum_{j,k} \mathbf{D}^k\mathbf{x}^k(j)\lambda_j^k+\mathbf{s}=\mathbf{b} & (\gamma) \\ & \mathbf{1}\boldsymbol{\lambda}^k=1, \quad \text{all } k & (\alpha_k) \\ & \mathbf{s}\geqslant \mathbf{0}, \boldsymbol{\lambda}^k \geqslant 0, \quad \text{all } k, & \end{array}\right\} \tag{6.23}$$

where γ and α_k are dual variables. The reduced costs associated with the variables of (6.23) are

$$\begin{Bmatrix} \gamma_m & \text{for } s_m \\ (\gamma \mathbf{D}^k - \mathbf{c}^k)\mathbf{x}^k(j) + \alpha_k & \text{for } \lambda_j^k \end{Bmatrix}.$$

Any variable with reduced cost greater than zero is a candidate for basis entry. For a given k, finding the variable λ_j^k having the largest reduced cost involves solving a subproblem of the form

$$\left.\begin{aligned} \min \quad & z^k(\gamma) = (\mathbf{c}^k - \gamma \mathbf{D}^k)\mathbf{x}^k \\ \text{s.t.} \quad & \mathbf{A}\mathbf{x}^k = \mathbf{r}^k \\ & \mathbf{0} \leqslant \mathbf{x}^k \leqslant \mathbf{u}^k, \end{aligned}\right\} \tag{6.24}$$

using an extreme point algorithm. Let $\mathbf{x}^k(j^*)$ denote an optimum extreme point for (6.24) with an optimal objective value of $z^k(\gamma)$. If $\alpha_k > z^k(\gamma)$, then $\lambda_{j^*}^k$ is a candidate for basis entry. Otherwise there is no extreme point of Λ^k that is a candidate for basis entry. Furthermore, $\mathbf{D}^k\mathbf{x}^k(j^*)$ and $\mathbf{c}^k\mathbf{x}^k(j^*)$ provide the column data for the variable $\lambda_{j^*}^k$.

Problem 6.23 is called the master program, and the K minimal cost network flow problems (6.24) are termed the subprograms. The master program is solved by the primal simplex method with the subprograms used to test for optimality and to select candidates for entering the basis of the master program.

ALG 6.2 PRICE-DIRECTIVE DECOMPOSITION ALGORITHM

0 *Initialization.* Find an initial feasible basis for (6.23) (the master program) and the corresponding dual variables. If a feasible basis is not readily available, artificial variables and a two-phase procedure may be used.

1 *Pricing.* Set

$$\psi_1 = \{\mathbf{x}^k(j) : \mathbf{x}^k(j) \text{ solves (6.24) and } \alpha_k > z^k(\gamma)\}$$

and

$$\psi_2 = \{s_m : \gamma_m > 0\}.$$

If $\psi_1 \cup \psi_2 = \phi$, terminate; otherwise select a candidate from $\psi_1 \cup \psi_2$ for basis entry.

2 *Pivot.* Generate the entering column, update this column, pivot in the master (i.e., update the basis inverse), update the dual variables, and return to Step 1.

We will now develop a lower bound on the solution of (6.23) that can be used in an attempt to speed termination. Consider the following proposition.

Proposition 6.6

Let $\hat{c}$ be the objective function value at any iteration of the simplex method applied to (6.23) with current dual variables given by γ and $\boldsymbol{\alpha}$. Let $\bar{\mathbf{x}}^1,\ldots,\bar{\mathbf{x}}^K$ denote any feasible solution for MP. If $\gamma \leqslant \mathbf{0}$, then

$$\sum_k \mathbf{c}^k\bar{\mathbf{x}}^k \geqslant \hat{c} + \sum_k \left[z^k(\gamma) - \alpha_k \right].$$

Proof. Since $\bar{\mathbf{x}}^1,\ldots,\bar{\mathbf{x}}^K$ is feasible, $\Sigma_k \mathbf{D}^k\bar{\mathbf{x}}^k \leqslant \mathbf{b}$. Since $\gamma \leqslant \mathbf{0}$,

$$\gamma \sum_k \mathbf{D}^k\bar{\mathbf{x}}^k \geqslant \gamma\mathbf{b}. \tag{6.25}$$

Since $z^k(\gamma)$ is optimal for γ, $z^k(\gamma) \leqslant (\mathbf{c}^k - \gamma\mathbf{D}^k)\bar{\mathbf{x}}^k$. Summing over k and rearranging, we obtain

$$\sum_k \mathbf{c}^k\bar{\mathbf{x}}^k \geqslant \sum_k \left[\gamma\mathbf{D}^k\bar{\mathbf{x}}^k + z^k(\gamma) \right]. \tag{6.26}$$

Substituting (6.25) into (6.26), we obtain

$$\sum_k \mathbf{c}^k\bar{\mathbf{x}}^k \geqslant \gamma\mathbf{b} \sum_k z^k(\gamma)$$

or

$$\begin{aligned} \sum_k \mathbf{c}^k\bar{\mathbf{x}}^k &\geqslant \gamma\mathbf{b} + \sum_k \alpha_k + \sum_k \left[z^k(\gamma) - \alpha_k \right] \\ &\geqslant \hat{c} + \sum_k \left[z^k(\gamma) - \alpha_k \right], \end{aligned}$$

since $\gamma\mathbf{b} + \Sigma_k \alpha_k = \hat{c}$. ■

A reasonable implementation of ALG 6.2 would not determine the complete set ψ_1 in the pricing operation. But to obtain the bound, $z^k(\gamma)$ for each $k = 1,\ldots,K$ must be determined. Rather than calculating this bound at every pass through ALG 6.2, we suggest that it be computed only when all subproblems are automatically solved in the pricing operation. With this implementation the bound is obtained almost free of charge.

The decomposition considered here placed the upper bound constraints $\mathbf{0} \leqslant \mathbf{x}^k \leqslant \mathbf{u}^k$ with the subproblems. It is possible to decompose so that $\mathbf{0} \leqslant \mathbf{x}^k \leqslant \mathbf{u}^k$ for all k is retained in the master program. The result is that the subproblems are then solvable as shortest path problems.

6.3 RESOURCE-DIRECTIVE DECOMPOSITION

The basic idea of the resource-directive approach is to distribute the mutual arc capacity among the commodities in such a way that the solutions of the K subproblems yield a solution to the coupled problem. At each iteration, an allocation is made and K minimal cost network flow problems are solved. The sum of the capacities allocated to an arc over all commodities is less than or equal to the arc capacity in the original problem. Hence the combined flow from the solutions of the subproblems provides a feasible flow for the original problem. Optimality is tested, and the procedure either terminates or a new arc-capacity allocation is developed.

After artificial variables have been added, MP becomes

$$\left.\begin{aligned} \min \quad & \sum_k \mathbf{c}^k\mathbf{x}^k + \gamma \sum_k \mathbf{1a}^k \\ \text{s.t.} \quad & \mathbf{A}\mathbf{x}^k + \mathbf{a}^k = r^k, \qquad \text{all } k \\ & \sum_k \mathbf{D}^k\mathbf{x}^k \leqslant \mathbf{b} \\ & \mathbf{0} \leqslant \mathbf{x}^k \leqslant \mathbf{u}^k, \mathbf{a}^k \geqslant \mathbf{0}, \end{aligned}\right\} \tag{6.27}$$

where γ is a large positive number, and $\mathbf{a}^k$ a vector of artificial variables. An equivalent statement of (6.27) is

$$\left.\begin{aligned} \min \quad & z(\mathbf{y}^1, \ldots, \mathbf{y}^K) \\ \text{s.t.} \quad & z(\mathbf{y}^1, \ldots, \mathbf{y}^K) = \sum_k z^k(\mathbf{y}^k) \\ & \sum_k \mathbf{D}^k\mathbf{y}^k = \mathbf{b} \\ & \mathbf{0} \leqslant \mathbf{y}^k \leqslant \mathbf{u}^k, \qquad \text{all } k, \end{aligned}\right\} \tag{6.28}$$

where

$$z^k(\mathbf{y}^k) = \min\{\mathbf{c}^k\mathbf{x}^k + \gamma\mathbf{1a}^k : \mathbf{A}\mathbf{x}^k + \mathbf{a}^k = \mathbf{r}^k, \mathbf{0} \leqslant \mathbf{x}^k \leqslant \mathbf{y}^k, \mathbf{a}^k \geqslant \mathbf{0}\},$$

$$z^k(\mathbf{y}^k) = \max\{\mathbf{r}^k\boldsymbol{\mu}^k - \mathbf{y}^k\boldsymbol{\nu}^k : \boldsymbol{\mu}^k\mathbf{A} - \boldsymbol{\nu}^k \leqslant \mathbf{c}^k, \boldsymbol{\mu}^k \leqslant \gamma\mathbf{1}, \boldsymbol{\nu}^k \geqslant \mathbf{0}\},$$

by duality theory.

We will now show that $z(\mathbf{y}^1,\ldots,\mathbf{y}^K)$ is convex.

Proposition 6.7

The function $z(\mathbf{y}^1,\ldots,\mathbf{y}^K)$ is convex over $\{\mathbf{y}^1,\ldots,\mathbf{y}^K : \mathbf{y}^1 \geqslant \mathbf{0},\ldots,\mathbf{y}^K \geqslant \mathbf{0}\}$.

Proof. Let $(\bar{\mathbf{y}}^1,\ldots,\bar{\mathbf{y}}^K)$ and $(\hat{\mathbf{y}}^1,\ldots,\hat{\mathbf{y}}^K)$ be chosen so that $\bar{\mathbf{y}}^1 \geqslant \mathbf{0},\ldots,\bar{\mathbf{y}}^K \geqslant \mathbf{0}$ and $\hat{\mathbf{y}}^1 \geqslant \mathbf{0},\ldots,\hat{\mathbf{y}}^K \geqslant \mathbf{0}$. Choose α such that $0 \leqslant \alpha \leqslant 1$. Then

$$z\left[\alpha\bar{\mathbf{y}}^1+(1-\alpha)\hat{\mathbf{y}}^1,\ldots,\alpha\bar{\mathbf{y}}^K+(1-\alpha)\hat{\mathbf{y}}^K\right]$$

$$=\sum_k z^k\left[\alpha\bar{\mathbf{y}}^k+(1-\alpha)\hat{\mathbf{y}}^k\right]$$

$$=\sum_k \max\left\{\mathbf{r}^k\boldsymbol{\mu}^k-\left[\alpha\bar{\mathbf{y}}^k+(1-\alpha)\hat{\mathbf{y}}^k\right]\boldsymbol{\nu}^k : \boldsymbol{\mu}^k\mathbf{A}-\boldsymbol{\nu}^k \leqslant \mathbf{c}^k, \boldsymbol{\mu}^k \leqslant \gamma\mathbf{1}, \boldsymbol{\nu}^k \geqslant \mathbf{0}\right\}$$

$$=\sum_k \max\left\{\alpha\left[\mathbf{r}^k\boldsymbol{\mu}^k-\bar{\mathbf{y}}^k\boldsymbol{\nu}^k\right]+(1-\alpha)\left[\mathbf{r}^k\boldsymbol{\mu}^k-\hat{\mathbf{y}}^k\boldsymbol{\nu}^k\right] : \boldsymbol{\mu}^k\mathbf{A}-\boldsymbol{\nu}^k \leqslant \mathbf{c}^k,\right.$$

$$\left.\cdot\boldsymbol{\mu}^k \leqslant \gamma\mathbf{1}, \boldsymbol{\nu}^k \geqslant \mathbf{0}\right\}$$

$$\leqslant \alpha\sum_k \max\left\{\mathbf{r}^k\boldsymbol{\mu}^k-\bar{\mathbf{y}}^k\boldsymbol{\nu}^k : \boldsymbol{\mu}^k\mathbf{A}-\boldsymbol{\nu}^k \leqslant \mathbf{c}^k, \boldsymbol{\mu}^k \leqslant \gamma\mathbf{1}, \boldsymbol{\nu}^k \geqslant \mathbf{0}\right\}$$

$$+(1-\alpha)\max\sum_k \left\{\mathbf{r}^k\boldsymbol{\mu}^k-\hat{\mathbf{y}}^k\boldsymbol{\nu}^k : \boldsymbol{\mu}^k\mathbf{A}-\boldsymbol{\nu}^k \leqslant \mathbf{c}^k, \boldsymbol{\mu}^k \leqslant \gamma\mathbf{1}, \boldsymbol{\nu}^k \geqslant \mathbf{0}\right\}$$

$$=\alpha z(\bar{\mathbf{y}}^1,\ldots,\bar{\mathbf{y}}^K)+(1-\alpha)z(\hat{\mathbf{y}}^1,\ldots,\hat{\mathbf{y}}^K).$$

Hence $z(\bar{\mathbf{y}}^1,\ldots,\bar{\mathbf{y}}^K)$ is convex. ■

Numerous approaches are available to solve the convex program, and we present two algorithms here.

6.3.1 Method of Tangential Approximation

Let

$$\Lambda^k=\left\{(\boldsymbol{\mu}^k,\boldsymbol{\nu}^k) : \boldsymbol{\mu}^k\mathbf{A}-\boldsymbol{\nu}^k \leqslant \mathbf{c}^k,\ \boldsymbol{\mu}^k \leqslant \gamma\mathbf{1},\ \boldsymbol{\nu}^k \geqslant \mathbf{0}, \text{ and } (\boldsymbol{\mu}^k,\boldsymbol{\nu}^k) \text{ an extreme point}\right\}.$$

Then (6.28) may be stated as follows:

$$\left.\begin{aligned}
\min \quad & \sum_k \sigma^k \\
\text{s.t.} \quad & \sigma^k \geqslant \mathbf{r}^k\boldsymbol{\mu}^k - \mathbf{y}^k\boldsymbol{\nu}^k, \qquad \text{all } (\boldsymbol{\mu}^k,\boldsymbol{\nu}^k)\varepsilon\Lambda^k \text{ and all } k \\
& \sum_k \mathbf{D}^k\mathbf{y}^k = \mathbf{b} \\
& \mathbf{0} \leqslant \mathbf{y}^k \leqslant \mathbf{u}^k, \qquad \text{all } k.
\end{aligned}\right\} \tag{6.29}$$

Suppose $\Delta^k \subset \Lambda^k$. Let $z(\Lambda)$ denote the optimal objective value of (6.29), and let $z(\Delta)$ denote the optimal objective value of (6.29) with Λ^k replaced by Δ^k. Then $z(\Delta) \leqslant z(\Lambda)$ provides a lower bound for (6.29). An upper bound is given by $\Sigma_k z^k(\mathbf{y}^k)$ for any $(\mathbf{y}^1,\ldots,\mathbf{y}^K)$ such that $\Sigma_k \mathbf{D}^k\mathbf{y}^k = \mathbf{b}$. The algorithm is now presented.

ALG 6.3 RESOURCE-DIRECTIVE DECOMPOSITION ALGORITHM USING TANGENTIAL APPROXIMATION

0 *Initialization.* Set $i \leftarrow 0$, and let $[\mathbf{y}_0^1,\ldots,\mathbf{y}_0^K]$ be any element of $([\mathbf{y}^1,\ldots,\mathbf{y}^K] : \Sigma_k \mathbf{D}^k\mathbf{y}^k = \mathbf{b}, \mathbf{0} \leqslant \mathbf{y}^k \leqslant \mathbf{u}^k, \text{ all } k\}$. Set $\Delta^k = \phi, \sigma^k \leftarrow -\infty$ for each k.

1 *Solve Subproblems* (*Determine Upper Bound*). Solve

$$z^k(\mathbf{y}_i^k) = \max\left\{\mathbf{r}^k\boldsymbol{\mu}^k - \mathbf{y}_i^k\boldsymbol{\nu}^k : \boldsymbol{\mu}^k\mathbf{A} - \boldsymbol{\nu}^k \leqslant \mathbf{c}^k, \boldsymbol{\mu}^k \leqslant \gamma\mathbf{1}, \boldsymbol{\nu}^k \geqslant \mathbf{0}\right\}$$

for each $k = 1,\ldots,K$, using an extreme point method.

2 *Check Optimality* (*Lower Bound = Upper Bound*?). Does $\Sigma_k \sigma_k = \Sigma_k z^k(\mathbf{y}_i^k)$? If so, terminate. The optimum is given by the solution to

$$z^k(\mathbf{y}_i^k) = \min\left\{\mathbf{c}^k\mathbf{x}^k + \gamma\mathbf{1}\mathbf{a}^k : \mathbf{A}\mathbf{x}^k + \mathbf{a}^k = \mathbf{r}^k, \mathbf{0} \leqslant \mathbf{x}^k \leqslant \mathbf{y}_i^k, \mathbf{a}^k \geqslant \mathbf{0}\right\}$$

for $k = 1,\ldots,K$. If not, add $(\boldsymbol{\mu}_i^k, \boldsymbol{\nu}_i^k)$ to Δ^k for each k.

3 *Solve Master Program.* Set $i \leftarrow i+1$. Let $\mathbf{y}_i^1,\ldots,\mathbf{y}_i^K$ denote the solution to

$$\begin{aligned}
\min \quad & \sum_k \sigma_k \\
\text{s.t.} \quad & \sigma_k \geqslant \mathbf{r}^k\bar{\boldsymbol{\mu}}^k - \mathbf{y}_i^k\bar{\boldsymbol{\nu}}^k \qquad \text{for each } k \text{ and all } (\bar{\boldsymbol{\mu}}^k,\bar{\boldsymbol{\nu}}^k)\varepsilon\Delta^k \\
& \sum_k \mathbf{D}^k\mathbf{y}_i^k = \mathbf{b} \\
& \mathbf{0} \leqslant \mathbf{y}_i^k \leqslant \mathbf{u}^k, \qquad \text{all } k,
\end{aligned}$$

and return to Step 1.

6.3.2 Subgradient Optimization

In this section we discuss the general subgradient algorithm for convex programs and indicate the specialization of this procedure for MP. Consider the nonlinear program

$$\left.\begin{array}{ll}\min & g(\mathbf{y}) \\ \text{s.t.} & \mathbf{y}\varepsilon\Gamma,\end{array}\right\} \tag{6.30}$$

where g is a real-valued function that is convex over the compact, convex, and nonempty set Γ. A vector $\boldsymbol{\eta}$ will be called a *subgradient* of $g(\mathbf{y})$ at $\bar{\mathbf{y}}$ if $g(\mathbf{y})-g(\bar{\mathbf{y}})\geqslant\boldsymbol{\eta}(\mathbf{y}-\bar{\mathbf{y}})$ for all $\mathbf{y}\varepsilon\Gamma$. Note that, if $g(\mathbf{y})$ is differentiable at $\bar{\mathbf{y}}$, the only subgradient of $\bar{\mathbf{y}}$ is the gradient. For points at which $g(\mathbf{y})$ is not differentiable, the subgradient is any linear support of g at $\bar{\mathbf{y}}$. Subgradients are illustrated in Figure 6.9. For any $\bar{\mathbf{y}}\varepsilon\Gamma$ we denote the set of all subgradients of g at $\bar{\mathbf{y}}$ by $\partial g(\bar{\mathbf{y}})$.

The subgradient algorithm makes use of an operation that we call the *projection operation*. The projection of a point $\mathbf{x}$ onto Γ, denoted by $P[\mathbf{x}]$, is defined to be any point $\bar{\mathbf{y}}\varepsilon\Gamma$ that is nearest Γ with respect to the Euclidean norm. Using the projection operation, we now present the subgradient algorithm in its most general form.

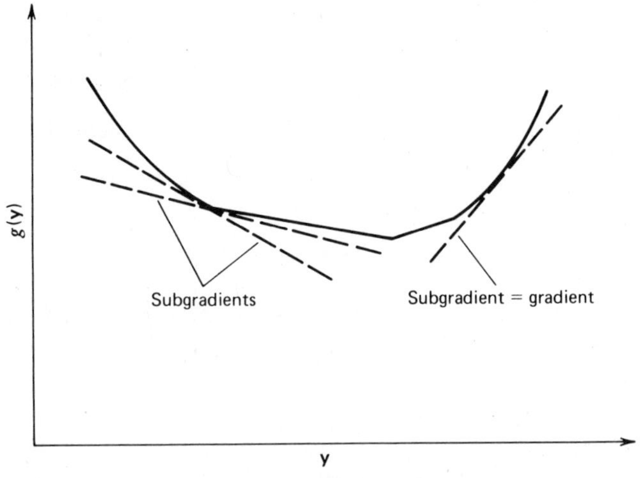

FIGURE 6.9 Illustration of subgradients.

ALG 6.4 SUBGRADIENT OPTIMIZATION ALGORITHM

0 *Initialization.* Let $\mathbf{y}_0$ be any element of Γ, select a set of step sizes $s_0, s_1, s_2, \ldots$, and set $i \leftarrow 0$.

1 *Find Subgradient.* Let $\boldsymbol{\eta}_i \varepsilon \partial g(\mathbf{y}_i)$. If $\boldsymbol{\eta}_i = \mathbf{0}$, terminate with $\mathbf{y}_i$ optimal.
2 *Move to New Point.* Set $\mathbf{y}_{i+1} \leftarrow P[\mathbf{y}_i - s_i \boldsymbol{\eta}_i]$, set $i \leftarrow i+1$, and return to 1.

Various proposals have been offered for the step sizes $s_0, s_1, \ldots$. They usually depend on a sequence of constants $\{\lambda_0, \lambda_1, \lambda_2, \ldots\}$ that satisfy the following conditions:

$$\lambda_i \geqslant 0, \qquad \text{all } i; \qquad \lim_{i \to \infty} \lambda_i = 0; \qquad \text{and}$$

$$\sum_i \lambda_i = \infty.$$

Given a sequence as described above, plus other similar assumptions, convergence can be guaranteed for the following step sizes:

(i) $s_i = \lambda_i$,

(ii) $s_i = \dfrac{\lambda_i}{\|\boldsymbol{\eta}_i\|^2}$,

(iii) $s_i = \dfrac{\lambda_i [\, g(\mathbf{y}_i) - g^* \,]}{\|\boldsymbol{\eta}_i\|^2}$,

where g^* denotes the optimal objective value. The available convergence results for each of these step sizes is given in Appendix C.

The subgradient algorithm as given (ALG 6.4) may be considered an exact technique. However, the termination criterion (see Step 1) is not computationally effective, and no effective optimality test is available. Furthermore, ALG 6.4 will terminate only if the solution to (6.30) is the unconstrained minimum of g over all of n-space. Consequently, ALG 6.4 must be modified for practical implementation, and the resulting algorithm must be viewed as a heuristic procedure.

Recall that the resource-directive problem formulation may be described as follows:

$$\left.\begin{aligned} \min \quad & z(\mathbf{y}^1, \ldots, \mathbf{y}^K) \\ \text{s.t.} \quad & z(\mathbf{y}^1, \ldots, \mathbf{y}^K) = \textstyle\sum_k z^k(\mathbf{y}^k) \\ & \textstyle\sum_k \mathbf{D}^k \mathbf{y}^k = \mathbf{b} \\ & \mathbf{0} \leqslant \mathbf{y}^k \leqslant \mathbf{u}^k, \qquad \text{all } k, \end{aligned}\right\} \tag{6.31}$$

where

$$z^k(\mathbf{y}^k) = \max\{\mathbf{r}^k \boldsymbol{\mu}^k - \mathbf{y}^k \boldsymbol{\nu}^k : \boldsymbol{\mu}^k \mathbf{A} - \boldsymbol{\nu}^k \leqslant \mathbf{c}^k, \boldsymbol{\mu}^k \leqslant \gamma \mathbf{1}, \boldsymbol{\nu}^k \geqslant \mathbf{0}\}.$$

To apply the subgradient algorithm to (6.31) one must have a means for determining a subgradient of $z(\mathbf{y}^1,\ldots,\mathbf{y}^K)$ at a given point. Consider the following proposition.

Proposition 6.8

Let $(\bar{\mathbf{y}}^1,\ldots,\bar{\mathbf{y}}^K) \geqslant \mathbf{0}$ be any allocation, and let $(\bar{\boldsymbol{\mu}}^k,\bar{\boldsymbol{\nu}}^k)$ denote the optimal solution to $z^k(\bar{\mathbf{y}}^k)$ for $k=1,\ldots,K$. Then $(-\bar{\boldsymbol{\nu}}^1,\ldots,-\bar{\boldsymbol{\nu}}^K)$ is a subgradient of z at $(\bar{\mathbf{y}}^1,\ldots,\bar{\mathbf{y}}^K)$.

Proof. Let $(\mathbf{y}^1,\ldots,\mathbf{y}^K)$ be any nonnegative allocation, and let $(\boldsymbol{\mu}^k,\boldsymbol{\nu}^k)$ denote the solution to $z^k(\mathbf{y}^k)$ for $k=1,\ldots,K$. Then

$$\begin{aligned} z(\mathbf{y}^1,\ldots,\mathbf{y}^K) - z(\bar{\mathbf{y}}^1,\ldots,\bar{\mathbf{y}}^K) &= \sum_k (\mathbf{r}^k\boldsymbol{\mu}^k - \mathbf{y}^k\boldsymbol{\nu}^k) - \sum_k (\mathbf{r}^k\bar{\boldsymbol{\mu}}^k - \bar{\mathbf{y}}^k\bar{\boldsymbol{\nu}}^k) \\ &\geqslant \sum_k (\mathbf{r}^k\bar{\boldsymbol{\mu}}^k - \mathbf{y}^k\bar{\boldsymbol{\nu}}^k) - \sum_k (\mathbf{r}^k\bar{\boldsymbol{\mu}}^k - \bar{\mathbf{y}}^k\bar{\boldsymbol{\nu}}^k) \\ &= \sum_k (-\bar{\boldsymbol{\nu}}^k)(\mathbf{y}^k - \bar{\mathbf{y}}^k). \end{aligned}$$

Hence $(-\bar{\boldsymbol{\nu}}^1,\ldots,-\bar{\boldsymbol{\nu}}^K)$ is a subgradient of $z(\mathbf{y}^1,\ldots,\mathbf{y}^K)$ at $(\bar{\mathbf{y}}^1,\ldots,\bar{\mathbf{y}}^K)$. ■

Therefore the subgradient is available from the solutions of the subproblems, $z^1(\mathbf{y}^1),\ldots,z^K(\mathbf{y}^K)$.

We now examine the projection operation when applied to (6.31). Given an allocation $(\bar{\mathbf{y}}^1,\ldots,\bar{\mathbf{y}}^K)$, we must solve the following problem:

$$\begin{aligned} &\min\left\{ \|(\mathbf{y}^1,\ldots,\mathbf{y}^K) - (\bar{\mathbf{y}}^1,\ldots,\bar{\mathbf{y}}^K)\| : \sum_k \mathbf{D}^k\mathbf{y}^k = \mathbf{b}, \mathbf{0} \leqslant \mathbf{y}^k \leqslant \mathbf{u}^k, \text{ all } k \right\} \\ &\equiv \min\left\{ \left(\sum_{j,k} (y_j^k - \bar{y}_j^k)^2 \right)^{1/2} : \sum_k \mathbf{D}^k\mathbf{y}^k = \mathbf{b}, \mathbf{0} \leqslant \mathbf{y}^k \leqslant \mathbf{u}^k, \text{ all } k \right\} \\ &\equiv \min\left\{ \sum_{j,k} (y_j^k - \bar{y}_j^k)^2 : \sum_k \mathbf{D}^k\mathbf{y}^k = \mathbf{b}, \mathbf{0} \leqslant \mathbf{y}^k \leqslant \mathbf{u}^k, \text{ all } k \right\}. \end{aligned}$$

But this program decomposes on j; therefore we must solve

$$\min\left\{ \sum_k (y_j^k - \bar{y}_j^k)^2 : \sum_k \mathrm{D}_{jj}^k y_j^k = b_j, 0 \leqslant y_j^k \leqslant u_j^k, \text{ all } k \right\} \qquad (6.32)$$

for each arc. Letting $z_k = D^k_{jj} y^k_j$, we find that (6.32) takes the form

$$\left.\begin{aligned} \min \quad & \tfrac{1}{2}\mathbf{z}\bar{\mathbf{D}}\mathbf{z} - \mathbf{a}\mathbf{z} \\ \text{s.t.} \quad & \sum_j z_j = b \\ & \mathbf{0} \leqslant \mathbf{z} \leqslant \mathbf{u}, \end{aligned}\right\} \qquad (6.33)$$

where $\bar{\mathbf{D}}$ is a diagonal matrix. An efficient algorithm for solving (6.33) is given in Appendix D.

We now present the subgradient algorithm for MP.

ALG 6.5 SUBGRADIENT ALGORITHM FOR MP

0 *Initialization.* Let $(\mathbf{v}^1, \ldots, \mathbf{v}^K)$ be the solution to $z^k(\mathbf{u}^k)$ for $k = 1, \ldots, K$, and set the lower bound $\text{LB} \leftarrow \Sigma_k z^k(\mathbf{u}^k)$. Set $(\mathbf{y}^1_0, \ldots, \mathbf{y}^K_0) \leftarrow P[\mathbf{v}^1, \ldots, \mathbf{v}^K]$. Select a set of step sizes $s_0, s_1, s_2, \ldots$. Choose $\epsilon > 0$ to be used to terminate if an ϵ-optimum is found. Select an integer $L > 0$ denoting the maximum number of allocations. Set $i \leftarrow 0$ and the upper bound $\text{UB} \leftarrow \infty$.

1 *Find Subgradient.* Let $(\boldsymbol{\mu}^k_i, \hat{\boldsymbol{\nu}}^k_i)$ solve

$$z^k(\mathbf{y}^k_i) = \max\{\mathbf{r}^k \boldsymbol{\mu}^k - \mathbf{y}^k_i \boldsymbol{\nu}^k : \boldsymbol{\mu}^k \mathbf{A} - \boldsymbol{\nu}^k \leqslant \mathbf{c}^k, \boldsymbol{\mu}^k \leqslant \gamma \mathbf{1}, \boldsymbol{\nu}^k \geqslant \mathbf{0}\},$$

and set $\text{UB} \leftarrow \Sigma_k z^k(\mathbf{y}^k_i)$.

2 *Check for ϵ-Optimum.* If $\text{UB} - \text{LB} \leqslant \epsilon(\text{LB})$, terminate with $(\mathbf{y}^1_i, \ldots, \mathbf{y}^K_i)$ a guaranteed ϵ-optimum.

3 *Projection.* Set

$$(\mathbf{y}^1_{i+1}, \ldots, \mathbf{y}^K_{i+1}) \leftarrow P[(\mathbf{y}^1_i, \ldots, \mathbf{y}^K_i) + s_i(\hat{\boldsymbol{\nu}}^1_i, \ldots, \hat{\boldsymbol{\nu}}^K_i)],$$

and set $i \leftarrow i + 1$.

4 *Check for Iteration Count.* If $i = L$, terminate with the best solution found; otherwise go to Step 1.

A step size that has been used successfully in practice is

$$s_i = \frac{\lambda_i[z(\mathbf{y}^1_i, \ldots, \mathbf{y}^K_i) - \bar{g}]}{\Sigma_k \boldsymbol{\nu}^k_i \boldsymbol{\nu}^k_i},$$

where $\lambda_0 = \cdots = \lambda_{63} = 2$, $\lambda_{64} = \cdots = \lambda_{95} = 1$, $\lambda_{96} = \cdots = \lambda_{111} = 0.5$, $\lambda_{112} = \cdots = \lambda_{119} = 0.25$, $\lambda_{120} = \cdots = \lambda_{123} = 0.125$, and $\lambda_{124} = \cdots = \lambda_{127} = 0.0625$, and $\bar{g}$ is taken to be the lower bound.

EXERCISES

6.1 Consider the multicommodity network flow problem described in Table 6.8 with the supplies and demands as given in Table E6.1*a*.

Table E6.1*a*

Node Number	Commodity 1 Supply	Commodity 1 Demand	Commodity 2 Supply	Commodity 2 Demand
1	2	0	2	0
2	2	0	2	0
3	2	0	2	0
4	0	2	0	2
5	0	2	0	2
6	0	2	0	2

(a) Generate the cycle matrix corresponding to the solution shown in Table E6.1*b*.

Table E6.1*b*

Arc Number	Commodity 1 Basic	Commodity 1 Flow	Commodity 2 Basic	Commodity 2 Flow
1	Yes	2	Yes	0
2	No	0	Yes	2
3	Yes	0	No	0
4	No	0	Yes	1
5	Yes	2	No	0
6	Yes	0	Yes	1
7	No	0	Yes	1
8	No	0	Yes	0
9	Yes	2	Yes	1

(b) Perform three pivots using the primal partitioning algorithm (ALG 6.1). Begin with the solution given in Table E6.1*b*.
(c) Perform three iterations, using the price-directive decomposition algorithm (ALG 6.2). Append artificials to the master program, and begin with an all-aritificial start.
(d) Perform three iterations using the resource-directive decomposition algorithm (ALG 6.3). Begin at Step 1, using a lower bound of zero and an allocation of $b_j/2$ for each arc $j=1,\ldots,9$.

6.2 Consider these problems:

$$\left.\begin{array}{ll} \min & [(\mathbf{y}-\bar{\mathbf{y}})(\mathbf{y}-\bar{\mathbf{y}})]^{1/2} \\ \text{s.t.} & \mathbf{y}\varepsilon\Gamma \end{array}\right\} P_1$$

and

$$\left.\begin{array}{ll} \min & (\mathbf{y}-\bar{\mathbf{y}})(\mathbf{y}-\bar{\mathbf{y}}) \\ \text{s.t.} & \mathbf{y}\varepsilon\Gamma \end{array}\right\} P_2.$$

Prove that $\mathbf{y}^*$ solves P_1 if and only if $\mathbf{y}^*$ solves P_2.

6.3 Present a generalization of MP in which the graphs associated with each commodity may be distinct.

6.4 Redevelop the price-directive decomposition algorithm when the capacity constraints $\mathbf{x}^k \leqslant \mathbf{u}^k$ are placed in the master program rather than the subproblem.

6.5 Develop an all-artificial start to be used with the primal partitioning algorithm. (Indicate the procedure used to obtain the K rooted spanning trees in the initial solution.)

6.6 The U.S. Air Force contracts with civilian carriers to fly specified daily routes among its bases to meet the demand for supplies and equipment. The number of pounds per day that must be shipped among eight bases are given in Table E6.6*a*.

Table E6.6*a*

			To					
From	FFO	TIK	WRB	SZL	BLV	LRF	BYH	CBM
Wright-Patterson (FFO)	–	3000	3000	2000	2000	3000	2000	1000
Tinker (TIK)	3300	–	3400	1000	1000	1000	1000	1000
Robins (WRB)	4000	5000	–	1000	1000	1000	1000	1000
Whiteman (SZL)	2000	3000	4000	–	0	0	100	200
Scott (BLV)	4000	5000	0	0	–	0	0	100
Little Rock (LRF)	3000	4000	5000	0	0	–	0	0
Blytheville (BYH)	1000	0	2000	0	0	100	–	100
Columbus (CBM)	200	200	200	0	0	100	0	–

This demand must be met by the three routes listed in Table E6.6*b*, which are flown daily by aircraft having a capacity of 25,000 pounds each.

Table E6.6*b*

Route Number	Origin	Destination	Distance (Miles)
1	TIK	SZL	312
	SZL	BLV	200
	BLV	FFO	324
	FFO	BYH	409
	BYH	CBM	189
2	TIK	LRF	326
	LRF	BYH	103
	BYH	CBM	189
	CBM	WRB	404
	WRB	CBM	404
3	WRB	BYH	465
	BYH	BLV	113
	BLV	TIK	473

We assume that units may be transferred from one aircraft to another without incurring extra cost. Present a multicommodity formulation of the problem of meeting the daily demands that minimizes the pound-miles for the system. Let the commodities correspond to the eight origin nodes.

6.7 Consider a conference where a number of people are to attend seminars for a 3-day period. Each person will attend one seminar on each of 3 days, and all seminars are offered on each day. No one should attend the same seminar twice, and there are limitations on the number attending each seminar. Individual preferences for each seminar have been obtained. Develop a multicommodity model that can be used to assign people to seminars subject to the limitations described. Let the decision variables x_{ij}^k be 1 if individual i is to attend seminar j on day k, and 0 otherwise.

NOTES AND REFERENCES

Section 6.1

The basic work leading to the primal partitioning procedure is traceable to Bennett [22], Kaul [116], Saigal [164], and Hartman and Lasdon [93, 94].

These ideas have been further developed in the following papers: Maier [140], Kennington [118], Helgason and Kennington [98], and Ali, Helgason, Kennington, and Lall [4]. Grigoriadis and White [90] have developed a dual partitioning approach. The example given in this section is due to Evans [54].

Section 6.2

Ford and Fulkerson [64] were the first to pose this approach for solving multicommodity network flow problems. It is purported that their work inspired the well-known Dantzig-Wolfe [46] decomposition technique. Tomlin [177] was the first to develop a code using this technique. Other studies involving this approach have been presented by Jarvis [107], Cremeans, Smith, and Tyndall [40], Swoveland [175, 176], Weigel and Cremeans [182], Wollmer [186], Jarvis and Keith [108], Chen and DeWald [32], Jarvis and Martinez [109], and Helgason [102].

Section 6.3

Robacker [156] was the first to suggest this approach for multicommodity problems. A more general discussion of the tangential approximation method is given in Geoffrion [73], and computational experience using this method can be found in Swoveland [175]. The subgradient approach was first applied to this problem class by Held, Wolfe, and Crowder [97]. Computational experience with this approach can be found in Refs. [4] and [119].

Exercises

The problem described in Exercise 6.7 was first formulated by Major R. W. Langley, USAF.

General Comments

In addition to the algorithms presented in this chapter, there have been numerous papers concerning special results for this class of problems. A survey of these results can be found in papers by Rothschild and Whinston [162] and Kennington [120]. Most of these results are limited to two-commodity problems or multicommodity transportation problems having at most two sources or two destinations. The interested reader is referred to the following papers: Hu [105], Rothschild and Whinston [160, 161], Sakarovitch [165], Evans [54, 58], and Evans, Jarvis, and Duke [57].

CHAPTER 7

The Simplex Method for the Network with Side Constraints Model

In this chapter we present a specialization of the primal simplex method for the *network problem with side constraints:*

$$\left.\begin{array}{lll} \min & \mathbf{cx}+\mathbf{dz} & \\ \text{s.t.} & \mathbf{Ax} & =\mathbf{r} \\ & \mathbf{Sx}+\mathbf{Pz} & =\mathbf{b} \\ & \mathbf{0}\leqslant\mathbf{x}\leqslant\mathbf{u},\ \mathbf{0}\leqslant\mathbf{z}\leqslant\mathbf{v} & \end{array}\right\}\quad \text{NPS},$$

where $\mathbf{A}$ is a node-arc incidence matrix for the graph $\mathcal{G}=[\mathcal{N},\mathcal{A}]$ having $\bar{I}$ nodes and $\bar{J}$ arcs with $\bar{I}\leqslant\bar{J}$, and $\mathbf{S}$ and $\mathbf{P}$ are arbitrary matrices. We make the following assumptions about NPS.

Assumptions

1 $\mathcal{G}$ is connected.

2 $[\mathbf{S} \mid \mathbf{P}]$ has full row rank.

3 Total supply equals total demand (i.e., $\mathbf{1r}=0$).

If $\mathcal{G}$ is not connected, then artificial arcs having zero bounds can be appended to NPS to satisfy the first assumption. If $[\mathbf{S} \,\vdots\, \mathbf{P}]$ does not have full row rank, then artificial variables having zero upper bounds can be appended to the system $\mathbf{Pz}$ to satisfy the second assumption. If $\mathbf{lr} > 0$, then the third assumption can always be satisfied by adding a dummy demand node along with the appropriate zero-cost arcs.

Since the system $\mathbf{Ax} = \mathbf{r}$ has rank $\bar{I} - 1$ (see Proposition 3.12), we add a root arc to NPS to obtain

$$\left.\begin{array}{ll} \min & \mathbf{cx} + \mathbf{dz} \\ \text{s.t.} & \mathbf{Ax} + \mathbf{e}^l a \qquad = \mathbf{r} \\ & \mathbf{Sx} + \mathbf{Pz} \qquad = \mathbf{b} \\ & 0 \leqslant a \leqslant 0, \ \mathbf{0} \leqslant \mathbf{x} \leqslant \mathbf{u}, \ \mathbf{0} \leqslant \mathbf{z} \leqslant \mathbf{v} \end{array}\right\} \quad \mathfrak{NPS}.$$

Then the constraint matrix

$$\bar{\mathbf{A}} = \left(\begin{array}{c|c|c} \mathbf{A} & & \mathbf{e}^l \\ \hline \mathbf{S} & \mathbf{P} & \end{array}\right)$$

has full row rank.

Note that NPS or $\mathfrak{NPS}$ may be viewed as a generalization of the multicommodity network flow problem, MP. The only difference is that for MP the graphs associated with the K commodities are not connected, and the side constraints for MP, that is, $\Sigma_k \mathbf{D}^k \mathbf{x}^k \leqslant \mathbf{b}$, have special structure, namely, each $\mathbf{D}^k$ is a diagonal matrix. By adding artificial arcs to a multicommodity problem to connect the individual graphs, any multicommodity problem can be placed in the form of NPS.

In this chapter we present a specialization of the primal simplex algorithm to solve $\mathfrak{NPS}$. This specialization simply partitions the basis so that a portion of the basis corresponds to a rooted spanning tree. Then all calculations involving this component of the basis are executed via labeling operations rather than by matrix multiplication.

7.1 BASIS CHARACTERIZATION

In this section we show that a basis for

$$\bar{\mathbf{A}} = \underbrace{}_{}\left(\begin{array}{c|c|c} \mathbf{A} & & \mathbf{e}^l \\ \hline \mathbf{S} & \mathbf{P} & \end{array}\right) \begin{array}{l} \}\bar{I} \\ \}\bar{K} \end{array}$$
$$\qquad \bar{J} \quad \bar{L} \quad 1$$

may be partitioned so that one of the components corresponds to a rooted spanning tree. Consider the following proposition.

Proposition 7.1

Every basis for $\overline{\mathbf{A}}$ may be placed in the following form:

$$\overline{\mathbf{B}}=\underbrace{}_{\bar{I}}\left(\begin{array}{c:c}\mathbf{B} & \mathbf{C}\\ \hdashline \mathbf{D} & \mathbf{F}\end{array}\right)\begin{array}{l}\}\bar{I}\\ \}\bar{K}\end{array},\qquad \text{columns: } \underbrace{\mathbf{B},\mathbf{D}}_{\bar{I}}\ \underbrace{\mathbf{C},\mathbf{F}}_{\bar{K}}$$

where $\mathbf{B}$ is a submatrix of $[\mathbf{A} \mid \mathbf{e}^l]$ and $\text{Det}(\mathbf{B})\neq 0$.

Proof. Clearly every basis for $\overline{\mathbf{A}}$ may be displayed as follows:

$$\overline{\mathbf{B}}=\underbrace{\left(\begin{array}{c}\mathbf{V}\\ \hdashline \mathbf{U}\end{array}\right)}_{\bar{I}+\bar{K}}\begin{array}{l}\}\bar{I}\\ \}\bar{K}\end{array}.$$

Suppose the rank of $\mathbf{V}<\bar{I}$; then the rows of $\mathbf{V}$ are not linearly independent, contradicting the linear independence of $\overline{\mathbf{B}}$. Therefore $\mathbf{V}$ can be partitioned into $[\mathbf{B} \mid \mathbf{C}]$ with $\text{Det}(\mathbf{B})\neq 0$. ■

Since $\mathbf{B}$ is a basis from $[\mathbf{A} \mid \mathbf{e}^l]$, by Proposition 3.16, the columns of $\mathbf{B}$ correspond to a rooted spanning tree, say $\mathfrak{T}_B$. We will show that all operations involving $\mathbf{B}$ or $\mathbf{B}^{-1}$ can be efficiently executed by labeling operations involving $\mathfrak{T}_B$.

Before developing the partitioned inverse of $\overline{\mathbf{B}}$, we show that a special matrix, called the *working basis*, is invertible.

Proposition 7.2

If

$$\overline{\mathbf{B}}=\left(\begin{array}{c:c}\mathbf{B} & \mathbf{C}\\ \hdashline \mathbf{D} & \mathbf{F}\end{array}\right)\begin{array}{l}\}\bar{I}\\ \}\bar{K}\end{array},\qquad \text{columns: } \underbrace{\mathbf{B},\mathbf{D}}_{\bar{I}}\ \underbrace{\mathbf{C},\mathbf{F}}_{\bar{K}}$$

is invertible and $\mathbf{B}$ is invertible, then $\mathbf{F}-\mathbf{D}\mathbf{B}^{-1}\mathbf{C}$ is invertible.

Proof. Let

$$\mathbf{T}=\left(\begin{array}{c:c}\mathbf{I} & -\mathbf{B}^{-1}\mathbf{C}\\ \hdashline & \mathbf{I}\end{array}\right).$$

Then $\text{Det}(\mathbf{T}) \neq 0$. Also,

$$\overline{\mathbf{B}}\mathbf{T} = \left(\begin{array}{c|c} \mathbf{B} & \\ \hline \mathbf{D} & \mathbf{F} - \mathbf{D}\mathbf{B}^{-1}\mathbf{C} \end{array}\right),$$

and $\text{Det}(\overline{\mathbf{B}}\mathbf{T}) = \text{Det}(\overline{\mathbf{B}})\,\text{Det}(\mathbf{T}) \neq 0$ since $\text{Det}(\overline{\mathbf{B}}) \neq 0$ and $\text{Det}(\mathbf{T}) \neq 0$. Therefore the $\overline{K} \times \overline{K}$ matrix $\mathbf{F} - \mathbf{D}\mathbf{B}^{-1}\mathbf{C}$ has full rank, implying that $\mathbf{F} - \mathbf{D}\mathbf{B}^{-1}\mathbf{C}$ is invertible. ■

We now state an important result that makes use of the inverse of the working basis.

Proposition 7.3

If

$$\overline{\mathbf{B}} = \underbrace{\left(\begin{array}{c|c} \mathbf{B} & \mathbf{C} \\ \hline \mathbf{D} & \mathbf{F} \end{array}\right)}_{\overline{I} \quad \overline{K}} \begin{array}{l} \}\overline{I} \\ \}\overline{K} \end{array}$$

is invertible and $\mathbf{B}$ is invertible, then

$$\overline{\mathbf{B}}^{-1} = \left(\begin{array}{c|c} \mathbf{B}^{-1} + \mathbf{B}^{-1}\mathbf{C}\mathbf{Q}^{-1}\mathbf{D}\mathbf{B}^{-1} & -\mathbf{B}^{-1}\mathbf{C}\mathbf{Q}^{-1} \\ \hline -\mathbf{Q}^{-1}\mathbf{D}\mathbf{B}^{-1} & \mathbf{Q}^{-1} \end{array}\right), \tag{7.1}$$

where $\mathbf{Q} = \mathbf{F} - \mathbf{D}\mathbf{B}^{-1}\mathbf{C}$.

Proof. This follows immediately from Proposition 7.2, block multiplication of $\overline{\mathbf{B}}$ by the matrix of (7.1), and the uniqueness of the inverse. ■

We will use the partitioning shown in (7.1) to specialize the simplex operations for $\mathcal{NPS}$.

7.2 DUAL VARIABLE CALCULATION

In this section we indicate how the structure of the basis $\overline{\mathbf{B}}$ may be used to efficiently calculate the dual variables $\boldsymbol{\pi} = \mathbf{c}^B\overline{\mathbf{B}}^{-1}$ to be used in the pricing operation (see ALG 2.1). Let $\boldsymbol{\pi}$ and $\mathbf{c}^B$ be partitioned to be compatible with

(7.1), that is, $\boldsymbol{\pi}=[\boldsymbol{\pi}^1 \,\vdots\, \boldsymbol{\pi}^2]$ and $\mathbf{c}^B=[\mathbf{c}^1 \,\vdots\, \mathbf{c}^2]$. Then

$$\left[\boldsymbol{\pi}^1 \,\vdots\, \boldsymbol{\pi}^2\right]=\left[\mathbf{c}^1 \,\vdots\, \mathbf{c}^2\right]\begin{pmatrix} \mathbf{B}^{-1}+\mathbf{B}^{-1}\mathbf{C}\mathbf{Q}^{-1}\mathbf{D}\mathbf{B}^{-1} & -\mathbf{B}^{-1}\mathbf{C}\mathbf{Q}^{-1} \\ -\mathbf{Q}^{-1}\mathbf{D}\mathbf{B}^{-1} & \mathbf{Q}^{-1} \end{pmatrix}$$
$$=\left[(\mathbf{c}^1+\mathbf{c}^1\mathbf{B}^{-1}\mathbf{C}\mathbf{Q}^{-1}\mathbf{D}-\mathbf{c}^2\mathbf{Q}^{-1}\mathbf{D})\mathbf{B}^{-1} \,\vdots\, (\mathbf{c}^2-\mathbf{c}^1\mathbf{B}^{-1}\mathbf{C})\mathbf{Q}^{-1}\right]. \tag{7.2}$$

Recall that ALG 3.2 can be used to obtain the $\bar{I}$-component vector $\mathbf{c}^1\mathbf{B}^{-1}$ via labeling operations on $\mathfrak{T}_B$ rather than by matrix operations. Therefore the calculations for (7.2) should be ordered according to the following algorithm.

ALG 7.1 DUAL VARIABLE CALCULATION

1 Set $\gamma^1 \leftarrow \mathbf{c}^1\mathbf{B}^{-1}$, using ALG 3.2.
2 Set $\gamma^2 \leftarrow \mathbf{c}^1+\gamma^1\mathbf{C}\mathbf{Q}^{-1}\mathbf{D}-\mathbf{c}^2\mathbf{Q}^{-1}\mathbf{D}$.
3 Set $\boldsymbol{\pi}^1 \leftarrow \gamma^2\mathbf{B}^{-1}$, using ALG 3.2.
4 Set $\boldsymbol{\pi}^2 \leftarrow (\mathbf{c}^2-\gamma^1\mathbf{C})\mathbf{Q}^{-1}$.

The potential savings of this algorithm over the standard calculation $\mathbf{c}^B\overline{\mathbf{B}}^{-1}$ are in Steps 1 and 3, where the structure of $\mathbf{B}$ is exploited.

7.3 COLUMN UPDATE

In this section we indicate how the structure of the basis $\overline{\mathbf{B}}$ may be used to efficiently calculate the updated column $\mathbf{y}=\overline{\mathbf{B}}^{-1}\overline{\mathbf{A}}(k)$ to be used in the ratio test (see ALG 2.1). Two cases will be considered.

Case 1 Suppose the entering column corresponds to an arc. Let $\mathbf{y}$ and $\overline{\mathbf{A}}(k)$ be partitioned to be compatible with (7.1), that is,

$$\mathbf{y}=\begin{bmatrix}\mathbf{y}^1\\ \mathbf{y}^2\end{bmatrix} \quad \text{and} \quad \overline{\mathbf{A}}(k)=\begin{pmatrix}\mathbf{A}(k)\\ \mathbf{S}(k)\end{pmatrix}.$$

Then

$$\begin{bmatrix}\mathbf{y}^1\\ \mathbf{y}^2\end{bmatrix}=\begin{pmatrix} \mathbf{B}^{-1}+\mathbf{B}^{-1}\mathbf{C}\mathbf{Q}^{-1}\mathbf{D}\mathbf{B}^{-1} & -\mathbf{B}^{-1}\mathbf{C}\mathbf{Q}^{-1} \\ -\mathbf{Q}^{-1}\mathbf{D}\mathbf{B}^{-1} & \mathbf{Q}^{-1} \end{pmatrix}\begin{pmatrix}\mathbf{A}(k)\\ \mathbf{S}(k)\end{pmatrix}$$
$$=\begin{bmatrix}\mathbf{B}^{-1}\left[\mathbf{A}(k)+\mathbf{C}\mathbf{Q}^{-1}\mathbf{D}\mathbf{B}^{-1}\mathbf{A}(k)-\mathbf{C}\mathbf{Q}^{-1}\mathbf{S}(k)\right]\\ \mathbf{Q}^{-1}\left[\mathbf{S}(k)-\mathbf{D}\mathbf{B}^{-1}\mathbf{A}(k)\right]\end{bmatrix}. \tag{7.3}$$

Before determining the order in which the calculations of (7.3) should be executed, we first examine an algorithm for solving the system $\mathbf{By}=\mathbf{d}$ where $\mathbf{d}$ is an arbitrary vector.

ALG 7.2 SOLVING $\boldsymbol{By}=\boldsymbol{d}$

0 *Initialization.* Let $\mathbf{B}$ be any basis from $[\mathbf{A} \mid \mathbf{e}^l]$, and let $\mathfrak{T}_B=[\hat{\mathfrak{N}},\hat{\mathfrak{A}}]$ denote the corresponding rooted spanning tree. Renumber the arcs so that the jth column of $\mathbf{B}$ corresponds to arc e_j.

1 *Find Endpoint.* Let n be any endpoint of the rooted tree $[\hat{\mathfrak{N}},\hat{\mathfrak{A}}]$. The root node is considered to be an endpoint if the only incident arc is the root arc. Let arc j denote the arc in $[\hat{\mathfrak{N}},\hat{\mathfrak{A}}]$ incident on node n.

2 *Calculate* y_j. If $F(j)=n$, $y_j \leftarrow d_j$ and $\delta \leftarrow 1$; otherwise $y_j \leftarrow d_j$ and $\delta \leftarrow -1$.

3 *Check for Termination.* If $n=l$, terminate; otherwise $\hat{\mathfrak{N}} \leftarrow \hat{\mathfrak{N}}-\{n\}$ and $\hat{\mathfrak{A}} \leftarrow \hat{\mathfrak{A}}-\{e_j\}$.

4 *Update d.* If $\delta=1$, $d_{T(j)} \leftarrow d_{T(j)}-y_j$; otherwise $d_{F(j)} \leftarrow d_{F(j)}+y_j$. Go to Step 1.

Step 1 of ALG 7.2 can be easily implemented by using the inverse or reverse thread (see Appendix B).

We now present an algorithm for performing the calculations of (7.3).

ALG 7.3 COLUMN UPDATE (CASE 1)

1 Set $\boldsymbol{\gamma}^1 \leftarrow \mathbf{B}^{-1}\mathbf{A}(k)$, using (3.5).

2 Set $\boldsymbol{\gamma}^2 \leftarrow \mathbf{A}(k)+\mathbf{CQ}^{-1}\mathbf{D}\boldsymbol{\gamma}^1-\mathbf{CQ}^{-1}\mathbf{S}(\mathrm{k})$.

3 Set $\mathbf{y}^1 \leftarrow \mathbf{B}^{-1}\boldsymbol{\gamma}^2$ by applying ALG 7.2.

4 Set $\mathbf{y}^2 \leftarrow \mathbf{Q}^{-1}[\mathbf{S}(\mathrm{k})-\mathbf{D}\boldsymbol{\gamma}^1]$.

As with ALG 7.1, the potential savings of this algorithm over the standard calculation $\mathbf{y}=\overline{\mathbf{B}}^{-1}\overline{\mathbf{A}}(k)$ are in Steps 1 and 3, where the structure of $\mathbf{B}$ is exploited.

Case 2 Suppose the entering column is not an arc. Let $\mathbf{y}$ and $\overline{\mathbf{A}}(k)$ again be partitioned to be compatible with (7.1), that is,

$$\mathbf{y}=\begin{bmatrix}\mathbf{y}^1\\ \hdashline \mathbf{y}^2\end{bmatrix} \quad \text{and} \quad \overline{\mathbf{A}}(k)=\begin{bmatrix}\mathbf{0}\\ \hdashline \mathbf{P}(k)\end{bmatrix}. \tag{7.4}$$

Then

$$\begin{bmatrix} \mathbf{y}^1 \\ \hline \mathbf{y}^2 \end{bmatrix} = \left(\begin{array}{c|c} \mathbf{B}^{-1}+\mathbf{B}^{-1}\mathbf{C}\mathbf{Q}^{-1}\mathbf{D}\mathbf{B}^{-1} & -\mathbf{B}^{-1}\mathbf{C}\mathbf{Q}^{-1} \\ \hline -\mathbf{Q}^{-1}\mathbf{D}\mathbf{B}^{-1} & \mathbf{Q}^{-1} \end{array}\right) \begin{bmatrix} \mathbf{0} \\ \hline \mathbf{P}(k) \end{bmatrix} \tag{7.5}$$

$$= \begin{bmatrix} -\mathbf{B}^{-1}\mathbf{C}\mathbf{Q}^{-1}\mathbf{P}(k) \\ \hline \mathbf{Q}^{-1}\mathbf{P}(k) \end{bmatrix}. \tag{7.6}$$

We now present an algorithm for performing the calculations of (7.6).

ALG 7.4 COLUMN UPDATE (CASE 2)

1 Set $\mathbf{y}^2 \leftarrow \mathbf{Q}^{-1}\mathbf{P}(k)$.
2 Set $\gamma \leftarrow \mathbf{C}\mathbf{y}^2$.
3 Set $\mathbf{y}^1 \leftarrow -\mathbf{B}^{-1}\gamma$ by applying ALG 7.2.

7.4 WORKING BASIS INVERSE UPDATE

In this section we present the formulae that can be used to update the working basis inverse after each simplex pivot. Let the basis for $\mathfrak{NPS}$ be displayed as follows:

$$\overline{\mathbf{B}} = \underbrace{\left(\begin{array}{c|c} \mathbf{B} & \mathbf{C} \\ \hline \mathbf{D} & \mathbf{F} \end{array}\right)}_{\bar{I}\quad\ \bar{K}} \begin{array}{l} \}\bar{I} \\ \}\bar{K} \end{array},$$

where the first $\bar{I}$ columns will be called *key columns*, the last $\bar{K}$ columns will be called *nonkey columns*, and Det(**B**)$\neq$0. Associated with every $\overline{\mathbf{B}}$ we define a transformation matrix **L** as follows:

$$\mathbf{L} = \left(\begin{array}{c|c} \mathbf{I} & -\mathbf{B}^{-1}\mathbf{C} \\ \hline & \mathbf{I} \end{array}\right).$$

Then

$$\overline{\mathbf{B}}\mathbf{L} = \left(\begin{array}{c|c} \mathbf{B} & \\ \hline \mathbf{D} & \mathbf{F}-\mathbf{D}\mathbf{B}^{-1}\mathbf{C} \end{array}\right),$$

and $\mathbf{Q}=\mathbf{F}-\mathbf{D}\mathbf{B}^{-1}\mathbf{C}$ is called the working basis. In this section we develop formulae for updating $\mathbf{Q}^{-1}$ after each simplex pivot.

Let $\overline{\mathbf{B}}_i$, $\mathbf{L}_i$, and $\mathbf{Q}_i$ denote the basis, transformation matrix, and working basis at iteration i. Then we wish to determine an expression for $\mathbf{Q}_{i+1}^{-1}$ in terms of $\mathbf{Q}_i^{-1}$.

As is done in Appendix E, in order to maintain the proper partitioning it may be necessary to interchange a nonkey column with a key column that is to leave the basis before pivoting. For this case two changes in $\mathbf{Q}^{-1}$ will be made to accomplish the update, that is,

$$\overline{\mathbf{B}}_{i+1}^{-1} = \mathbf{E}\overline{\mathbf{B}}_i^{-1},$$

where $\mathbf{E}$ may be either an elementary column matrix or a permutation matrix. Let $\mathbf{E}$ be partitioned to be compatible with $\overline{\mathbf{B}}$, that is,

$$\mathbf{E} = \begin{pmatrix} \mathbf{E}_1 & \mathbf{E}_2 \\ \mathbf{E}_3 & \mathbf{E}_4 \end{pmatrix} \begin{matrix} \}\bar{I} \\ \}\bar{K} \end{matrix}.$$
$$\qquad \underbrace{}_{\bar{I}} \quad \underbrace{}_{\bar{K}}$$

Then from (E.8) of Appendix E

$$\mathbf{Q}_{i+1}^{-1} = (\mathbf{E}_4 - \mathbf{E}_3\mathbf{B}^{-1}\mathbf{C})\mathbf{Q}_i^{-1}. \tag{7.7}$$

In determining the updating formulae, we must examine two major cases with subcases.

Case 1 The leaving column is nonkey. For this case $\mathbf{E}$ takes the form

$$\mathbf{E} = \begin{pmatrix} \mathbf{I} & \mathbf{E}_2 \\ & \mathbf{E}_4 \end{pmatrix},$$

and (7.7) reduces to $\mathbf{Q}_{i+1}^{-1} = \mathbf{E}_4\mathbf{Q}_i^{-1}$, where $\mathbf{E}_4$ is an elementary column matrix.

Case 2 The leaving column is a key column. Then the leaving column is an arc in the spanning tree $\mathfrak{T}_B$. If this arc is removed from $\mathfrak{T}_B$, some other arc must replace it to maintain the structure of $\overline{\mathbf{B}}$. Therefore we attempt to switch the leaving column with some nonkey column. Suppose the leaving arc is in the lth column of $\mathbf{B}$. Let $\boldsymbol{\beta}^l$ denote the lth row of $\mathbf{B}^{-1}$, and let $\boldsymbol{\gamma} = \boldsymbol{\beta}^l\mathbf{C}$. Then, if $\gamma_j \neq 0$, the jth column of $\mathbf{C}$ can be interchanged with the lth column of $\mathbf{B}$, and the resulting matrix will be nonsingular. We now consider two subcases.

Subcase 2a $\boldsymbol{\gamma} \neq \mathbf{0}$. Suppose $\gamma_j \neq 0$. Consider the following permutation matrix:

$$\mathbf{E} = \left[\begin{array}{c|c|c|c|c} \mathbf{I} & & & & \\ \hline & & & 1 & \\ \hline & & \mathbf{I} & & \\ \hline & 1 & & & \\ \hline & & & & \mathbf{I} \end{array}\right] \begin{array}{l} \\ \leftarrow \text{row } l \\ \\ \leftarrow \text{row } \bar{I}+j \\ \\ \end{array}$$

$$\uparrow \text{column } l \qquad \uparrow \text{column } \bar{I}+j$$

Then

$$\mathbf{E}_3 = \left[\begin{array}{c|c|c} & & \\ \hline & 1 & \\ \hline & & \end{array}\right] \leftarrow \text{row } j$$

$$\uparrow \text{column } l$$

and

$$\mathbf{E}_4 = \left[\begin{array}{c|c|c} \mathbf{I} & & \\ \hline & & \\ \hline & & \mathbf{I} \end{array}\right].$$

$$\uparrow \text{column } j$$

Therefore $\mathbf{E}_4 - \mathbf{E}_3\mathbf{B}^{-1}\mathbf{C}$ reduces to

$$\mathbf{E}_5 = \left[\begin{array}{c|c|c} \mathbf{I} & & \\ \hline & -\boldsymbol{\beta}^j\mathbf{C} & \\ \hline & & \mathbf{I} \end{array}\right],$$

and $\mathbf{Q}_{i+1}^{-1} = \mathbf{E}_5\mathbf{Q}_i^{-1}$. Once the leaving column has been made nonkey, Case 1 may be applied.

Subcase 2b $\boldsymbol{\gamma} = \mathbf{0}$. For this case a direct pivot is made, and $\mathbf{Q}_{i+1}^{-1} = \mathbf{Q}_i^{-1}$.

7.5 EXAMPLE

In this section we demonstrate ALG 2.2 with an example. Consider the minimum cost network flow problem, shown in Table 7.1. with these side constraints:

$$\left\{\begin{array}{lllll} 10x_1 & -2x_3 & +3x_4 - 2x_5 & \leqslant 16 \\ x_1 & +4x_3 & \qquad\quad +x_5 & \leqslant 10 \end{array}\right\}.$$

Table 7.1 Node Requirements

Node	Supply	Demand
1	10	0
2	0	0
3	0	0
4	0	10

Arc j	u_j	c_j	$F(j)$	$T(j)$
1	12	0	1	2
2	18	10	1	3
3	5	2	2	3
4	12	1	2	4
5	1	3	3	2
6	16	4	3	4

The network is illustrated in Figure 7.1. After adding slacks and a root arc, the example may be written as follows:

$$
\begin{array}{lcccccccccccc}
\text{min} & & 10x_2 & +2x_3 & +x_4 & +3x_5 & +4x_6 & & & & & & \\
\text{s.t.} & x_1 & +x_2 & & & & & & + & x_7 & = & 10 \\
 & -x_1 & & +x_3 & +x_4 & -x_5 & & & & & = & 0 \\
 & & -x_2 & -x_3 & & +x_5 & +x_6 & & & & = & 0 \\
 & & & & -x_4 & & -x_6 & & & & = & -10 \\
 & 10x_1 & & -2x_3 & +3x_4 & -2x_5 & & +z_1 & & & = & 16 \\
 & x_1 & & +4x_3 & & +x_5 & & & +z_2 & & = & 10
\end{array}
$$

$$0<x_1<12,\ 0<x_2<18,\ 0<x_3<5,\ 0<x_4<12,\ 0<x_5<1,\ 0<x_6<16,\ 0<z_1<\infty,\ 0<z_2<\infty,\ 0<x_7<0.$$

We will now apply ALG 2.2 to this problem, using the specializations developed in this chapter.

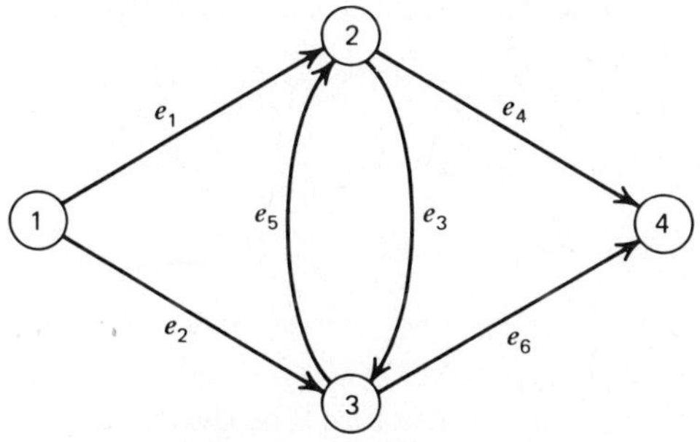

FIGURE 7.1 Sample network.

0 (Initialization). Consider the following initial basis:

$$\bar{\mathbf{B}}=\begin{array}{c} \begin{matrix} x_7 & x_2 & x_5 & x_6 & x_1 & x_3 \end{matrix} \\ \left[\begin{array}{cccc|cc} 1 & 1 & & & 1 & \\ & & -1 & & -1 & 1 \\ & -1 & 1 & 1 & & -1 \\ & & & -1 & & \\ \hline & & -2 & & 10 & -2 \\ & & 1 & & 1 & 4 \end{array}\right] \end{array}=\left[\begin{array}{c|c} \mathbf{B} & \mathbf{C} \\ \hline \mathbf{D} & \mathbf{F} \end{array}\right].$$

By Proposition 7.3

$$\mathbf{Q}^{-1}=\left[\begin{array}{c|c} \frac{1}{12} & \frac{1}{15} \\ \hline & \frac{1}{5} \end{array}\right], \qquad \mathbf{B}^{-1}=\left[\begin{array}{c|c|c|c} 1 & 1 & 1 & 1 \\ \hline & -1 & -1 & -1 \\ \hline & -1 & & \\ \hline & & & -1 \end{array}\right],$$

and

$$\bar{\mathbf{B}}^{-1}=\left[\begin{array}{c|c|c|c|c|c} 1 & 1 & 1 & 1 & & \\ \hline & -\frac{9}{10} & -1 & -1 & -\frac{1}{12} & -\frac{1}{15} \\ \hline & -\frac{7}{10} & & & -\frac{1}{12} & \frac{2}{15} \\ \hline & & & -1 & & \\ \hline & -\frac{1}{10} & & & \frac{1}{12} & \frac{1}{15} \\ \hline & \frac{1}{5} & & & & \frac{1}{5} \end{array}\right].$$

$\mathcal{T}_B$ is shown in Figure 7.2.

The current solution is

$$\left[x_1^B \;\; x_2^B \;\; x_3^B \;\; x_4^N \;\; x_5^B \;\; x_6^B \;\; x_7^B \;\; z_1^N \;\; z_2^N\right] = \left[2 \;\; 8 \;\; 2 \;\; 0 \;\; 0 \;\; 10 \;\; 0 \;\; 0 \;\; 0\right].$$

1 (Pricing). We apply ALG 7.1 to obtain the dual variables.
a. We use ALG 3.2 to obtain γ^1, as shown in Figure 7.3.

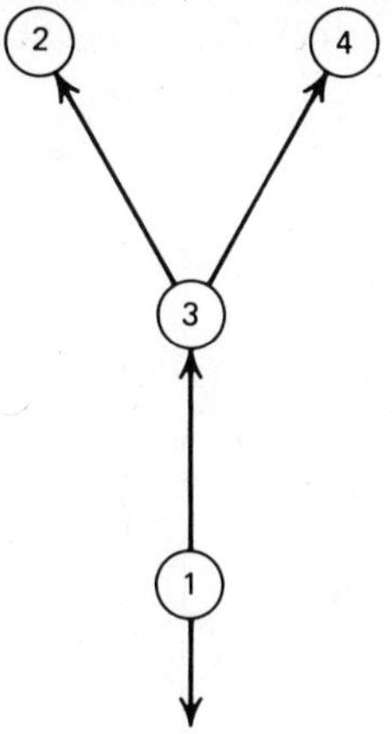

FIGURE 7.2 Initial basis.

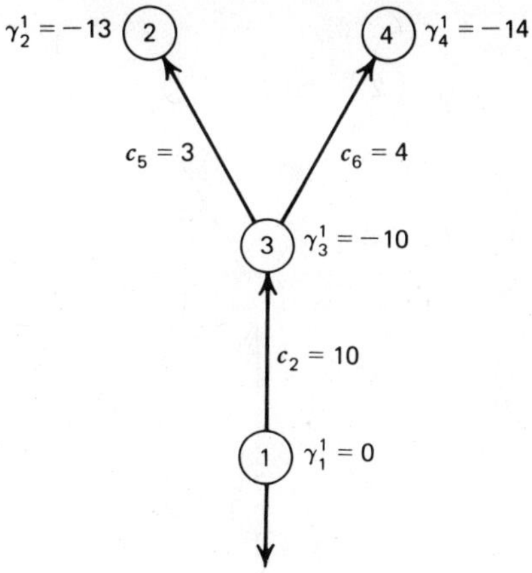

FIGURE 7.3 Calculation of $\boldsymbol{\gamma}^1$ for pricing.

b. $\boldsymbol{\gamma}^2 \leftarrow \mathbf{c}^1 + \boldsymbol{\gamma}^1\mathbf{C}\mathbf{Q}^{-1}\mathbf{D} - \mathbf{c}^2\mathbf{Q}^{-1}\mathbf{D}.$

$$\gamma^2 = [0 \quad 10 \quad 3 \quad 4] + [0 \quad -13 \quad -10 \quad -14]\left[\begin{array}{c|c} 1 & 0 \\ \hline -1 & 1 \\ \hline 0 & -1 \\ \hline 0 & 0 \end{array}\right]$$

$$\cdot \left[\begin{array}{c|c} \frac{1}{12} & \frac{1}{15} \\ \hline & \frac{1}{5} \end{array}\right]\left[\begin{array}{c|c|c|c} - & - & \begin{matrix}-2\\1\end{matrix} & - \end{array}\right]$$

$$-[0 \quad 2]\left[\begin{array}{c|c} \frac{1}{12} & \frac{1}{15} \\ \hline & \frac{1}{5} \end{array}\right]\left[\begin{array}{c|c|c|c} - & - & \begin{matrix}-2\\1\end{matrix} & - \end{array}\right],$$

$$\gamma^2 = [0 \quad 10 \quad 3 \quad 4] + [0 \quad 0 \quad -\tfrac{57}{30} \quad 0] - [0 \quad 0 \quad \tfrac{2}{5} \quad 0]$$

$$= [0 \quad 10 \quad \tfrac{7}{10} \quad 4].$$

c. $\boldsymbol{\pi}^1 \leftarrow \boldsymbol{\gamma}^2\mathbf{B}^{-1}$. This calculation is shown in Figure 7.4.

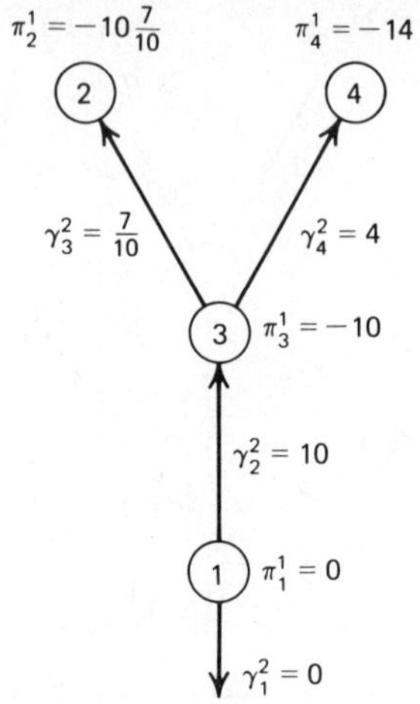

FIGURE 7.4 Calculation of $\boldsymbol{\pi}^1$.

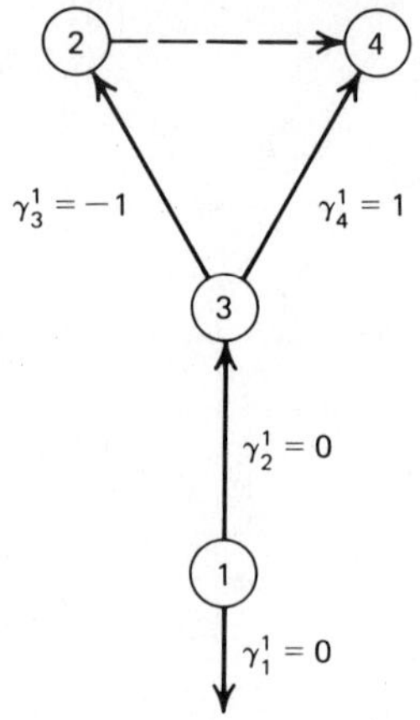

FIGURE 7.5 Ratio test $\boldsymbol{\gamma}^1$ calculation.

d. $\boldsymbol{\pi}^2 \leftarrow (\mathbf{c}^2 - \boldsymbol{\gamma}^1\mathbf{C})\mathbf{Q}^{-1}$.

$$\boldsymbol{\pi}^2 = \left\{ [0 \quad 2] - [0 \quad -13 \quad -10 \quad -14] \begin{bmatrix} 1 & 0 \\ -1 & 1 \\ 0 & -1 \\ 0 & 0 \end{bmatrix} \right\} \begin{bmatrix} \frac{1}{12} & \frac{1}{15} \\ & \frac{1}{5} \end{bmatrix}$$

$$= \left[-1\tfrac{1}{12} \quad \tfrac{26}{15} \right].$$

We now calculate the reduced costs using the dual variables $[\boldsymbol{\pi}^1 \,\vdots\, \boldsymbol{\pi}^2]$. For x_4: $\pi_2^1 - \pi_4^1 + 3\pi_1^2 - c_4 = -10\frac{7}{10} - (-14) + 3(-1\frac{1}{12}) > 0 \Rightarrow x_4 \varepsilon \psi^1$. Let $x_k = x_4$ and $\delta = 1$.

2 (Ratio Test). We apply ALG 7.3 to obtain the updated column.

a. $\boldsymbol{\gamma}^1 \leftarrow \mathbf{B}^{-1}\mathbf{A}(4)$. This calculation is shown in Figure 7.5.

b. $\boldsymbol{\gamma}^2 \leftarrow \mathbf{A}(4) + \mathbf{C}\mathbf{Q}^{-1}\mathbf{D}\boldsymbol{\gamma}^1 - \mathbf{C}\mathbf{Q}^{-1}\mathbf{S}(4)$.

$$\boldsymbol{\gamma}^2 = \begin{bmatrix} 0 \\ 1 \\ 0 \\ -1 \end{bmatrix} + \begin{bmatrix} 1 & 0 \\ -1 & 1 \\ 0 & -1 \\ 0 & 0 \end{bmatrix} \begin{bmatrix} \frac{1}{12} & \frac{1}{15} \\ & \frac{1}{5} \end{bmatrix} \begin{bmatrix} & & -2 & \\ & & 1 & \end{bmatrix} \begin{bmatrix} 0 \\ 0 \\ -1 \\ 1 \end{bmatrix}$$

$$- \begin{bmatrix} 1 & 0 \\ -1 & 1 \\ 0 & -1 \\ 0 & 0 \end{bmatrix} \begin{bmatrix} \frac{1}{12} & \frac{1}{15} \\ & \frac{1}{5} \end{bmatrix} \begin{pmatrix} 3 \\ 0 \end{pmatrix} = \begin{bmatrix} 0 \\ 1 \\ 0 \\ -1 \end{bmatrix} + \begin{bmatrix} \frac{1}{10} \\ -\frac{3}{10} \\ \frac{1}{5} \\ 0 \end{bmatrix} - \begin{bmatrix} \frac{1}{4} \\ -\frac{1}{4} \\ 0 \\ 0 \end{bmatrix} = \begin{bmatrix} -\frac{3}{20} \\ \frac{19}{20} \\ \frac{1}{5} \\ -1 \end{bmatrix}.$$

c. $\mathbf{y}^1 \leftarrow \mathbf{B}^{-1}\gamma^2$.

$$\mathbf{y}^1 = \begin{bmatrix} -\frac{3}{20} \\ 0 \\ 0 \\ 0 \end{bmatrix} + \begin{bmatrix} \frac{19}{20} \\ -\frac{19}{20} \\ -\frac{19}{20} \\ 0 \end{bmatrix} + \begin{bmatrix} \frac{1}{5} \\ -\frac{1}{5} \\ 0 \\ 0 \end{bmatrix} + \begin{bmatrix} -1 \\ 1 \\ 0 \\ 1 \end{bmatrix}$$

$$= \begin{bmatrix} 0 \\ -\frac{3}{20} \\ -\frac{19}{20} \\ 1 \end{bmatrix}.$$

The graphical computation of the vector components is illustrated in Figure 7.6.

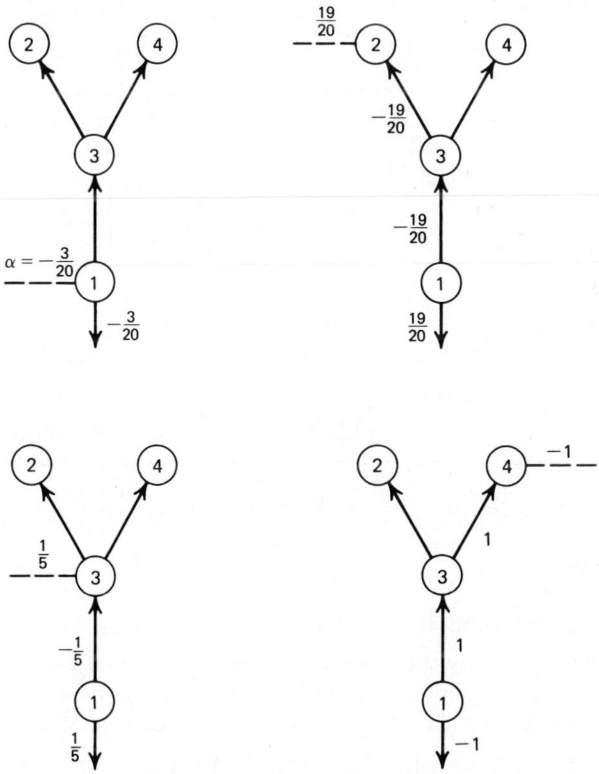

FIGURE 7.6 Calculations of $\mathbf{y}^1$.

d. $\mathbf{y}^2 \leftarrow \mathbf{Q}^{-1}[\mathbf{S}(4) - \mathbf{D}\gamma^1]$.

$$\mathbf{y}^2 = \left[\begin{array}{c:c} \frac{1}{12} & \frac{1}{15} \\ \hdashline & \frac{1}{5} \end{array}\right] \left\{ \left(\begin{array}{c} 3 \\ \hdashline 0 \end{array}\right) - \left(\begin{array}{c:c:c:c} & & -2 & \\ \hdashline & & 1 & \end{array}\right) \left[\begin{array}{r} 0 \\ \hdashline 0 \\ \hdashline -1 \\ \hdashline 1 \end{array}\right] \right\} = \left[\begin{array}{c} \frac{3}{20} \\ \hdashline \frac{1}{5} \end{array}\right].$$

$$\therefore \left[\begin{array}{c} \mathbf{y}^1 \\ \hline \mathbf{y}^2 \end{array}\right] = \left[\begin{array}{r} 0 \\ \hdashline -\frac{3}{20} \\ \hdashline -\frac{19}{20} \\ \hdashline 1 \\ \hline \frac{3}{20} \\ \hdashline \frac{1}{5} \end{array}\right], \qquad \bar{\mathbf{B}}^{-1}\left(\frac{r}{b}\right) = \left[\begin{array}{c} 0 \\ \hdashline 8 \\ \hdashline 0 \\ \hdashline 10 \\ \hline 2 \\ \hdashline 2 \end{array}\right].$$

$$\therefore \Delta_1 = \min\left\{ \tfrac{10}{1}, \frac{2}{\frac{3}{20}}, \frac{2}{\frac{1}{5}} \right\} = 10,$$

$$\Delta_2 = \min\left\{ \frac{18-8}{\frac{3}{20}}, \frac{1-0}{\frac{19}{20}} \right\} = \tfrac{20}{19},$$

$$\Delta = \min\left\{ 10, \tfrac{20}{19}, 12 \right\} = \tfrac{20}{19}.$$

The leaving variable will be x_5. This completes one pricing operation and one ratio test for the sample problem.

EXERCISES

7.1 Prove Proposition 7.3.

7.2 Consider the network problem shown in Table E7.2. Suppose there are these additional side constraints: (i) the sum of flow in arcs 3 and 4 cannot exceed 8, and (ii) the flow on arc 5 is never smaller than the flow on arc 3.

Table E7.2 Problem Description

Node	Supply	Demand
1	10	0
2	0	0
3	0	0
4	0	10

Arc j	$F(j)$	$T(j)$	c_j	u_j
1	1	2	1	8
2	1	3	2	6
3	2	3	0	13
4	2	4	10	5
5	3	2	−5	8
6	3	4	2	5

(a) Use ALG 7.1 to determine the dual variables associated with the following feasible solution: $[x_1^B \;\; x_2^B \;\; x_3^N \;\; x_4^B \;\; x_5^N \;\; x_6^N] = [5 \;\; 5 \;\; 0 \;\; 5 \;\; 0 \;\; 5]$.

(b) For your solution in (a) use ALG 7.3 to determine the updated columns associated with arc 6.

(c) Solve this problem with the specialized simplex algorithm, beginning with the initial solution given in (a).

7.3 Define a data structure to implement the primal simplex algorithm for $\mathcal{NPS}$.

7.4 Beginning with an all-artificial start, make two simplex pivots with the following network problem with side constraints:

$$
\begin{aligned}
\min \quad & 2x_1 && && +4x_3 && \\
\text{s.t.} \quad & x_1 && +x_2 && && \leqslant 5 \\
& -x_1 && && +x_3 && \leqslant -2 \\
& && -x_2 && -x_3 && \leqslant -2 \\
& 2x_1 && +x_2 && && \leqslant 7 \\
& -x_1 && +2x_2 && +x_3 && \geqslant 0 \\
& 0 \leqslant x_1 \leqslant 4,\; 0 \leqslant x_2 \leqslant 1,\; 0 \leqslant x_3 \leqslant 3.
\end{aligned}
$$

7.5 Beginning with the solution $[x_1 \quad x_2 \quad x_3]=[3 \quad 1 \quad 1]$, make one simplex pivot with the following network problem with side constraints:

$$
\begin{array}{llllll}
\max & 2x_1 & & +4x_3 & & \\
\text{s.t.} & x_1 & +x_2 & & \leqslant & 5 \\
& -x_1 & & +x_3 & \leqslant & -2 \\
& & -x_2 & -x_3 & \leqslant & -2 \\
& 2x_1 & +x_2 & & \leqslant & 7 \\
& -x_1 & +2x_2 & +x_3 & \geqslant & 0 \\
& \multicolumn{5}{l}{0 \leqslant x_1 \leqslant 4,\ 0 \leqslant x_2 \leqslant 1,\ 0 \leqslant x_3 \leqslant 3.}
\end{array}
$$

NOTES AND REFERENCES

The results of this chapter may be traced to the work of Bennett [22], Kaul [116], Hartman and Lasdon [93], and Chen and Saigal [33]. Similar ideas have been investigated by Graves and McBride [89]. These basic ideas have been generalized by Glover and Klingman [83] and specialized by Glover, Karney, Klingman, and Russell [85].

CHAPTER 8

The Convex Cost Network Flow Problem

In this chapter we present several algorithms for solving the *convex cost network flow problem*:

$$\left.\begin{array}{ll}\min & g(\mathbf{x}) \\ \text{s.t.} & \mathbf{A}\mathbf{x}=\mathbf{r} \\ & \mathbf{0}\leqslant\mathbf{x}\leqslant\mathbf{u}\end{array}\right\}\text{NLP},$$

where **A** is a node-arc incidence matrix for the graph $\mathcal{G}=[\mathcal{N},\mathcal{A}]$, $\mathbf{1r}=0$, and $g(\cdot)$ is a convex function. The NLP is a specially structured nonlinear program and may be solved by any of a host of nonlinear techniques. In this chapter we present specializations of some of these techniques that exploit the underlying graphical structure. These specializations completely eliminate the need for matrix operations and execute all operations directly on $\mathcal{G}$.

8.1 PIECEWISE LINEAR APPROXIMATION

In this section we consider the case where $g(\mathbf{x})$ is separable, that is, $g(\mathbf{x})=\Sigma_j g_j(x_j)$. It can easily be shown that each component function g_j must be convex because of the assumed convexity of g. One approach to the solution of NLP is to approximate each component function $g_j(x_j)$ with a continuous piecewise linear function and solve the derived problem. If the approximations used satisfy certain conditions, the derived problem can be solved as a minimal cost network flow model.

Suppose K linear segments are to be used to approximate $g_j(x_j)$. Let the interval $[0,u_j]$ be partitioned into K segments, each of length v_j^k for $k=1,\ldots,K$, so that $u_j=\Sigma_k v_j^k$. Let $u_j^0=0$, and let u_j^k denote the right endpoint for the kth segment, so that $u_j^k=\Sigma_{p=1}^{p=k} v_j^p$. We may define a partitioning of the flow on arc e_j corresponding to the partitioning of $[0,u_j]$ by defining $x_j^k=\max\{0,\min\{x_j-u_j^{k-1},u_j^k-u_j^{k-1}\}\}$ for $k=1,\ldots,K$. If unit cost c_j^k is assigned to flow x_j^k, the function $\hat{g}_j(x_j)=g_j(0)+\Sigma_k c_j^k x_j^k$ provides a continuous piecewise linear approximation to $g_j(x_j)$. Since $g_j(x_j)$ must be convex, it is desirable that the approximation $\hat{g}_j(x_j)$ retain this property. Convexity for $\hat{g}_j(x_j)$ will hold if and only if $c_j^1 \leqslant c_j^2 \leqslant \cdots \leqslant c_j^K$.

Suppose that K linear segments are used in approximating $g_j(x_j)$ for each arc e_j and that the partitions and unit costs have been assigned. Then the derived optimization problem may be given as follows:

$$\min \quad \sum_{j,k} c_j^k x_j^k \tag{8.1}$$

$$\text{s.t.} \quad \sum_k \mathbf{A}\mathbf{x}^k=\mathbf{r} \tag{8.2}$$

$$\mathbf{0} \leqslant \mathbf{x}^k \leqslant \mathbf{v}^k \tag{8.3}$$

$$x_j^{k-1}=u_j^{k-1} \qquad \text{if } x_j^k>0. \tag{8.4}$$

Note that the constant $\Sigma_j g_j(0)$ has been omitted from (8.1). The effect of the approximation has been to replace each arc in $\mathcal{G}$ by K arcs. If $c_j^1 \leqslant c_j^2 \leqslant \cdots \leqslant c_j^K$ for each $e_j \varepsilon \mathcal{G}$, it is easy to see that constraint (8.4) may be omitted. One should also note that the pricing routine of the network algorithm can be specialized to account for multiple arcs. In other words, if an arc $e_{k_1}=(i,j)$ with zero flow prices out unfavorably, all other arcs $e_{k_l}=(i,j)$ with unit cost no less than that of e_{k_1} will also price out unfavorably. In general, for a K-part segmentation of an original arc, at most two segments need be priced.

We now turn to the problem of determining the unit costs c_j^k. As an example, we use the function $g_j(x_j)=x_j^3+1000$, illustrated in Figure 8.1. For the moment we assume that a partitioning of each $[0,u_j]$ using K segments has already been chosen. One method for obtaining c_j^k is to simply use a chordal approximation, that is, $c_j^k=[g_j(u_j^k)-g_j(u_j^{k-1})]/v_j^k$. A three-part segmentation using a chordal approximation to the function of Figure 8.1 is illustrated in Figure 8.2. It can be shown that this approximation will automatically produce the desired property, $c_j^1 \leqslant c_j^2 \leqslant \cdots \leqslant c_j^K$. However, unless a very fine partitioning of $[0,u_j]$ can be allowed, this

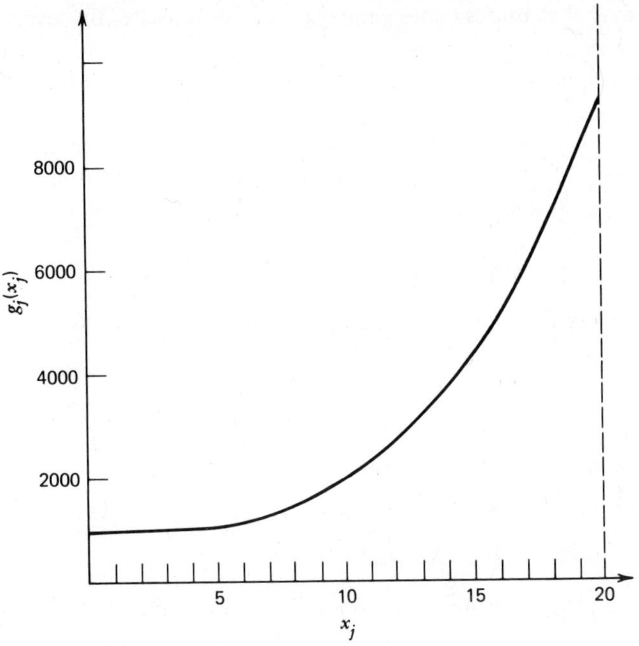

FIGURE 8.1 Illustration of $g_j(x_j) = x_j^3 + 1000$ over (0, 20).

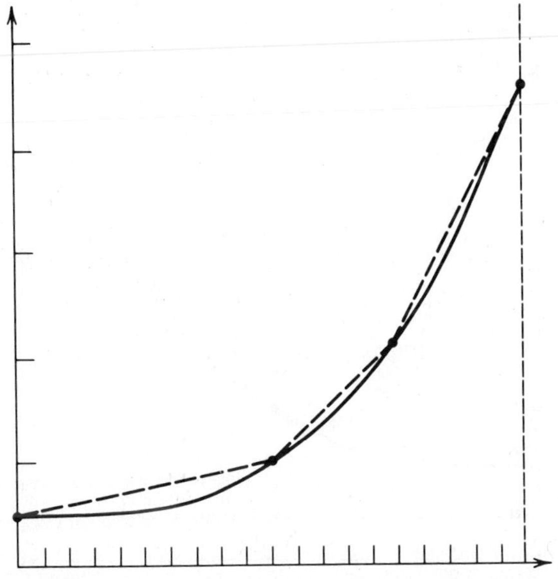

FIGURE 8.2 Illustration of three-part chordal approximation.

method may be undesirable since $g_j(x_j)$ is consistently overestimated, except at $u_j^0, u_j^1, \ldots, u_j^K$.

Another technique that may be applied to determine the unit costs is based on the idea of a least squares fit. Assuming again that the partitioning of $[0, u_j]$ has already been chosen, we may determine the unit costs recursively. (To simplify the notation we suppress the subscript j.) We deduce $\beta^1 = g(0)$, and let $\beta^{k+1} = \beta^k + c^k(u^k - u^{k-1})$ for $k = 1, \ldots, K-1$, so that

$$\hat{g}(x) = \beta^k + c^k(x - u^{k-1}) \qquad \text{for } u^{k-1} \leqslant x \leqslant u^k.$$

We select c^k as the value that minimizes the convex function

$$f(c^k) = \int_{u^{k-1}}^{u^k} [g(t) - \hat{g}(t)]^2 \, dt.$$

Setting $df(c^k)/dc^k$ to zero, we obtain

$$c^k = \frac{-3\beta^k}{2(u^k - u^{k-1})} + \frac{3}{(u^k - u^{k-1})^3} \int_{u^{k-1}}^{u^k} (t - u^{k-1}) g(t) \, dt.$$

Of course, c^1 must be determined first, c^2 next, and so on. A three-part

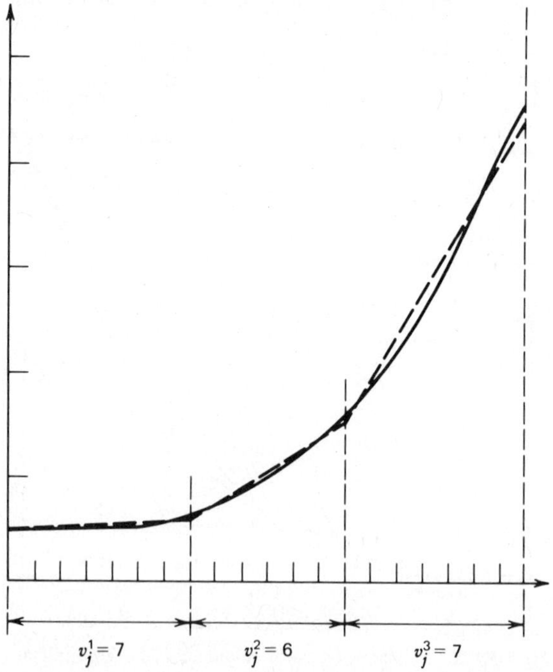

FIGURE 8.3 A three-part least squares linear approximation.

segmentation using this approach for the function of Figure 8.1 is illustrated in Figure 8.3.

The least squares method given here will not guarantee that $c^1 \leqslant c^2 \leqslant \cdots \leqslant c^K$. This can be achieved by careful choice of the number of segments and the positioning of $u^1, \ldots, u^K$. We believe that these choices are best made by careful consideration of and experimentation with the functions involved in the particular application at hand. If the functions involved contain linear portions, it is obviously desirable to use approximations that agree with these linear portions.

8.2 FRANK-WOLFE METHOD

In 1956 Marguerite Frank and Philip Wolfe proposed a method for solving nonlinear programs having a convex differentiable cost function and linear constraints. For our specialization of this technique we assume that g is differentiable over $\{\mathbf{x}:\mathbf{0} \leqslant \mathbf{x} \leqslant \mathbf{u}\}$, and we let $\nabla g(\mathbf{y})$ denote the gradient of g evaluated at $\mathbf{y}$. The Frank-Wolfe algorithm is a feasible direction method (see Chapter 2) and may be simply stated as follows: *Given a feasible flow at iteration k for* NLP, *say* $\mathbf{y}_k$, *one finds an improving feasible flow direction by solving the linear network flow problem min*$\{\nabla g(\mathbf{y}_k)\mathbf{x}:\mathbf{A}\mathbf{x}=\mathbf{r}, \mathbf{0} \leqslant \mathbf{x} \leqslant \mathbf{u}\}$. As a by-product of the solution of the linear network problem, one also obtains a lower bound.

The following well-known result, which will be given without proof, states that a convex function evaluated at a point never lies below a tangent plane passed through any other point.

Proposition 8.1

If $g(x)$ is convex and has continuous first derivatives, then $g(\mathbf{x}_1) \geqslant g(\mathbf{x}_2) + \nabla g(\mathbf{x}_2)(\mathbf{x}_1 - \mathbf{x}_2)$ for any two points $\mathbf{x}_1$ and $\mathbf{x}_2$.

This result is used to obtain a lower bound. Consider the following.

Proposition 8.2

Let $\mathbf{x}^*$ denote an optimum for NLP, and let $\mathbf{y}$ be any feasible flow. Let $\mathbf{z}$ solve $\min\{\nabla g(\mathbf{y})\mathbf{x}:\mathbf{A}\mathbf{x}=\mathbf{r}, \mathbf{0} \leqslant \mathbf{x} \leqslant \mathbf{u}\}$. Then $g(\mathbf{x}^*) \geqslant g(\mathbf{y}) + \nabla g(\mathbf{y})(\mathbf{z}-\mathbf{y})$.

Proof. By Proposition 8.1 we have

$$g(\mathbf{x}^*) \geqslant g(\mathbf{y}) + \nabla g(\mathbf{y})(\mathbf{x}^* - \mathbf{y}) = g(\mathbf{y}) + \nabla g(\mathbf{y})\mathbf{x}^* - \nabla g(\mathbf{y})\mathbf{y}.$$

But $\nabla g(\mathbf{y})\mathbf{x}^* \geqslant \nabla g(\mathbf{y})\mathbf{z}$. Then $g(\mathbf{x}^*) \geqslant g(\mathbf{y}) + \nabla g(\mathbf{y})(\mathbf{z}-\mathbf{y})$. ∎

Using the lower bound provided by Proposition 8.2, we may state the Frank-Wolfe algorithm as follows.

ALG 8.1 FRANK-WOLFE ALGORITHM

0 *Initialization.* Let $\mathbf{y}_0$ be any feasible flow. Set $k \leftarrow 0, \beta \leftarrow -\infty$, and choose the termination parameter, $\epsilon > 0$. (*Note*: β is the lower bound.)

1 *Solve Network Subproblem, Update Bound, Check for Termination.* Let $\mathbf{z}_k$ denote the solution to $\min\{\nabla g(\mathbf{y}_k)\mathbf{x} : \mathbf{Ax} = \mathbf{r}, \mathbf{0} \leqslant \mathbf{x} \leqslant \mathbf{u}\}$. Set $\beta \leftarrow \max\{\beta, g(\mathbf{y}_k) + \nabla g(\mathbf{y}_k)(\mathbf{z}_k - \mathbf{y}_k)\}$. If $g(\mathbf{y}_k) - \beta < \varepsilon$, terminate with $\mathbf{y}_k$ as an ε-optimum; otherwise $k \leftarrow k+1$.

2 *Line Search.* Let $\mathbf{y}_k$ be the flow on the line segment between $\mathbf{y}_{k-1}$ and $\mathbf{z}_{k-1}$ having the smallest objective function value, and go to Step 1.

At least three variations of this basic algorithm have been developed. The method of parallel tangents (PARTAN) integrates movements in the flow direction $(\mathbf{y}_k - \mathbf{y}_{k-2})$ with the basic Frank-Wolfe steps. This algorithm is described below.

ALG 8.2 FRANK-WOLFE WITH PARTAN

0 *Initialization.* Same as ALG 8.1.

1 *Solve Network Subproblem, Update Bound, Check for Termination.* Same as ALG 8.1.

2 *Line Search.* Let ω be the flow on the line segment between $\mathbf{y}_{k-1}$ and $\mathbf{z}_{k-1}$ having the smallest objective function value. If $k=1, \mathbf{y}_1 \leftarrow \omega$, and go to Step 1.

3 *PARTAN Step.* Let $\mathbf{y}_k$ be the feasible flow having the smallest objective function value on the half-ray from $\mathbf{y}_{k-2}$ through ω, and go to Step 1.

Philip Wolfe later presented a modification of the original Frank-Wolfe algorithm that incorporated what he termed *away steps*. The modified algorithm is described below.

ALG 8.3 FRANK-WOLFE ALGORITHM WITH AWAY STEPS

0 *Initialization.* Same as ALG 8.1.

1 *Find Toward Step Direction, Update Bound, Check for Termination.* Same as ALG 8.1 with $k \leftarrow k+1$ deleted.

2 *Find Away Step Direction.* Let ω denote the solution to $\max\{\nabla g(\mathbf{y}_k)\mathbf{x} : \mathbf{A}\mathbf{x}=\mathbf{r}, \mathbf{0} \leqslant \mathbf{x} \leqslant \mathbf{u}, x_j = 0 \text{ if } (y_k)_j = 0 \text{ or } u_j\}$. Set $\mathbf{d}^2 \leftarrow \mathbf{y}_k - \omega$.

3 *Find Maximum Movement in Away Direction.* Let α^2 denote the solution to $\max_{\alpha \geqslant 0}\{\alpha : \mathbf{0} \leqslant \omega + \alpha \mathbf{d}^2 \leqslant \mathbf{u}\}$. If $\alpha^2 \leqslant 1$, go to Step 5.

4 *Select Direction.* If $|\nabla g(\mathbf{y}_k)\mathbf{z}_k| > |\nabla g(\mathbf{y}_k)\omega|$, go to Step 5; otherwise go to Step 6.

5 *Line Search with Toward Direction.* Let $\mathbf{y}_{k+1}$ be the flow on the line segment between $\mathbf{y}_k$ and $\mathbf{z}_k$ having the smallest objective value. Set $k \leftarrow k+1$, and go to Step 1.

6 *Line Search with Away Direction.* Let α^* denote the solution to $\min_{1 \leqslant \alpha \leqslant \alpha^2}\{g(\mathbf{y}_k + \alpha \mathbf{d}^2)\}$. Set $k \leftarrow k+1, \mathbf{y}_k \leftarrow \mathbf{y}_{k-1} + \alpha^* \mathbf{d}^2$, and go to Step 1.

Another proposed enhancement of the basic Frank-Wolfe algorithm involves replacing Step 2 of ALG 8.1 with the following:

Search. Let (λ^*, α^*) denote the solution to

$$\min \quad g\left(\mathbf{y}_0\lambda + \sum_{\lambda=0}^{\lambda=k+1} \mathbf{z}_i \alpha_i\right)$$

$$\text{s.t.} \quad \lambda + \sum_i \alpha_i = 1$$

$$\lambda, \alpha_i \geqslant 0, \qquad \text{for all } i.$$

Set $\mathbf{y}_k \leftarrow \lambda^* \mathbf{y}_0 + \Sigma_i \alpha_i^* \mathbf{z}_i$, and go to Step 1.

This replaces the simple line search with a search over the convex hull of all feasible points found at that stage in the solution process. This technique was originally suggested by Charles Holloway.

8.3 ZOUTENDIJK'S METHOD OF FEASIBLE DIRECTIONS

In this section we present a specialization of the work of G. Zoutendijk for the convex cost network flow problem, NLP. We assume that $g(\mathbf{x})$ is differentiable over $\{\mathbf{x} : \mathbf{0} \leqslant \mathbf{x} \leqslant \mathbf{u}\}$. Given a feasible flow $\mathbf{y}$, any feasible flow direction $\mathbf{d}$ must satisfy the following:

$$\left\{\begin{array}{ll} d_j \geqslant 0, & \text{for all } e_j \text{ such that } y_j = 0 \\ d_j \leqslant 0 & \text{for all } e_j \text{ such that } y_j = u_j \end{array}\right\}.$$

Since $\mathbf{d}$ is a feasible flow direction, $\mathbf{A}(\mathbf{y}+\alpha\mathbf{d})$ must equal $\mathbf{r}$ for some $\alpha > 0$. Since $\mathbf{Ay}=\mathbf{r}, \mathbf{Ad}=0$. Thus an appropriate direction-finding program is the following:

$$\begin{aligned} \min \quad & \nabla g(\mathbf{y})\mathbf{d} & & (8.5) \\ \text{s.t.} \quad & \mathbf{Ad}=\mathbf{0} & & (8.6) \\ & d_j \geqslant 0 \quad \text{for all } j \text{ such that } y_j = 0 & & (8.7) \\ & d_j \leqslant 0 \quad \text{for all } j \text{ such that } y_j = u_j & & (8.8) \\ & \|\mathbf{d}\| = 1. & & (8.9) \end{aligned}$$

Because of the nonlinear constraint (8.9), program (8.5)–(8.9) is difficult to solve. Hence a relaxation of (8.5)–(8.9), wherein (8.9) is replaced by

$$-\mathbf{1} \leqslant \mathbf{d} \leqslant \mathbf{1},$$

is usually substituted. The relaxed problem is simply a minimal cost network flow problem. If an improving feasible flow direction for (8.5)–(8.9) exists, an optimum solution to the relaxed problem will be an improving feasible flow direction, although not necessarily the best local flow direction.

8.4 CONVEX SIMPLEX METHOD

The convex simplex method of Willard Zangwill may be viewed as a generalization of the linear simplex method. It adheres to the simplex strategy of partitioning the variables into basic and nonbasic sets, but differs from the linear simplex method in that the nonbasic variables are allowed to assume values other than their upper and lower bounds. Like the linear simplex method, the convex simplex method allows a single nonbasic variable to change flow at each iteration. For this presentation we assume that the gradient of g exists over $\{\mathbf{x}:\mathbf{0} \leqslant \mathbf{x} \leqslant \mathbf{u}\}$.

Suppose we partition the flow conservation equations $\mathbf{Ax}=\mathbf{r}$ into $[\mathbf{B} \vdots \mathbf{N}]\mathbf{x}=\mathbf{r}$, where $\mathbf{B}$ is a basis. Likewise, we partition $\mathbf{x}$ into $[\mathbf{x}^B \vdots \mathbf{x}^N]$ and $\mathbf{u}$ into $[\mathbf{u}^B \vdots \mathbf{u}^N]$. Then NLP may be stated as follows:

$$\begin{aligned} \min \quad & g([\mathbf{x}^B \vdots \mathbf{x}^N]) \\ \text{s.t.} \quad & \mathbf{x}^B = \mathbf{B}^{-1}\mathbf{r} - \mathbf{B}^{-1}\mathbf{N}\mathbf{x}^N \\ & \mathbf{0} \leqslant \mathbf{x}^B \leqslant \mathbf{u}^B \\ & \mathbf{0} \leqslant \mathbf{x}^N \leqslant \mathbf{u}^N. \end{aligned}$$

After substituting for $\mathbf{x}^B$ we obtain

$$\begin{aligned} \min \quad & f(\mathbf{x}^N) \\ \text{s.t.} \quad & \mathbf{0} \leqslant \mathbf{B}^{-1}\mathbf{r} - \mathbf{B}^{-1}\mathbf{N}\mathbf{x}^N \leqslant \mathbf{u}^B \\ & \mathbf{0} \leqslant \mathbf{x}^N \leqslant \mathbf{u}^N, \end{aligned}$$

where $f(\mathbf{x}^N) = g([\mathbf{B}^{-1}\mathbf{r} - \mathbf{B}^{-1}\mathbf{N}\mathbf{x}^N \,|\, \mathbf{x}^N])$. By making an analogous partioning of ∇g, that is, $\nabla g(\mathbf{x}) = [\nabla g^B(\mathbf{x}) \,|\, \nabla g^N(\mathbf{x})]$, we obtain

$$\nabla f(\mathbf{x}^N) = \nabla g^N(\mathbf{x}) - \nabla g^B(\mathbf{x})\mathbf{B}^{-1}\mathbf{N}. \tag{8.10}$$

Hence, for direction $\mathbf{d}$, $\nabla f(\mathbf{x}^N)d/\|d\|$ is the directional derivative in nonbasic space. Therefore, if the ith component of $\nabla f(\mathbf{x}^N)$ is less than zero (greater than zero), then $\mathbf{e}^i$ $(-\mathbf{e}^i)$ is an improving direction in nonbasic space. Note that (8.10) assumes the same form as the components of the reduced cost (i.e., $\mathbf{c}^N - \mathbf{c}^B\mathbf{B}^{-1}\mathbf{N}$ from Chapter 2), with the vectors $\mathbf{c}^N$ and $\mathbf{c}^B$ replaced by $\nabla g^N(\mathbf{x})$ and $\nabla g^B(\mathbf{x})$, respectively.

Given the direction of change (i.e., the nonbasic variables to change), a one-dimensional search is required to determine the magnitude of the change. An actual simplex pivot is performed only if the resulting change forces one of the basic variables to zero or upper bound.

Let $\bar{I}$ denote the row dimension of $\mathbf{A}$. Since $\mathbf{A}$ has rank $\bar{I}-1$, we add an artificial variable to some row (node), say row l, to obtain the following system:

$$\mathbf{A}\mathbf{x} + \mathbf{e}^l a = \mathbf{r}. \tag{8.11}$$

The variable a does not appear in the objective function, and the upper bound for a is set to zero. Let $\mathbf{B}$ be any basis for (8.11), and let $\mathcal{T}_B$ be the corresponding rooted spanning tree (see Proposition 3.16).

We now show that the components of (8.10) may be calculated using $\mathcal{T}_B$. Let e_k be any arc not in $\mathcal{T}_B$, and let $P = \{s_1, e_{j_1}, s_2, e_{j_2}, \ldots, s_n, e_{j_n}, s_{n+1}\}$ denote the path linking $F(k)$ to $T(k)$. Let the corresponding orientation sequence be denoted by $O(k)$. It was shown in Section 3.3 that for the minimal cost network flow problem the reduced cost for any nonbasic arc e_k may be computed in one of two ways. One may compute

$$\sum_{i=1}^{i=n} c_{j_i} O_i(k) - c_k$$

or use the following relationship to compute the dual variables:

$$\pi_l = 0, \qquad \pi_{F(p)} - \pi_{T(p)} = c_p, \qquad \text{for all } e_p \,\varepsilon\, \mathcal{T}_B,$$

and from the dual variables compute

$$\pi_{F(k)} - \pi_{T(k)} - c_k.$$

Either of these approaches may be extended for the convex simplex algorithm, and we choose the former for this presentation. We now present the algorithm for the special case in which $g(\mathbf{x})$ is separable.

ALG 8.4 CONVEX SIMPLEX ALGORITHM

0 *Initialization.* Let $\bar{\mathbf{x}}$ be a feasible flow, and determine the basis tree $\mathcal{T}_B$. Let $\epsilon > 0$ be a termination tolerance.

1 *Pricing.* For $e_m \notin \mathcal{T}_B$, let $P = \{s_1, e_{j_1}, s_2, e_{j_2}, \ldots, s_n, e_{j_n}, s_{n+1}\}$ denote the path in $\mathcal{T}_B$ linking $F(\mathrm{m})$ to $T(\mathrm{m})$, let $C(\mathrm{m})$ denote the cycle formed by P together with e_m, and let $O(\mathrm{m})$ denote the corresponding orientation sequence. Let

$$\bar{c}_m = \sum_{1=1}^{i=n} O_i(m) \frac{\partial g(\bar{\mathbf{x}})}{\partial x_{j_i}}.$$

Let

$$\psi_1 = \{ e_m : x_m < u_m \quad \text{and} \quad \bar{c}_m > \epsilon \}$$

and

$$\psi_2 = \{ e_m : x_m > 0 \quad \text{and} \quad \bar{c}_m < -\epsilon \}.$$

If $\psi_1 \cup \psi_2 = \phi$, terminate with $\bar{x}$ an optimum; otherwise select $e_k \varepsilon \psi_1 \cup \psi_2$ and set

$$\delta \leftarrow \begin{cases} +1 & \text{if } e_k \varepsilon \psi_1, \\ -1 & \text{if } e_k \varepsilon \psi_2. \end{cases}$$

2 *Ratio Test.* Set

$$\Delta_1 \leftarrow \min_{O_i(k)=\delta} \{ \bar{x}_{j_i}, \infty \},$$

$$\Delta_2 \leftarrow \min_{-O_i(k)=\delta} \{ u_{j_i} - \bar{x}_{j_i}, \infty \},$$

$$\Delta \leftarrow \min\{\Delta_1, \Delta_2\}.$$

If $\Delta = 0$, go to Step 5.

3 *Check for Pivot.* Define the arc-length vector $\mathbf{y}$ as follows:

$$y_j \leftarrow \begin{cases} O_i(k) & \text{if } e_j = e_{j_i} \varepsilon C(k), \\ 0 & \text{otherwise.} \end{cases}$$

Set

$$\gamma \leftarrow -\delta \sum_{i=1}^{i=n} O_i(k)\left[\frac{\partial g(\bar{\mathbf{x}}-\Delta\delta\mathbf{y})}{\partial x_j}\right].$$

If $\gamma < 0$, set $\bar{\mathbf{x}} \leftarrow \bar{\mathbf{x}} - \Delta\delta\mathbf{y}$ and go to Step 5.

4 *Line Search (Bisecting)*

a. Set $\alpha_0 \leftarrow 0$ and $\alpha_1 \leftarrow \Delta$.

b. Set $\alpha \leftarrow (\alpha_0 + \alpha_1)/2$.

c. Set

$$\gamma \leftarrow -\delta \sum_{i=1}^{i=n} O_i(k)\left[\frac{\partial g(\bar{\mathbf{x}}-\alpha\delta\mathbf{y})}{\partial x_j}\right].$$

d. If $|\gamma| < \epsilon$, set $\bar{\mathbf{x}} \leftarrow \bar{\mathbf{x}} - \alpha\delta\mathbf{y}$ and go to Step 1.

e. If $\gamma > 0$, set $\alpha_1 \leftarrow \alpha$ and go to b; otherwise set $\alpha_0 \leftarrow \alpha$ and go to b.

5 *Pivot.* The flow on some arc in $C(k)$ has become zero or has attained its upper bound. Remove one such arc from $\mathcal{T}_B$, add the arc e_k to form a new basis tree, and go to Step 1.

Any line search may be substituted for Step 4. We present the bisecting search to illustrate how this may be implemented once $C(k)$ is known. Consider the sample problem shown in Table 8.1.

Table 8.1 Problem Description

Node	Supply	Demand
1	4	0
2	2	0
3	4	0
4	0	5
5	0	0
6	0	5

Arc j	*OBJ* Function	Capacity	$F(j)$	$T(j)$
e_1	x_1^2	3	1	2
e_2	$3x_2^2$	5	3	2
e_3	$2x_3^2$	5	1	4
e_4	$5x_4^2$	7	2	5
e_5	$9x_5^2$	9	3	6
e_6	$10x_6^2$	9	5	4
e_7	$15x_7^2$	11	5	6

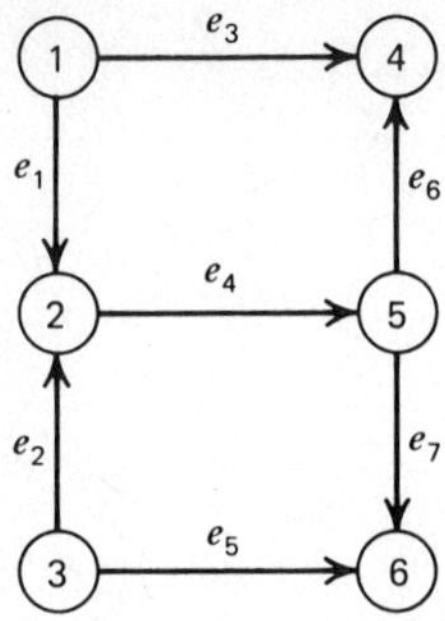

FIGURE 8.4 Sample network.

The network is illustrated in Figure 8.4.

0 (Initialization). Let $[x_1^B \;\; x_2^B \;\; x_3^N \;\; x_4^B \;\; x_5^N \;\; x_6^B \;\; x_7^B] = [2 \;\; 0 \;\; 2 \;\; 4 \;\; 4 \;\; 3 \;\; 1]$ and $\epsilon = 1$.

1 (Pricing). Let $e_m = e_3 = (1,4)$. Pricing is shown graphically in Figure 8.5.

$$\bar{c}_3 = (+1)(4) + (+1)(40) + (+1)(60) + (-1)(8) = 96.$$

$\therefore x_3 \epsilon \psi_1$. Let $e_k = e_3$ and $\delta = 1$.

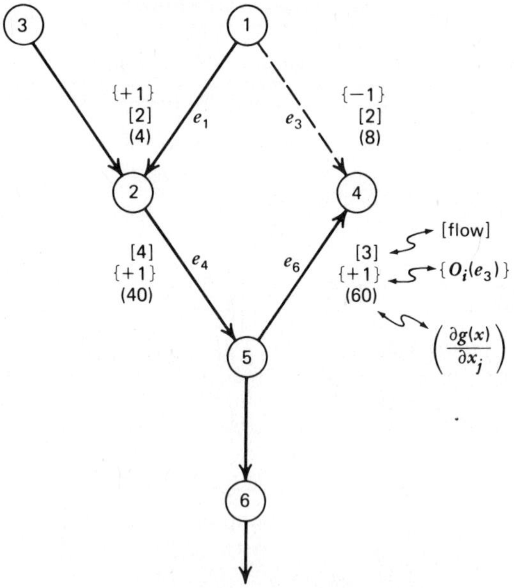

FIGURE 8.5 Pricing at iteration 1.

2 (Ratio Test). Set

$$\Delta_1 = \min\{2, 4, 3\} = 2,$$
$$\Delta_2 = \min\{5 - 2\} = 3,$$
$$\Delta = \min\{2, 3\} = 2.$$

3 (Check for Pivot).

$$\mathbf{y} = \begin{bmatrix} +1 \\ 0 \\ -1 \\ +1 \\ 0 \\ +1 \\ 0 \end{bmatrix} \begin{matrix} e_1 \\ e_2 \\ e_3 \\ e_4 \\ e_5 \\ e_6 \\ e_7 \end{matrix}$$

$$\gamma = -1[(+1)(0) + (+1)(20) + (+1)(20) + (-1)(16)] = -24.$$

This calculation is shown graphically in Figure 8.6.

$$[x_1^B \quad x_2^B \quad x_3^N \quad x_4^B \quad x_5^N \quad x_6^B \quad x_7^B] = [0 \quad 0 \quad 4 \quad 2 \quad 4 \quad 1 \quad 1].$$

5 (Pivot). Since e_1 has flow equal to zero, remove e_1.

1 (Pricing at Iteration 2). Let $e_m = e_5 = (3, 6)$. Pricing is shown graphically in Figure 8.7.

$$\bar{c}_5 = (+1)(0) + (+1)(20) + (+1)(30) + (-1)(72) = -22.$$

Therefore $e_5 \epsilon \psi_2$. Let $e_k = e_5$ and $\delta = -1$.

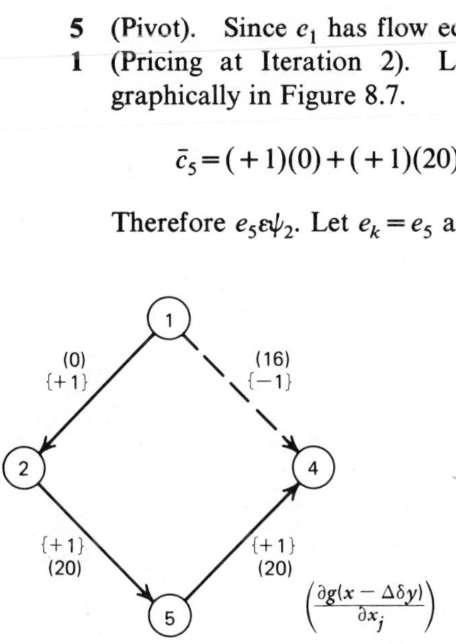

FIGURE 8.6 Calculation of γ at iteration 1.

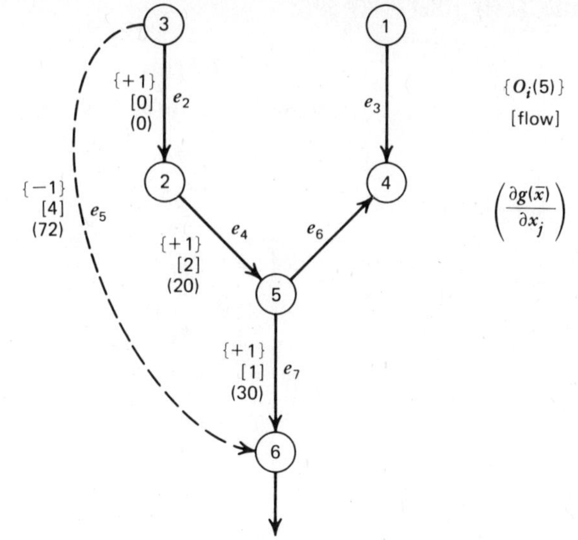

FIGURE 8.7 Pricing at iteration 2.

2 (Ratio Test). Set

$$\Delta_1 = \min\{4\} = 4,$$
$$\Delta_2 = \min\{5-0, 7-2, 11-1\} = 5,$$
$$\Delta = \min\{4, 5\} = 4.$$

3 (Check for Pivot).

$$\mathbf{y} = \begin{bmatrix} 0 \\ +1 \\ 0 \\ +1 \\ -1 \\ 0 \\ +1 \end{bmatrix} \begin{matrix} e_1 \\ e_2 \\ e_3 \\ e_4 \\ e_5 \\ e_6 \\ e_7 \end{matrix}$$

$$\gamma = -(-1)[(+1)(24) + (+1)(60) + (+1)(150) + (-1)(0)] = +234.$$

This calculation is shown graphically in Figure 8.8.

4 (Line Search).

a. $\alpha_0 = 0, \alpha_1 = 4.$

b. $\alpha = (0+4)/2 = 2.$

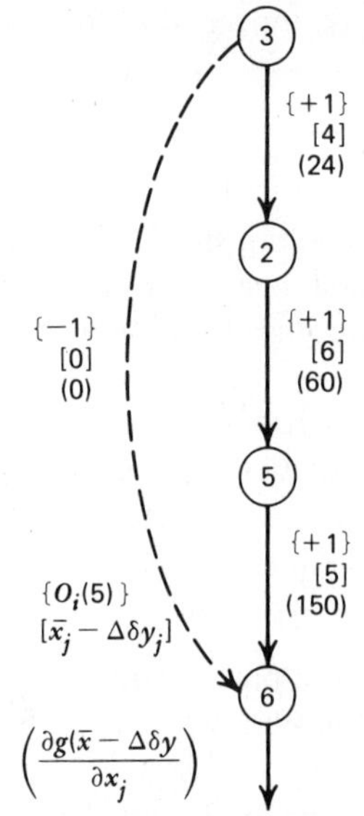

FIGURE 8.8 Calculation of γ at iteration 2.

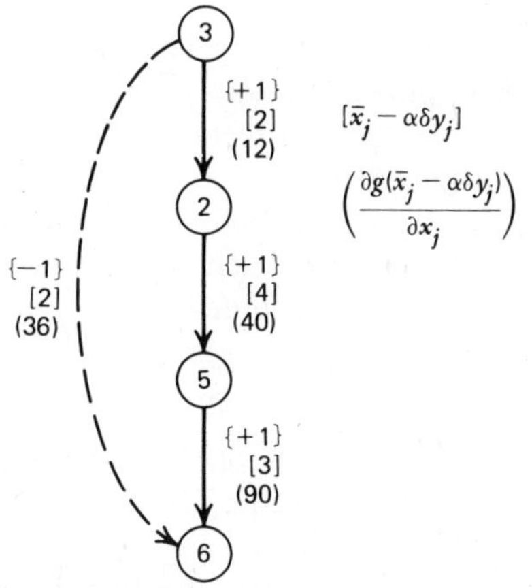

FIGURE 8.9 Line search at iteration 2.

c. $\gamma = -(-1)[(+1)(12) + (+1)(40) + (+1)(90) + (-1)(36)] = +106.$

This calculation is shown graphically in Figure 8.9.

d. No.

e. $\alpha_1 \leftarrow 2$.

We terminate this example at this point even though optimality has not been attained.

EXERCISES

8.1 Show that, if g is convex and $g(\mathbf{x}) = \Sigma_j g_j(x_j)$, then each g_j is convex.

8.2 Formulate a more general approximation method based on least squares whereby the quantities $\mathbf{u}^1, \ldots, \mathbf{u}^K$, are also selected. Discuss the practical application of the method.

8.3 Select u^1 and u^2 so that a three-part segmentation for the function $g(\mathbf{x}) = |1 - x|, 0 \leqslant x \leqslant 2$, using the least squares idea, yields $c^3 < c^2$.

8.4 Show that $f(c^k) = \int_{u^{k-1}}^{u^k} [g(t) - \hat{g}(t)]^2 \, dt$ is convex.

8.5 Show that the chordal approximation for a convex function is convex.

8.6 Using only the definition of convexity, prove that the following functions are convex:

(a) $g(x) = x^2$.
(b) $g(x) = e^x$.

8.7 It has been shown that the problem of finding steady state flows in a water pipe distribution network can be modeled as a separable convex cost network flow problem. Consider the water distribution network illustrated in Figure 8.10.
After a ground node and arcs connecting the ground node to each of the reservoirs have been added, the nonlinear network program shown in Table E8.7 is obtained.

(a) Show that the objective function is convex.

(b) Let the right endpoints for arc 1 be given by $u_1^1 = 2, u_1^2 = 5$, and $u_1^3 = 8$. Using these endpoints, determine a three-part chordal approximation and a three-part least squares approximation for arc 1. Plot your approximations.

(c) Beginning with the initial solution given above, execute two passes through the Frank-Wolfe algorithm (ALG 8.1).

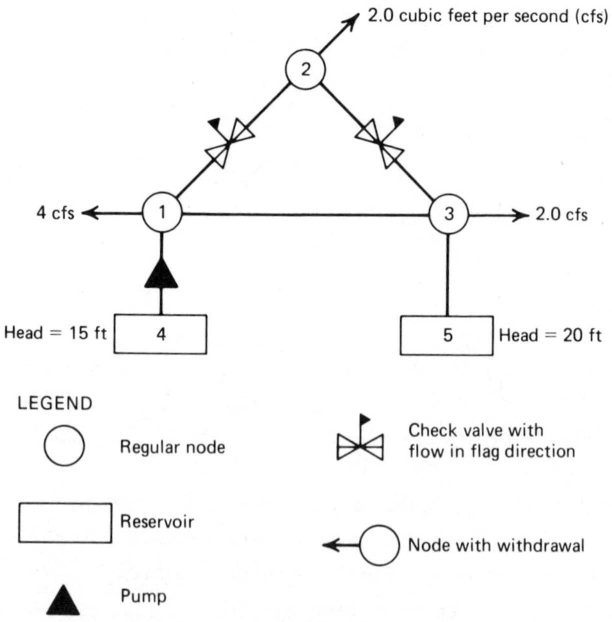

FIGURE 8.10 Water distribution network.

Table E8.7

Node	Supply	Demand
1	0	4
2	0	2
3	0	2
4	0	0
5	0	0
6	8	0

Arc	"From" Node	"To" Node	Objective Function	Initial Flows
1	1	2	$52x_1^{2.85}$	1
2	1	3	$68x_2^{2.85}$	0
3	3	1	$58x_3^{2.85}$	1
4	3	2	$52x_4^{2.85}$	1
5	4	1	$10x_5^3+5x_5^2-100x_5$	4
6	5	3	$25x_6^{2.85}$	4
7	6	4	$-15x_7$	4
8	6	5	$-20x_8$	4

(d) Using the initial solution given above, determine the feasible direction obtained by applying Zoutendijk's method.

(e) Using the initial solution given above, execute two passes through the convex simplex algorithm (ALG 8.4).

8.8 Design a data structure for implementation of (*a*) the Frank-Wolfe algorithm (ALG 8.1) and (*b*) the convex simplex algorithm (ALG 8.4) (i.e., indicate the number of node-length and arc-length arrays required and the information to be stored in each such array).

8.9 Generalize the convex simplex algorithm for the nonseparable case, and perform two passes of your algorithm on the problem shown in Table E8.9.

The objective function is given by $g(x_1,x_2,x_3,x_4,x_5)=(x_1+x_2)^2+x_3x_4+x_5^3$.

8.10 Consider the network shown in Table E8.10. Suppose we wish to solve this linear network problem by using Zoutendijk's method of feasible directions. Develop the direction-finding program associated with the current solution, and transform the problem so that it takes the NP form.

Table E8.9

Node	Supply	Demand
1	0	5
2	0	2
3	0	3
4	10	0

Arc	"From" Node	"To" Node	Capacity, u_j
1	4	1	4
2	4	2	7
3	1	2	2
4	3	1	8
5	2	3	5

Table E8.10

Node Number	Requirement
1	10
2	0
3	0
4	−10

Arc Number, j	"From" Node, $F(j)$	"To" Node, $T(j)$	Unit Cost, c_j	Arc Capacity, u_j	Current Solution, y_j
1	1	2	10	15	10
2	2	3	10	10	10
3	3	4	20	12	10
4	1	3	1	5	0
5	3	2	2	10	0
6	2	4	3	6	0

NOTES AND REFERENCES

Three basic types of problems are modeled as convex cost network flow problems: equilibrium problems, production-distribution problems, and communication network design problems. Equilibrium models appear in studies involving urban transportation systems (such as Abdulaal and LeBlanc [2], Charnes and Cooper [31], Dafermos [42,43], Ferland [61], Florian [62], Florian, Nguyen, and Soumis [63], Jewell [111], Jorgensen [114], LeBlanc, Morlok, and Pierskalla [134], Leventhal, Nemhauser, and Trotter [137], Nguyen [146,147], and Tomlin [178]), pipe network systems (such as Collins, Cooper, Helgason, Kennington, and LeBlanc [35] and Hall [92]), and electrical systems (such as Charnes and Cooper [29], Cooper and Kennington [37], Dennis [50], and Hu [106]). Production-distribution problems that have been modeled as convex cost network flow problems can be found in papers by Cooper and LeBlanc [38], LeBlanc and Cooper [133], LeBlanc, Abdulaal, and Helgason [136], Rosenthal [158], Sharp, Snyder, and Greene [167], and Williams [184]. Applications involving communication network design are given in Cantor and Gerla [27] and Fratta, Gerla, and Kleinrock [69].

Section 8.1

Professor Mike Collins of Southern Methodist University first suggested to one of the authors the least squares idea for determining the linear approximations. Other discussions of this basic idea can be found in Refs. [26], [39], [48], [150], and [152].

Section 8.2

The Frank-Wolfe algorithm originally appeared in Frank and Wolfe [68]. The various enhancements of this procedure can be traced to Refs. [104], [139], and [185].

Section 8.3

This section is a specialization of the work presented in Zoutendijk [189]. Beale [21] was the first to apply a feasible flow direction algorithm (in the spirit of that described by Zoutendijk) to the convex cost network flow problem. Extensions of Beale's work and other related approaches are given in Ferland [61], Hu [106], Klein [121], Menon [143], and Weintraub [183].

Section 8.4

The Convex Simplex Algorithm was developed by Zangwill [187]. A specialization of Zangwill's algorithm to the network structure can be found in Nguyen [147] and Helgason and Kennington [100].

General Remarks

A new feasible direction algorithm for nonlinear networks that makes use of second-order information has been developed by Dembo and Klincewicz [49].

APPENDIX **A**

Characterization of a Tree

Proposition

For a graph $\mathcal{T}=[\mathcal{N},\mathcal{A}]$ having at least one node, the following are equivalent:

(i) $\mathcal{T}$ is a tree.

(ii) For every distinct pair of nodes (p,q) of $\mathcal{N}$, there is a unique path in $\mathcal{T}$ that links p to q.

(iii) $\mathcal{T}$ has one less arc than node and is connected.

(iv) $\mathcal{T}$ has one less arc than node and is acyclic.

Proof. Let $\mathcal{T}=[\mathcal{N},\mathcal{A}]$ be a graph having at least one node.

Proof of (*i*) *implies* (*ii*). Assume that $\mathcal{T}$ is a tree, and let (p,q) be any nodes in $\mathcal{N}$ with $p \neq q$. Suppose there are two distinct paths, $P=\{s_1, e_{j_1}, s_2, e_{j_2}, \ldots, s_n, e_{j_n}, s_{n+1}\}$ and $\bar{P}=\{\bar{s}_1, \bar{e}_{j_1}, \bar{s}_2, \bar{e}_{j_2}, \ldots, \bar{s}_t, \bar{e}_{j_t}, \bar{s}_{t+1}\}$ that link p to q. Let i be the least nonnegative integer such that $s_i=\bar{s}_i$ and $s_{i+1} \neq \bar{s}_{i+1}$. Such an i exists since $s_1=\bar{s}_1$ and, if no such i existed, P and $\bar{P}$ would not be distinct. Let m be the least positive integer such that $m>i$ and, for some k, $s_m=\bar{s}_k$. Such an m exists since the last elements of P and $\bar{P}$ are identical. Define the finite sequence

$$C=\left\{s_i, e_{j_i}, s_{i+1}, e_{j_{i+1}}, \ldots, s_m, \bar{e}_{j_{k-1}}, \bar{s}_{k-1}, \bar{e}_{j_{k-2}}, \bar{s}_{k-2}, \ldots, \bar{s}_i\right\}.$$

By construction, C is a cycle. This contradicts the assumption that $\mathcal{T}$ is a tree and hence acyclic. Therefore P and $\bar{P}$ cannot be distinct.

Proof of (ii) implies (iii) (Proof by induction). Assume (ii). Then $\mathcal{T}$ is obviously connected. Let n be the number of nodes of $\mathcal{T}$. Suppose $n=2$. Since there is a unique path linking these two nodes, $\mathcal{T}$ has one arc. Thus the proposition holds for $n=2$.

We now assume that, for any integer $n \geqslant 2$, (iii) holds for all m satisfying $1 \leqslant m \leqslant n$, and we show that (iii) holds for $n+1$. Let $\mathcal{T}=[\mathcal{N},\mathcal{A}]$ satisfy (ii) and have $n+1$ nodes. Let e_r be any element of $\mathcal{A}$. Consider the following node sets:

$$\mathcal{M}^1=\{i : i\,\varepsilon\,\mathcal{N} \text{ and } e_r \text{ is not in the path linking } F(r) \text{ to } i\},$$

$$\mathcal{M}^2=\{i : i\,\varepsilon\,\mathcal{N} \text{ and } e_r \text{ is not in the path linking } T(r) \text{ to } i\}.$$

Since every path has at least one arc, $F(r)$ and $T(r)$ are not members of either $\mathcal{M}^1$ or $\mathcal{M}^2$.

We now show that any other node of $\mathcal{N}$ is in $\mathcal{M}^1 \cup \mathcal{M}^2$. Let t be any member of $\mathcal{N}-\{F(r),T(r)\}$. By (ii) there is a unique path linking $F(r)$ to t, say $P=\{s_1,e_{j_1},s_2,e_{j_2},\ldots,s_k,e_{j_k},s_{k+1}\}$ of length k. If e_r is not an arc of P, $t\,\varepsilon\,\mathcal{M}^1$. If e_r is an arc of P, it must be the first one, since the nodes of a path are distinct. The sequence $Q=\{s_3,e_{j_3},s_4,e_{j_4},\ldots,s_{k+1}\}$ is a path linking $T(r)$ to t. Furthermore, e_r is not an arc of Q since by Proposition 3.1 the arcs of P are distinct. Thus $t\,\varepsilon\,\mathcal{M}^2$.

We now show that $\mathcal{M}^1$ and $\mathcal{M}^2$ are disjoint. Assume, to the contrary, that there is some p such that $p\,\varepsilon\,\mathcal{M}^1 \cap \mathcal{M}^2$. Thus there is a path P linking $F(r)$ to p and a path $\bar{P}$ linking $T(r)$ to p, with e_r not an arc of P or $\bar{P}$. Let v be the length of $\bar{P}$. Consider the sequence $Q=\{F(r),e_r,\bar{s}_1,\bar{e}_{j_1},\bar{s}_2,\bar{e}_{j_2},\ldots,\bar{s}_{v+1}\}$. By construction Q is a path linking $F(r)$ to p. But Q is distinct from P since e_r is not an arc of P. This contradicts the assumption that $\mathcal{T}$ has property (ii). Hence $\mathcal{N}^1=\mathcal{M}^1\cup\{F(r)\}$ and $\mathcal{N}^2=\mathcal{M}^2\cup\{T(r)\}$ are disjoint and nonempty and contain all the nodes of $\mathcal{N}$.

Consider a similar partitioning of the arcs as follows. Let

$$\mathcal{A}^1=\{e_j : e_j\,\varepsilon\,\mathcal{A} \text{ and } e_j \text{ is an arc of the path linking } F(r) \text{ to } i \text{ for some } i\,\varepsilon\,\mathcal{M}^1\},$$

$$\mathcal{A}^2=\{e_j : e_j\,\varepsilon\,\mathcal{A} \text{ and } e_j \text{ is an arc of the path linking } T(r) \text{ to } i \text{ for some } i\,\varepsilon\,\mathcal{M}^2\}.$$

Arc e_r is not a member of $\mathcal{A}^1$ or $\mathcal{A}^2$. We now show that any other arc of $\mathcal{A}$ is in $\mathcal{A}^1\cup\mathcal{A}^2$. Let e_u be any member of $\mathcal{A}-\{e_r\}$. If $F(u)=F(r)$, then $e_u\,\varepsilon\,\mathcal{A}^1$. If $F(u)\neq F(r)$, there is a unique path $P=\{s_1,e_{j_1},\ldots,s_w,e_{j_w},s_{w+1}\}$ in $\mathcal{T}$ linking $F(r)$ to $F(u)$. If e_r is an arc of P, it must be the first arc since the nodes of a path are distinct. We now consider four cases.

Case 1 Suppose neither e_u nor e_r is in P. Let $Q=\{s_1,e_{j_1},\ldots,s_{w+1},e_u,T(u)\}$. Then Q is a path linking $F(r)$ to $T(u)$ not including e_r. Hence $T(u)\,\varepsilon\,\mathfrak{M}^1$ and $e_u\,\varepsilon\,\mathcal{A}^1$.

Case 2 Suppose e_u is in P and e_r is not in P. Then $F(u)\,\varepsilon\,\mathfrak{M}^1$ and $e_u\,\varepsilon\,\mathcal{A}^1$.

Case 3 Suppose e_r is in P and e_u is not in P. Let $Q=\{s_2,e_{j_2},\ldots,s_{w+1},e_u,T(u)\}$. Then Q is a path linking $T(r)$ to $T(u)$ not including e_r. Hence $T(u)\,\varepsilon\,\mathfrak{M}^2$ and $e_u\,\varepsilon\,\mathcal{A}^2$.

Case 4 Suppose both e_u and e_r are in P. If $T(u)=T(r)$, then $e_u\,\varepsilon\,\mathcal{A}^2$. If $T(u)\neq T(r)$, let $Q=\{s_2,e_{j_2},\ldots,s_{w+1}\}$. Then Q is a path linking $T(r)$ to $F(u)$ not including e_r. Hence $F(u)\,\varepsilon\,\mathfrak{M}^2$ and $e_u\,\varepsilon\,\mathcal{A}^2$.

Therefore $e_u\,\varepsilon\,\mathcal{A}^1\cup\mathcal{A}^2$.

We now show that $\mathcal{A}^1$ and $\mathcal{A}^2$ are disjoint. Assume, to the contrary, that there is some arc e_j such that $e_j\,\varepsilon\,\mathcal{A}^1\cap\mathcal{A}^2$. Then there exist nodes $p\,\varepsilon\,\mathfrak{M}^1$ and $q\,\varepsilon\,\mathfrak{M}^2$, and paths $P=\{s_1,e_{j_1},s_2,e_{j_2},\ldots,s_t\}$ and $\bar{P}=\{\bar{s}_1,\bar{e}_{j_1},\bar{s}_2,\bar{e}_{j_2},\ldots,\bar{s}_v\}$, linking $F(r)$ to p and $T(r)$ to q, respectively, with e_j an arc of both P and $\bar{P}$. Thus there exist positive integers i and k such that $s_i=\bar{s}_k=F(j)$. Define the sequences $Q=\{s_1,e_{j_1},s_2,e_{j_2},\ldots,s_i\}$ and $\bar{Q}=\{\bar{s}_1,\bar{e}_{j_1},\bar{s}_2,\bar{e}_{j_2},\ldots,\bar{s}_k\}$. Then Q links $F(r)$ to $F(j)$ and $\bar{Q}$ links $T(r)$ to $F(j)$. By the definition of $\mathfrak{M}^1$ and $\mathfrak{M}^2$, e_r is not an arc of P or $\bar{P}$ and thus is not an arc of Q or $\bar{Q}$. Thus $F(j)\,\varepsilon\,\mathfrak{M}^1$ and $F(j)\,\varepsilon\,\mathfrak{M}^2$, contradicting the previous result that $\mathfrak{M}^1\cap\mathfrak{M}^2=\phi$. Thus $\mathcal{A}^1\cap\mathcal{A}^2=\phi$ and $\mathcal{A}=\mathcal{A}^1\cup\mathcal{A}^2\cup\{e_r\}$.

By construction, $\mathfrak{T}^1=[\mathfrak{M}^1,\mathcal{A}^1]$ and $\mathfrak{T}^2=[\mathfrak{M}^2,\mathcal{A}^2]$ are both subgraphs of $\mathfrak{T}$ having $\mathfrak{M}^1\cap\mathfrak{M}^2=\phi$ and $\mathcal{A}^1\cap\mathcal{A}^2=\phi$. We now show that $\mathfrak{T}^1$ and $\mathfrak{T}^2$ are connected. Let (p,q) be a distinct pair of nodes from $\mathfrak{M}^1$. We now consider three cases.

Case 1 Suppose $p=F(r)$. Then $q\,\varepsilon\,\mathfrak{M}^1$. By the definition of $\mathfrak{M}^1$, there is a path in $\mathfrak{T}$ linking p to q and this path must be in $\mathfrak{T}^1$.

Case 2 Suppose $q=F(r)$. Then $p\,\varepsilon\,\mathfrak{M}^1$. By the same argument used above, there is a path in $\mathfrak{T}^1$ linking p to q.

Case 3 Suppose $p\neq F(r)$ and $q\neq F(r)$. By the definition of $\mathfrak{M}^1$, there exist paths $P=\{s_1,e_{j_1},s_2,e_{j_2},\ldots,s_t\}$ and $\bar{P}=\{\bar{s}_1,\bar{e}_{j_1},\bar{s}_2,\bar{e}_{j_2},\ldots,\bar{s}_v\}$ in $\mathfrak{T}$ linking $F(r)$ to p and $F(r)$ to q, respectively. P and $\bar{P}$ must also be paths in $\mathfrak{T}^1$. Let k be the largest positive integer such that $s_k=\bar{s}_k$. Such a k exists since $s_1=\bar{s}_1$, $s_t=p$, $\bar{s}_v=q$, and $p\neq q$. Consider the sequence $Q=\{s_t,e_{j_{t-1}},s_{t-1},e_{j_{t-2}},s_{t-2},\ldots,s_k,\bar{e}_{j_k},\bar{s}_{k+1},\bar{e}_{j_{k+1}},\ldots,\bar{s}_v\}$. By construction, Q is a path linking p to q and must be a path in $\mathfrak{T}^1$.

Therefore $\mathfrak{T}^1$ is connected. By similar arguments $\mathfrak{T}^2$ is connected.

Since $\mathcal{T}$ has property (ii), $\mathcal{T}^1$ and $\mathcal{T}^2$ must have property (ii). Since $\mathcal{T}^1$ and $\mathcal{T}^2$ have fewer than $n+1$ nodes but have at least one node, the induction hypothesis applies to each. Thus $\mathcal{T}^1$ and $\mathcal{T}^2$ each have one less arc than node. Let the order of $\mathfrak{M}^1$ and $\mathfrak{M}^2$ be given by n_1 and n_2. Since $\mathfrak{M}^1$ and $\mathfrak{M}^2$ are disjoint and contain all the nodes of $\mathcal{T}$, $n_1+n_2=n+1$. Since $\mathcal{Q}^1$, $\mathcal{Q}^2$, and $\{e_r\}$ are mutually disjoint and contain all the arcs of $\mathcal{T}$, $(n_1-1)+(n_2-1)+1=n$. Thus the induction is complete.

Proof of (*iii*) *implies* (*iv*). Assume (iii), and let the number of nodes in $\mathcal{T}$ be n. We need only show that $\mathcal{T}$ is acyclic. Assume, to the contrary, that we can form a cycle C of length m from $\mathcal{T}$. Then C has m distinct nodes and m distinct arcs. Let $\mathcal{N}^1$ be the set of nodes from C, and let $\mathcal{Q}^1$ be the set of arcs from C. Then $\mathcal{N}-\mathcal{N}^1\neq\phi$, since $m=n$ contradicts the assumption that $\mathcal{T}$ has one less arc than node. Let $\mathcal{N}^2$ be the set of nodes from $\mathcal{N}-\mathcal{N}^1$ that are linked to a node of $\mathcal{N}^1$ by a path of length 1. $\mathcal{N}^2\neq\phi$ since otherwise $\mathcal{T}$ would not be connected. Let $\mathcal{Q}^2$ be the set of all arcs that are elements of a path of length 1 linking a node of $\mathcal{N}^2$ to a node of $\mathcal{N}^1$. $\mathcal{Q}^2$ must contain at least as many arcs as it does nodes, and none are arcs of $\mathcal{Q}^1$. As long as $\mathcal{N}-\cup_{j=1,\ldots,i}\mathcal{N}^j\neq\phi$, we can continue this process, defining $\mathcal{N}^{i+1}$ to be the set of all nodes in $\mathcal{N}-\cup_{j=1,\ldots,i}\mathcal{N}^j$ that are linked to a node of $\mathcal{N}^i$ by a path of length 1 and defining $\mathcal{Q}^{i+1}$ to be the set of arcs that are in a path of length 1 linking a node of $\mathcal{N}^{i+1}$ to a node of $\mathcal{N}^i$. As before, $\mathcal{Q}^{i+1}$ must contain as many arcs as nodes, and none is an arc of $\cup_{j=1,\ldots,i}\mathcal{Q}^i$. Since $\mathcal{N}$ is finite, this process will stop. When the process stops, we will have identified as many arcs as nodes, contradicting the assumption that $\mathcal{T}$ has one less arc than node. Hence $\mathcal{T}$ is acyclic.

Proof of (*iv*) *implies* (*i*). Assume (iv). We need only show that $\mathcal{T}$ is connected. Let $\mathfrak{M}$ be the set of all maximal connected subgraphs of $\mathcal{T}$. Let $p\in\mathcal{N}$. If no arc of $\mathcal{Q}$ is incident on p, $\mathcal{T}^1=[\{p\},\phi]\in\mathfrak{M}$. Otherwise some arc, say e_j, is incident on p, so that $\mathcal{T}^2=[\{F(j),T(j)\},\{e_j\}]$ will be a subgraph of some member of $\mathfrak{M}$. Hence every node of $\mathcal{N}$ will be in some member of $\mathfrak{M}$. The same reasoning used with $\mathcal{T}^2$ implies that each arc of $\mathcal{Q}$ is in some member of $\mathfrak{M}$. All members of $\mathfrak{M}$ must be mutually disjoint with respect to both nodes and arcs; otherwise they would not be maximal subgraphs. Since $\mathcal{T}$ is acyclic, each member of $\mathfrak{M}$ is acyclic and thus must be a tree. Suppose $\mathfrak{M}$ has at least two elements. Then the number of arcs in some member of $\mathfrak{M}$ is two less than the number of nodes. But by the above arguments the number of arcs in the members of $\mathfrak{M}$ equals one less than the number of nodes. Therefore $\mathfrak{M}$ has only one member, $\mathcal{T}$ itself. Hence $\mathcal{T}$ is connected and is a tree. ■

APPENDIX **B**

Data Structures for Network Programs

This appendix presents a concise summary of detailed information for the class of data structures that employ a label known as the thread. It presents the most efficient data structures within this class along with detailed algorithms for implementing the primal simplex method on a graph using each of these data structures.

B.1 LABELS FOR ROOTED TREES

Let $\mathcal{T}=[\mathcal{N},\mathcal{A}]$ be a rooted tree with root node l. By Proposition 3.6 there is a unique path linking any node $i \neq l$ to node l, and we denote this path by $P(i)$. Node i will be called a *successor* of node n, if n is in $P(i)$. We denote the set of successors of node n by $U(n)$ and the *number of successors* of node n by t_n.

We define a label for $\mathcal{T}$ to be a mapping with domain $\mathcal{N}$. The *distance label*, denoted by d_i, is given as follows:

$$d_i = \begin{cases} 0 & \text{if } i = l, \\ \text{length of } P(i) & \text{otherwise.} \end{cases}$$

The *predecessor label*, denoted by p_i, is given by

$$p_i = \begin{cases} 0 & \text{if } i = l, \\ P_3(i) & \text{otherwise.} \end{cases}$$

For any one-to-one mapping from $\mathcal{N}$ onto $\mathcal{N}$, say s_i, we define the family of maps by the recursion

$$\begin{aligned} s^1(i) &= s_i, \\ s^{j+1}(i) &= s^j(s_i). \end{aligned}$$

Then s_i is called a *thread label* if $U(i)=\{s^j(i):j=1,\ldots,t_i\}$, when $t_i \neq 0$. For a given rooted tree, many such maps can typically be defined. Given a thread label, the *preorder distance label*, denoted by g_i, is a mapping from $\mathcal{N}$ onto $\mathcal{N}$ such that

$$g_i = \begin{cases} 1, & i = l, \\ j+1, & i = s^j(l). \end{cases}$$

Given a thread label, the *last successor label*, denoted by n_i, is given by

$$n_i = \begin{cases} i & \text{if } U(i)=\phi, \\ s^j(i) & \text{otherwise, where } s^j(i) \in U(i) \text{ and } s^{j+1}(i) \notin U(i). \end{cases}$$

To illustrate these mappings, Table B.1 gives the labels for the rooted tree of Figure B.1.

The data structures that follow all represent $\mathcal{T}$ using the predecessor and the thread labels plus various combinations of the other labels. Note

Table B.1 Labels for The Rooted Tree of Figure B.1

Node, i	Distance, d_i	Predecessor, p_i	Thread, s_i	Last Successor, n_i	Number of Successor, t_i	Preorder Distance, g_i
1	1	11	2	10	10	2
2	2	1	3	10	9	3
3	3	2	4	7	5	4
4	4	3	5	4	1	5
5	4	3	6	6	2	6
6	5	5	7	6	1	7
7	4	3	8	7	1	8
8	3	2	9	8	1	9
9	3	2	10	10	2	10
10	4	9	11	10	1	11
11	0	0	1	10	11	1

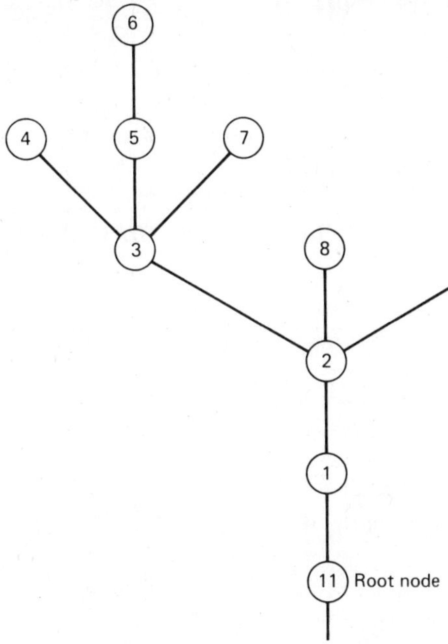

FIGURE B.1 Sample rooted tree.

that each label used in the data structure requires a node-length array. Furthermore, in general it is true that an efficient implementation using a data structure with $K+1$ node-length arrays will result in faster solution times than one with only K such arrays. Hence, in the absence of budgetary and other design restrictions, the appropriate data structure for a given problem is a function of the core storage available. We now present the data structures and corresponding algorithms for implementation using two, three, and four node-length arrays for representing $\mathfrak{T}$.

Suppose $\mathfrak{T}$ has $\bar{I}$ nodes. Then $\mathfrak{T}$ has $\bar{I}-1$ arcs and the root arc. Therefore the pertinent information about the arcs is also carried in node-length arrays, where for $i \neq l$ the information concerning the arc connecting nodes i and p_i is associated with node i. Suppose arc e_k connects nodes i and p_i. To facilitate the computations, we make use of an *oriented arc identifier*, m_i, which is defined to be k if $e_k=(p_i,i)$ and $-k$ if $e_k=(i,p_i)$. The flow on e_k is denoted by α_i. To implement the pricing operation, it is desirable to maintain the values of the dual variables. Thus three additional node-length arrays are required, which may also be considered as labels.

B.2 PREDECESSOR/THREAD

The following data structure, first proposed by Glover, Klingman, and Stutz, uses five node-length arrays plus a one-bit flag for each node. Given that the entering arc is $e_\phi = (u, v)$, the ratio test is given by ALG-B.1, where it is assumed that all flags are initially zero.

ALG B.1 RATIO TEST (PREDECESSOR/THREAD)

0 *Initialization*
  a. [Initialize γ_1 and γ_2] If $\alpha_\phi = 0$, $\gamma_1 \leftarrow -1$ and $\gamma_2 \leftarrow 1$; otherwise $\gamma_1 \leftarrow 1$ and $\gamma_2 \leftarrow -1$.
  b. [Set Δ to bound] $\Delta \leftarrow \mathrm{u}_\phi$.

1 *Set Flags for All Nodes on Path from u to Root*
  a. [Begin at u] $i \leftarrow u$.
  b. [Set flag at i] $f_i \leftarrow 1$.
  c. [Last node in path] If $i = l$, go to Step 2; otherwise $i \leftarrow p_i$ and go to b.

2 *Determine Maximum Allowable Flow Change on Path from v to First Flagged Node*
  a. [Begin at v] $i \leftarrow v$.
  b. [Flag set at i?] If $f_i = 1$, $w \leftarrow i$ and go to Step 3.
  c. [Set flag at i] $f_i \leftarrow 1$.
  d. [Increase or decrease flow?] If $m_i \gamma_2 > 0$, go to f.
  e. [Increasing flow] If $u_{|m_i|} - \alpha_i > \Delta$, $i \leftarrow p_i$ and go to b; otherwise $\Delta \leftarrow u_{|m_i|} - \alpha_i$, $\mu \leftarrow 2$, $\mathcal{K} \leftarrow i$, $i \leftarrow p_i$ and go to b.
  f. [Decreasing flow] If $\alpha_i > \Delta$, $i \leftarrow p_i$ and go to b; otherwise $\Delta \leftarrow \alpha_i$, $\mu \leftarrow 2$, $\mathcal{K} \leftarrow i$, $i \leftarrow p_i$, and go to b.

3 *Determine Maximum Allowable Flow Change on Path from u to w.*
  a. [Begin at u] $i \leftarrow u$.
  b. [End of cycle?] If $i = w$, terminate.
  c. [Increase or decrease flow] If $m_i \gamma_1 > 0$, go to e.
  d. [Increasing flow] If $u_{|m_i|} - \alpha_i > \Delta$, $i \leftarrow p_i$ and go to b; otherwise $\Delta \leftarrow u_{|m_i|} - \alpha_i$, $\mu \leftarrow 1$, $\mathcal{K} \leftarrow i$, $i \leftarrow p_i$, and go to b.
  e. [Decreasing flow] If $\alpha_i > \Delta$, $i \leftarrow p_i$ and go to b; otherwise, $\Delta \leftarrow \alpha_i$, $\mu \leftarrow 1$, $\mathcal{K} \leftarrow i$, $i \leftarrow p_i$, and go to b.

At the termination of ALG-B.1, the path from u to v is given by the path from u to w together with the path from v to w. The maximum allowable flow change Δ has been determined, and flags are set for nodes on the paths from u to v and from w to l. The updating algorithm may be stated as follows.

ALG B.2 UPDATE (PREDECESSOR/THREAD)

1 *Determine whether a Full Update is Required.* If $\Delta \neq u_\phi$, go to Step 3.

2 *Update Flows and Reset Flags on Path from u to v*
  a. [Set up for loop] $\delta_1 \leftarrow u$, $\delta_2 \leftarrow v$, and $k \leftarrow 1$.
  b. [Initialize parameters] $\delta \leftarrow \delta_k$ and $\gamma \leftarrow \gamma_k \Delta$.
  c. [Node equal w?] If $\delta = w$, go to g.
  d. [Increase or decrease flow?] If $m_\delta > 0$, $\beta \leftarrow 1$; otherwise $\beta \leftarrow -1$.
  e. [Update flow and reset flag] $\alpha_\delta \leftarrow \alpha_\delta - \gamma\beta$ and $f_\delta \leftarrow 0$.
  f. [Follow predecessor] $\delta \leftarrow p_\delta$, and go to c.
  g. [Path complete?] If $k = 1$, $k \leftarrow 2$ and go to b; otherwise go to Step 9.

3 *Initialization*
  a. [Set dual variable change] $\sigma \leftarrow c_\phi$.
  b. [Entering variable at bound?] If $\alpha_\phi = u_\phi$, $\Delta' \leftarrow u_\phi - \Delta$, $\sigma \leftarrow -\sigma$; otherwise $\Delta' \leftarrow \Delta$.
  c. [Initialize q, q', and λ] If $\mu = 1$, $q \leftarrow u$, $q' \leftarrow v$, $\lambda \leftarrow -\phi$, $\sigma \leftarrow -\sigma$; otherwise $q \leftarrow v$, $q' \leftarrow u$, and $\lambda \leftarrow \phi$.
  d. [Initialize i, j, and γ] $i \leftarrow q$, $j \leftarrow p_q$, and $\gamma \leftarrow \gamma_\mu \Delta$.
  e. [Save predecessor of $\mathcal{K}$] $\mathcal{K}' \leftarrow p_{\mathcal{K}}$.

4 *Update Dual, Flow, and Arc Identification, and Reset Flag at i*
  a. [Update dual] $\pi_i \leftarrow \pi_i + \sigma$.
  b. [Save flow] $\alpha' \leftarrow \alpha_i$.
  c. [Update flow] $\alpha_i \leftarrow \Delta'$.
  d. [Save direction] If $m_i > 0$, $\beta \leftarrow 1$; otherwise $\beta \leftarrow -1$.
  e. [Save arc] $m \leftarrow |m_i|$.
  f. [Update arc identification] $m_i \leftarrow \lambda$.
  g. [Reset flag] $f_i \leftarrow 0$.

5 *Determine Last Successor of i and Update Dual Variables*
  a. [Begin at i] $z \leftarrow i$ and $x \leftarrow s_i$.
  b. [Is x a successor?] If $x = l$ or $f_{p_x} = 1$, go to Step 6.
  c. [Update dual variable] $\pi_x \leftarrow \pi_x + \sigma$.
  d. [Follow thread] $z \leftarrow x$, $x \leftarrow s_x$, and go to b.

6 *Find Reverse Thread for Node i*
  a. [Begin at j] $r \leftarrow j$.
  b. [Is r reverse thread?] If $s_r = i$, go to Step 7.
  c. [Follow thread] $r \leftarrow s_r$, and go to b.

7 *Update Predecessor and Thread*
  a. [Final update?] If $i = \mathcal{K}$, go to i.
  b. [Determine updated flow for j] $\Delta' \leftarrow \alpha' - \gamma\beta$.

c. [Determine arc identification for j] $\lambda \leftarrow -\beta m$.
d. [Update threads of r and z] $s_r \leftarrow x$ and $s_z \leftarrow j$.
e. [Save i] $r \leftarrow i$.
f. [Update i] $i \leftarrow j$.
g. [Update j] $j \leftarrow p_j$.
h. [Update predecessor of i] $p_i \leftarrow r$, and go to Step 4.
i. [Update threads of r and z] $s_r \leftarrow x$ and $s_z \leftarrow s_{q'}$.
j. [Update thread of q'] $s_{q'} \leftarrow q$.
k. [Update predecessor of q] $p_q \leftarrow q'$.

8 *Update Flows and Reset Flags on Path from q' to $\mathcal{K}'$* Same as Step 2 with Part a replaced by
a. [Set up for loop] If $\mu = 1$, $\delta_1 \leftarrow \mathcal{K}'$, $\delta_2 \leftarrow q'$, and $k \leftarrow 1$; otherwise $\delta_1 \leftarrow q'$, $\delta_2 \leftarrow \mathcal{K}'$, and $k \leftarrow 1$.

9 *Reset Flags on Path from w to the Root*
a. [Begin at w] $i \leftarrow w$.
b. [Reset flag at i] $f_i \leftarrow 0$.
c. [End of path?] If $i = l$, terminate; otherwise $i \leftarrow p_i$, and go to b.

B.3 PREDECESSOR/THREAD/DISTANCE

By augmenting the preceding data structure with the distance label, one can eliminate the flagging operations over the parts of the tree in which no other operations are performed, that is, Step 1 of ALG-B.1 and Step 9 of ALG-B.2 can be eliminated. The ratio test can be performed simultaneously with the determination of the path from u to v. In addition, the nested subscripting in Step 5b of ALG-B.2 is not required. To gain these simplifications, however, one must pay a penalty—the distance label must be updated. Fortunately, the set of nodes whose distance label may change after each pivot is identical to the set of nodes whose corresponding dual variable may change. Therefore the update for the distance label is integrated into Steps 4 and 5 of ALG-B.2. The distance label was first used by Srinivasan and Thompson. Letting $e_\phi = (u, v)$ denote the entering arc, one can determine the leaving arc as follows.

ALG B.3 RATIO TEST (PREDECESSOR/THREAD/DISTANCE)

0 Initialization. Same as ALG-B.1, Step 0, plus
c. [Initialize δ_1 and δ_2] $\delta_1 \leftarrow u$ and $\delta_2 \leftarrow v$.

1 *Determine Maximum Allowable Flow Change on the Partial Path from u to v Having Distance Labels Greater than $MIN(d_u, d_v)$*

a. [Begin at node having largest distance label]

$$\text{If } d_u - d_v \begin{cases} <0, & d \leftarrow d_v - d_u,\ \mu \leftarrow 2. \\ =0, & \text{go to Step 2.} \\ >0, & d \leftarrow d_u - d_v,\ \mu \leftarrow 1. \end{cases}$$

b. [Set δ, γ, and counter] $\delta \leftarrow \delta_\mu$, $\gamma \leftarrow \gamma_\mu$, and $k \leftarrow 1$.
c. [Increase or decrease flow?] If $m_\delta \gamma > 0$, go to e.
d. [Increasing flow] If $u_{|m_\delta|} - \alpha_\delta > \Delta$, $\delta \leftarrow p_\delta$ and go to f; otherwise $\Delta \leftarrow u_{|m_\delta|} - \alpha_\delta$, $\mathcal{K} \leftarrow \delta$, $\delta \leftarrow p_\delta$, and go to f.
e. [Decreasing flow] If $\alpha_\delta > \Delta$, $\delta \leftarrow p_\delta$; otherwise $\Delta \leftarrow \alpha_\delta$, $\mathcal{K} \leftarrow \delta$, and $\delta \leftarrow p_\delta$.
f. [Distance at δ equals $MIN(d_u, d_v)$] If $k < d$, $k \leftarrow k+1$, and go to c.
g. [Reset δ_μ] $\delta_\mu \leftarrow \delta$.

2 *Determine Maximum Allowable Flow Change on the Remainder of the Path from u to v*

a. [End of path?] If $\delta_1 = \delta_2$, $w \leftarrow \delta_1$ and terminate; otherwise $k \leftarrow 1$.
b. [Load δ] $\delta \leftarrow \delta_k$.
c. [Increase or decrease flow?] If $m_\delta \gamma_k > 0$, go to e.
d. [Increasing flow] If $u_{|m_\delta|} - \alpha_\delta > \Delta$, $\delta_k \leftarrow p_\delta$, and go to f; otherwise $\Delta \leftarrow u_{|m_\delta|} - \alpha_\delta$, $\mu \leftarrow k$, $\mathcal{K} \leftarrow \delta$, $\delta_k \leftarrow p_\delta$, and go to f.
e. [Decreasing flow] If $\alpha_\delta > \Delta$, $\delta_k \leftarrow p_\delta$; otherwise $\Delta \leftarrow \alpha_\delta$, $\mu \leftarrow k$, $\mathcal{K} \leftarrow \delta$, $\delta_k \leftarrow p_\delta$.
f. [Both sides examined?] If $k = 1$, $k \leftarrow 2$ and go to b; otherwise go to a.

At the termination of ALG-B.3, the maximum allowable flow change is Δ and the path from u to v is given by the path from u to w together with the path from v to w.

The updating algorithm is given as follows.

ALG B.4 UPDATE (PREDECESSOR/THREAD/DISTANCE)

1 *Determine Whether a Full Update is Required.* If $\Delta \neq u_\phi$, go to Step 3.

2 *Update Flows on Path from u to v.* Same as ALG-B.2, Step 2, with Part e replaced by

e. [Update flow] $\alpha_\delta \leftarrow \alpha_\delta - \gamma\beta$.

Also, Part g is replaced by

g. [Path complete?] If $k = 1$, $k \leftarrow 2$ and go to b; otherwise terminate.

3 *Initialization.* Same as ALG-B.2, Step 3, plus

f. [Save distance label of q'] $\mathcal{L} \leftarrow d_{q'} + 1$.

4 *Update Dual, Flow, Arc Identification, and Distance at i.* Same as ALG-B.2, Step 4, with Part g replaced by

g. [Save distance at i and difference of $\mathcal{L}$ and d_i] $\mathcal{L}' \leftarrow d_i$ and $L \leftarrow \mathcal{L} - d_i$.

h. [Update distance of i] $d_i \leftarrow \mathcal{L}$.

5 *Determine Last Successor of i; Update Dual Variables and Distance Labels*

a. [Begin at i] $z \leftarrow i$ and $x \leftarrow s_i$.

b. [Is x a successor?] If $d_x \leqslant \mathcal{L}'$, go to Step 6.

c. [Update dual variable and distance at x] $\pi_x \leftarrow \pi_x + \sigma$ and $d_x \leftarrow d_x + L$.

d. [Follow thread] $z \leftarrow x$, $x \leftarrow s_x$, and go to b.

6 *Find Reverse Thread for Node i.* Same as ALG-B.2, Step 6.

7 *Update Predecessor and Thread.* Same as ALG-B.2, Step 7, with Part h replaced by

h. [Update $\mathcal{L}$ and the predecessor of i] $p_i \leftarrow r$, $\mathcal{L} \leftarrow \mathcal{L} + 1$, and go to Step 4.

8 *Update Flows on Path from q' to $\mathcal{K}'$.* Same as Step 2 with Part a replaced by

a. [Set up for loop] If $\mu = 1$, $\delta_1 \leftarrow \mathcal{K}'$, $\delta_2 \leftarrow q'$, and $k \leftarrow 1$; otherwise $\delta_1 \leftarrow q'$, $\delta_2 \leftarrow \mathcal{K}'$, and $k \leftarrow 1$.

B.4 PREDECESSOR/THREAD/NUMBER OF SUCCESSORS

By replacing the distance label by the number of successors label, one can retain the benefits of the data structure of Section B.3 while simultaneously reducing the computational burden required for the update. For each pivot the only nodes whose labels, t_i, may change are those on the path from u to v in the basis tree, $\mathcal{T}$, that is, the update for t_i can be integrated into Steps 4 and 8 of ALG-B.2. The major savings result from the elimination of Steps 4g and 4h of ALG-B.4. The use of the number of successors to replace the distance label was first suggested by Glover and Klingman.

Given that the entering arc is $e_\phi = (u, v)$, the ratio test is as follows.

ALG B.5 RATIO TEST (PREDECESSOR/THREAD/NUMBER OF SUCCESSORS)

0 *Initialization.* Same as ALG-B.1, Step 0.

1 *Determine Maximal Allowable Flow Change on the Path from u to v*

a. [Set δ_1 and δ_2] $\delta_1 \leftarrow u$ and $\delta_2 \leftarrow v$.
b. [Select node with smallest subtree] If $t_{\delta_1} > t_{\delta_2}$, $k \leftarrow 2$; otherwise $k \leftarrow 1$.
c. [Set δ] $\delta \leftarrow \delta_k$.
d. [Increase or decrease flow?] If $m_\delta \gamma_k > 0$, go to f.
e. [Increasing flow] If $u_{|m_\delta|} - \alpha_\delta > \Delta$, $\delta_k \leftarrow p_\delta$ and go to g; otherwise $\Delta \leftarrow u_{|m_\delta|} - \alpha_\delta$, $\mu \leftarrow k$, $\mathcal{K} \leftarrow \delta$, $\delta_k \leftarrow p_\delta$, and go to g.
f. [Decreasing flow] If $\alpha_\delta > \Delta$, $\delta_k \leftarrow p_\delta$; otherwise $\Delta \leftarrow \alpha_\delta$, $\mu \leftarrow k$, $\mathcal{K} \leftarrow \delta$, $\delta_k \leftarrow p_\delta$.
g. [End of path] If $\delta_1 = \delta_2$, $w \leftarrow \delta_1$ and terminate; otherwise go to b.

At the termination of ALG-B.5, the maximum allowable flow change is Δ and the path from u to v is given by the path from u to w together with the path from v to w.

The updating algorithm is given as follows.

ALG B.6 UPDATE (PREDECESSOR/THREAD/NUMBER OF SUCCESSORS)

1 *Determine Whether a Full Update is Required.* If $\Delta \neq u_\phi$, go to Step 3.

2 *Update Flows on Path from u to v.* Same as ALG-B.2, Step 2, with Part e replaced by
e. [Update flow] $\alpha_\delta \leftarrow \alpha_\delta - \gamma\beta$.
Also, Part g is replaced by
g. [Path complete?] If $k=1$, $k \leftarrow 2$ and go to b; otherwise terminate.

3 *Initialization.* Same as ALG-B.2, Step 3, plus
f. [Save t_q] $L \leftarrow t_q$.
g. [Update t_q] $t_q \leftarrow t_{\mathcal{K}}$.

4 *Update Dual, Flow, Arc Identification, and Number of Successors at i.* Same as ALG-B.2, Step 4, with Part g replaced by
g. [Last update?] If $j = \mathcal{K}'$, go to 5.
h. [Save number of successors at j] $L' \leftarrow t_j$.
i. [Update number of successors at j] $t_j \leftarrow t_q - L$.
j. [Save L' in L] $L \leftarrow L'$.

5 *Determine Last Successor of i and Update Dual Variables*
a. [Initialize z, x and counter] $z \leftarrow i$, $x \leftarrow s_i$, and $c \leftarrow 2$.
b. [Is x a successor?] If $c > t_i$, to to Step 6.
c. [Update dual variable] $\pi_x \leftarrow \pi_x + \sigma$.
d. [Follow thread] $z \leftarrow x$, $x \leftarrow s_x$, $c \leftarrow c+1$, and go to b.

6 *Find Reverse Thread for Node i.* Same as ALG-B.2, Step 6.

7 *Update Predecessor and Thread.* Same as ALG-B.2, Step 7.

8 *Update Flows and Number of Successors on Path from q' to $\mathcal{K}'$*

a. [Set up for loop] If $\mu=1$, $\delta_1 \leftarrow \mathcal{K}'$, $\delta_2 \leftarrow q'$, $\lambda \leftarrow -1$, and $k \leftarrow 1$; otherwise $\delta_1 \leftarrow q'$, $\delta_2 \leftarrow \mathcal{K}'$, $\lambda \leftarrow 1$, and $k \leftarrow 1$.

b. [Initialize parameters] $\delta \leftarrow \delta_k$ and $\gamma \leftarrow \gamma_k \Delta$.

c. [Node equal w?] If $\delta = w$, go to h.

d. [Increase or decrease flow?] If $m_\delta > 0$, $\beta \leftarrow 1$; otherwise $\beta \leftarrow -1$.

e. [Update flow] $\alpha_\delta \leftarrow \alpha_\delta - \gamma\beta$.

f. [Update number of successors] $t_\delta \leftarrow t_\delta + \lambda t_q$.

g. [Follow predecessor] $\delta \leftarrow p_\delta$, and go to c.

h. [Path complete?] If $k=1$, $k \leftarrow 2$, $\lambda \leftarrow -\lambda$, and go to b; otherwise terminate.

B.5 PREDECESSOR/THREAD/PREORDER DISTANCE

Another three-label data structure has been proposed by Bradley, Brown, and Graves. For this data structure the preorder distance replaces the distance or the number of successors label. This data structure yields a simpler cycle trace than either ALG-B.3 or ALG-B.5 but requires a more difficult update than either ALG-B.4 or ALG-B.6. Given that the entering arc is $e_\phi = (u, v)$, the ratio test is as follows.

ALG B.7 RATIO TEST (PREDECESSOR/THREAD/PREORDER DISTANCE)

0 *Initialization*

a. [Initialize δ_1, δ_2] If $g_u < g_v$, $\tau \leftarrow g_u$, $\cdot$ $\delta_1 \leftarrow v$, $\delta_2 \leftarrow u$, $k^* \leftarrow 1$; otherwise $\tau \leftarrow g_v$, $\delta_1 \leftarrow u$, $\delta_2 \leftarrow v$, $k^* \leftarrow -1$.

b. [Initialize γ_1, γ_2] If $\alpha_\phi = 0$, $\gamma_1 \leftarrow k^*$, $\gamma_2 \leftarrow -k^*$; otherwise $\gamma_1 \leftarrow -k^*$, $\gamma_2 \leftarrow k^*$.

c. [Set Δ to bound and initialize k] $\Delta \leftarrow u_\phi$, $k \leftarrow 1$.

1 *Determine Maximum Allowable Flow Change on the Path from u to v*

a. [Set δ and γ] $\delta \leftarrow \delta_k$, $\gamma \leftarrow \gamma_k$.

b. [Increase or decrease flow?] If $m_\delta \gamma > 0$, go to d.

c. [Increasing flow] If $u_{|m_\delta|} - \alpha_\delta > \Delta$, go to e; otherwise $\Delta \leftarrow u_{|m_\delta|} - \alpha_\delta$, $\mu \leftarrow k$, $\mathcal{K} \leftarrow \delta$, and go to e.

d. [Decreasing flow] If $\alpha_\delta \geq \Delta$, go to e; otherwise $\Delta \leftarrow \alpha_\delta$, $\mu \leftarrow k$, and $\mathcal{K} \leftarrow \delta$.

e. [Follow predecessor] $\delta \leftarrow p_\delta$.
f. [Check for termination]

$$\text{If } g_\delta \begin{cases} >\tau, & \text{go to b,} \\ =\tau, & w \leftarrow \delta \text{ and go to g,} \\ <\tau, & \tau \leftarrow g_\delta,\ k \leftarrow 2, \text{ and go to a.} \end{cases}$$

g. [Save μ] $\mu^* \leftarrow \mu$.
h. [Reset μ] If $k^* = -1$, terminate, otherwise $\mu \leftarrow 3 - \mu$ and terminate.

At the termination of ALG-B.7, the maximum allowable flow change is Δ and the path from u to v is given by the path from u to w together with the path from v to w.

The update is given as follows.

ALG B.8 UPDATE (PREDECESSOR/THREAD/PREORDER DISTANCE)

1 *Determine Whether a Full Update is Required.* Same as ALG-B.4, Step 1.
2 *Update Flows on Path from u to v.* Same as ALG-B.4, Step 2.
3 *Initialization.* Same as ALG-B.2, Step 3, plus
  f. [Check for leaving edge on left stem] If $\mu^* \neq 2$, go to g; otherwise $r^* \leftarrow k'$ and go to Step 4.
  g. [Is q' the root?] If $p_{q'} \neq 0$, go to h; otherwise $r^* \leftarrow q'$ and go to Step 4.
  h. [Begin at $p_{q'}$] $r^* \leftarrow p_{q'}$.
  i. [Is r^* a reverse thread?] If $s_r^* = q'$, go to Step 4; otherwise go to j.
  j. [Follow thread] $r^* \leftarrow s_r^*$ and go to i.
4 *Update Dual, Flow, and Arc Identification at i.* Same as ALG-B.2, Step 4, with Part g replaced by
  g. [Save preorder distance of i] $L' \leftarrow g_i$.
5 *Determine Last Successor of i and Update Dual Variables*
  a. [Begin at i] $z \leftarrow i$ and $x \leftarrow s_i$.
  b. [Is x a successor?] If $g_{p_x} < L'$, go to Step 6.
  c. [Update dual] $\pi_x \leftarrow \pi_x + \sigma$.
  d. [Follow thread] $z \leftarrow x$, $x \leftarrow s_x$.
  e. [Is x the root?] If $p_x = 0$, go to Step 6; otherwise go to b.
6 *Find Reverse Thread for Node i.* Same as ALG-B.2, Step 6.
7 *Update Predecessor and Thread.* Same as ALG-B.2, Step 7.

8 *Update Flows on Path from q' to $\mathcal{K}'$.* Same as ALG-B.4, Step 8.
9 *Update Preorder Distance Labels*
a. [Begin at r^*] $i \leftarrow r^*$.
b. [Initialize preorder distance] $g^* \leftarrow g_i$.
c. [Follow thread] $i \leftarrow s_i$.
d. [Check for termination] If $g_i = 1$, terminate; otherwise $g^* \leftarrow g^* + 1$, $g_i \leftarrow g^*$, and go to c.

B.6 PREDECESSOR/THREAD/NUMBER OF SUCCESSORS/LAST SUCCESSOR

If a basic arc is removed from $\mathcal{T}$, then $\mathcal{T}$ is partitioned into two trees, say $\mathcal{T}_1$ and $\mathcal{T}_2$, where the root l is in $\mathcal{T}_1$. The dual variables associated with the nodes of either $\mathcal{T}_1$ or $\mathcal{T}_2$ must be updated. Since a search would have been required to determine the size of $\mathcal{T}_1$ or $\mathcal{T}_2$, the preceding algorithms always updated the dual variables in $\mathcal{T}_2$. Augmenting the data structure of Section 4 with the number of successors allows one to select the smaller tree for dual variable updating. Following the work of Glover and Klingman, we transform $\mathcal{T}_1$ and $\mathcal{T}_2$ into independently labeled trees, update the dual variables in the tree with fewest nodes, and reconnect $\mathcal{T}_2$ to $\mathcal{T}_1$ with arc e_ϕ. An implementation of this procedure may allow the root to change from pivot to pivot (called rerooting) or may require the root to remain fixed. Our implementation requires that the root remain fixed throughout the procedure.

The ratio test is identical to ALG-B.5, and the update is as follows.

ALG B.9 UPDATE (PREDECESSOR/THREAD/NUMBER OF SUCCESSORS/LAST SUCCESSOR)

1 *Determine Whether a Full Update is Required.* If $\Delta \neq u_\phi$, $\mathcal{K}' \leftarrow p_{\mathcal{K}}$ and go to Step 3.
2 *Update Flows on Path from u to v.* Same as ALG-B.2, Step 2, with Part e replaced by
e. [Update flow] $\alpha_\delta \leftarrow \alpha_\delta - \gamma\beta$.
Also, part g is replaced by
g. [Path complete?] If $k=1$, $k \leftarrow 2$ and go to b; otherwise terminate.
3 *Update Thread of $\mathcal{T}_1$*
a. [Begin at $\mathcal{K}'$] $i \leftarrow \mathcal{K}'$.
b. [Thread of i equal $\mathcal{K}$?] If $s_i = \mathcal{K}$, $y \leftarrow i$, $s_y \leftarrow s_{n_{\mathcal{K}}}$, and go to Step 4.

c. [Follow thread] $i \leftarrow n_{s_i}$ and go to b.

4 *Update Number of Successors, Last Successor, and Flow on Path from $\mathcal{K}'$ to w*

a. [Initialize parameters] $i \leftarrow \mathcal{K}'$ and $\gamma \leftarrow \gamma_\mu \Delta$.

b. [Does last successor require updating?] If $n_i = n_{\mathcal{K}}$, $n_i \leftarrow y$; otherwise go to f.

c. [Node i equal node w?] If $i = w$, go to Step 5; otherwise $t_i \leftarrow t_i - t_{\mathcal{K}}$.

d. [Update flow] If $m_i > 0$, $\alpha_i \leftarrow \alpha_i - \gamma$; otherwise $\alpha_i \leftarrow \alpha_i + \gamma$.

e. [Follow predecessor] $i \leftarrow p_i$ and go to b.

f. [Node i equal node w?] If $i = w$, go to Step 6; otherwise $t_i \leftarrow t_i - t_{\mathcal{K}}$.

g. [Update flow] If $m_i > 0$, $\alpha_i \leftarrow \alpha_i - \gamma$; otherwise $\alpha_i \leftarrow \alpha_i + \gamma$.

h. [Follow predecessor] $i \leftarrow p_i$ and go to f.

5 *Update Number of Successors on the Path from w to the Root if Necessary*

a. [Is node i the root?] If $i = l$, go to Step 6; otherwise $i \leftarrow p_i$.

b. [Does last successor require updating?] If $n_i = n_{\mathcal{K}}$, $n_i \leftarrow y$ and go to a.

6 *Update for $\mathcal{T}_2$*

a. [Save thread of $n_{\mathcal{K}}$] $s' \leftarrow s_{n_{\mathcal{K}}}$.

b. [Update thread of $n_{\mathcal{K}}$] $s_{n_{\mathcal{K}}} \leftarrow \mathcal{K}$.

7 *Update Dual Variables on Smaller Tree*

a. [Determine smaller tree] If $t_{\mathcal{K}} < \frac{1}{2} t_l$, $i \leftarrow \mathcal{K}$, $L \leftarrow \mathcal{K}$; otherwise $i \leftarrow l$, $L \leftarrow l$, and $\sigma \leftarrow -\sigma$.

b. [Update dual variable] $\pi_i \leftarrow \pi_i + \sigma$.

c. [Follow thread] $i \leftarrow s_i$.

d. [At root?] If $i \neq L$, go to b.

e. [Restore thread of $n_{\mathcal{K}}$] $s_{n_{\mathcal{K}}} \leftarrow s'$.

f. [Restore thread of y] $s_y \leftarrow \mathcal{K}$.

8 *Initialization for Update of $\mathcal{T}_2$*

a. [Entering variable at bound?] If $\alpha_\phi = u_\phi$, $\Delta' \leftarrow u_\phi - \Delta$; otherwise $\Delta' \leftarrow \Delta$.

b. [Initialize q, q', and λ] If $\mu = 1$, $q \leftarrow u$, $q' \leftarrow v$, $\lambda \leftarrow -\phi$; otherwise $q \leftarrow v$, $q' \leftarrow u$, and $\lambda \leftarrow \phi$.

c. [Initialize i, j, and γ] $i \leftarrow q$, $j \leftarrow p_q$, and $\gamma \leftarrow \gamma_\mu \Delta$.

d. [Save t_q] $L \leftarrow t_q$.

e. [Update t_q] $t_q \leftarrow t_{\mathcal{K}}$.

9 *Update Flow, Arc Identification, and Number of Successors at i*

a. [Save flow] $\alpha' \leftarrow \alpha_i$.

b. [Update flow] $\alpha_i \leftarrow \Delta'$.

c. [Save direction] If $m_i > 0$, $\beta \leftarrow 1$; otherwise $\beta \leftarrow -1$.
d. [Save arc] $m \leftarrow |m_i|$.
e. [Update arc identification] $m_i \leftarrow \lambda$.
f. [Last update?] If $j = \mathcal{K}'$, go to Step 10.
g. [Save number of successors at j] $L' \leftarrow t_j$.
h. [Update number of successors at j] $t_j \leftarrow t_q - L$.
i. [Save L' in L] $L \leftarrow L'$.

10 *Find Reverse Thread for Node i.* Same as ALG-B.2, Step 6, with "go to Step 7" replaced by "go to Step 11" in Part b.

11 *Save Reverse Thread If Necessary.* [Node j equal $\mathcal{K}$?] If $j = \mathcal{K}$, go to Step 12; otherwise $r' \leftarrow r$ and $i' \leftarrow i$.

12 *Update Predecessor and Thread of $\mathfrak{T}_2$.* Same as ALG-B.2, Step 7, with Parts d, h, and i replaced by

d. [Update threads of r and n_i] $s_r \leftarrow s_{n_i}$ and $s_{n_i} \leftarrow j$.
h. [Update predecessor of i] $p_i \leftarrow r$ and go to Step 9.
i. [Update threads of r and n_i] $s_r \leftarrow s_{n_i}$ and $s_{n_i} \leftarrow s_{q'}$.

13 *Update Last Node in $\mathfrak{T}_2$*

a. [No change if $\mathcal{K}$ equals q] If $\mathcal{K} = q$, go to Step 14.
b. [Does $n_{\mathcal{K}}$ change?] If $n_{\mathcal{K}} \neq n_{i'}$, $i \leftarrow i'$ and go to c; otherwise $i \leftarrow i'$ and $n_{\mathcal{K}} \leftarrow r'$.
c. [Update n_i] $n_i \leftarrow r'$.
d. [End of path?] If $i = q$, go to Step 14; otherwise $i \leftarrow p_i$ and go to c.

14 *Update Flow, Number of Successors, and Last Successor on the Path from q' to w*

a. [Initialize parameters] $i \leftarrow q'$ and $\gamma \leftarrow -\gamma_\mu \Delta$.
b. [Does last successor require updating?] If $n_i = q'$, $n_i \leftarrow n_q$; otherwise go to f.
c. [Node i equal node w?] If $i = w$, go to Step 15; otherwise $t_i \leftarrow t_i + t_q$.
d. [Update flow] If $m_i > 0$, $\alpha_i \leftarrow \alpha_i - \gamma$; otherwise $\alpha_i \leftarrow \alpha_i + \gamma$.
e. [Follow predecessor] $i \leftarrow p_i$ and go to b.
f. [Node i equal node w?] If $i = w$, terminate; otherwise $t_i \leftarrow t_i + t_q$.
g. [Update flow] If $m_i > 0$, $\alpha_i \leftarrow \alpha_i - \gamma$; otherwise $\alpha_i \leftarrow \alpha_i + \gamma$.
h. [Follow predecessor] $i \leftarrow p_i$ and go to f.

15 *Update Number of Successors on the Path from w to the Root if Necessary*

a. [Is node i the root?] If $i = l$, terminate; otherwise $i \leftarrow p_i$.
b. [Does last successor require updating?] If $n_i = q'$, $n_i \leftarrow n_q$ and go to a; otherwise terminate.

NOTES AND REFERENCES

Most of the ideas presented in this appendix are due to Fred Glover and Dar Klingman and their colleagues at the University of Texas. These ideas were first compiled in Ref. [3], which draws freely upon the following: Bradley, Brown, and Graves [25], Glover, Klingman, and Stutz [79], Glover, Karney, Klingman, and Napier [77], Glover, Karney, and Klingman [76], and Srinivasan and Thompson [171, 172].

APPENDIX C

Convergence of Subgradient Optimization Algorithm

Consider the nonlinear program

$$\left.\begin{array}{ll}\min & g(\mathbf{y}) \\ \text{s.t.} & \mathbf{y}\varepsilon\Gamma\subset R^n\end{array}\right\} P,$$

where Γ is compact, convex, and nonempty and g is continuous and convex over Γ. Letting $s_0, s_1, s_2, \ldots$ denote the step sizes and $P[\mathbf{y}]$ denote the projection operator, we may state the subgradient algorithm as follows.

SUBGRADIENT ALGORITHM

0 *Initialization.* Let $\mathbf{y}_0 \varepsilon \Gamma$, and set $i \leftarrow 0$.

1 *Obtain Subgradient.* Let $\boldsymbol{\eta}_i \varepsilon \partial g(\mathbf{y}_i)$. If $\boldsymbol{\eta}_i = \mathbf{0}$, terminate with $\mathbf{y}_i$ optimal [$\partial g(\mathbf{y}_i)$ is the set of all subgradients at $\mathbf{y}_i$].

2 *Move to New Point.* Set $\mathbf{y}_{i+1} \leftarrow P[\mathbf{y}_i - s_i \boldsymbol{\eta}_i]$, $i \leftarrow i+1$, and go to Step 1.

Before specifying a particular form for $s_0, s_1, s_2, \ldots$, we develop three results that will be useful for proving the convergence results. Consider the following propositions.

Proposition C.1

Let $\mathbf{y} \varepsilon \Gamma$, and let $\mathbf{x} \varepsilon R^n$. Then $(\mathbf{x}-P[\mathbf{x}])(\mathbf{y}-P[\mathbf{x}]) \leqslant 0$.

Proof. Choose α so that $0<\alpha<1$. Since Γ is convex, $\alpha\mathbf{y}+(1-\alpha)P[\mathbf{x}] \varepsilon \Gamma$. By the definition of $P[\mathbf{x}]$, $\|\mathbf{x}-P[\mathbf{x}]\| \leqslant \|\mathbf{x}-(\alpha\mathbf{y}+(1-\alpha)P[\mathbf{x}])\|$. Thus

$$\begin{aligned}\|\mathbf{x}-P[\mathbf{x}]\|^2 &\leqslant \|\mathbf{x}-(\alpha\mathbf{y}+(1-\alpha)P[\mathbf{x}])\|^2\\ &= \|\mathbf{x}-P[\mathbf{x}]-\alpha(\mathbf{y}-P[\mathbf{x}])\|^2\\ &= \|\mathbf{x}-P[\mathbf{x}]\|^2+\alpha^2\|\mathbf{y}-P[\mathbf{x}]\|^2-2\alpha(\mathbf{x}-P[\mathbf{x}])(\mathbf{y}-P[\mathbf{x}]).\end{aligned}$$

Then $(\mathbf{x}-P[\mathbf{x}])(\mathbf{y}-P[\mathbf{x}]) \leqslant \|\mathbf{y}-P[\mathbf{x}]\|\alpha/2$. Since α can be taken arbitrarily close to 0,

$$(\mathbf{x}-P[\mathbf{x}])(\mathbf{y}-P[\mathbf{x}]) \leqslant 0.$$

■

Proposition C.2

Let $\mathbf{x}, \mathbf{y} \varepsilon R^n$. Then $\|P[\mathbf{x}]-P[\mathbf{y}]\| \leqslant \|\mathbf{x}-\mathbf{y}\|$.

Proof.

Case 1 Suppose $P[\mathbf{x}]=P[\mathbf{y}]$. Then

$$\|P[\mathbf{x}]-P[\mathbf{y}]\|=0 \leqslant \|\mathbf{x}-\mathbf{y}\|.$$

Case 2 Suppose $P[\mathbf{x}] \neq P[\mathbf{y}]$. Since $P[\mathbf{x}] \varepsilon \Gamma$ and $P[\mathbf{y}] \varepsilon \Gamma$, from Proposition C.1 we have that

$$(\mathbf{x}-P[\mathbf{x}])(P[\mathbf{y}]-P[\mathbf{x}]) \leqslant 0$$

and

$$(\mathbf{y}-P[\mathbf{y}])(P[\mathbf{x}]-P[\mathbf{y}]) \leqslant 0.$$

We may rewrite the above inequalities as

$$\mathbf{x}(P[\mathbf{y}]-P[\mathbf{x}])-P[\mathbf{x}]P[\mathbf{y}]+\|P[\mathbf{x}]\|^2 \leqslant 0$$

and

$$\mathbf{y}(P[\mathbf{x}]-P[\mathbf{y}])-P[\mathbf{y}]P[\mathbf{x}]+\|P[\mathbf{y}]\|^2 \leqslant 0.$$

Adding these inequalities, we obtain

$$(\mathbf{x}-\mathbf{y})(P[\mathbf{y}]-P[\mathbf{x}])+\|P[\mathbf{y}]-P[\mathbf{x}]\|^2 \leqslant 0.$$

From the Cauchy-Schwartz inequality,

$$-(\mathbf{x}-\mathbf{y})(P[\mathbf{y}]-P[\mathbf{x}]) \leqslant \|\mathbf{x}-\mathbf{y}\|\,\|P[\mathbf{y}]-P[\mathbf{x}]\|.$$

Thus

$$\|P[\mathbf{y}]-P[\mathbf{x}]\|^2 \leqslant \|\mathbf{x}-\mathbf{y}\|\,\|P[\mathbf{y}]-P[\mathbf{x}]\|.$$

Since $P[\mathbf{x}] \neq P[\mathbf{y}]$,

$$\|P[\mathbf{x}]-P[\mathbf{y}]\| \leqslant \|\mathbf{x}-\mathbf{y}\|.$$

■

Proposition C.3

If $\boldsymbol{\eta}_i \neq \mathbf{0}$,

$$\|\mathbf{y}_{i+1}-\mathbf{y}\|^2 \leqslant \|\mathbf{y}_i-\mathbf{y}\|^2+s_i^2\|\boldsymbol{\eta}_i\|^2+2s_i\boldsymbol{\eta}_i(\mathbf{y}-\mathbf{y}_i) \qquad \text{for any } \mathbf{y}\,\varepsilon\,\Gamma.$$

Proof. Let i be any iteration of the subgradient algorithm. Suppose $\boldsymbol{\eta}_i \neq \mathbf{0}$. Let $\mathbf{y}$ be any element of Γ.

By Proposition C.2

$$\|P[\mathbf{y}_i-s_i\boldsymbol{\eta}_i]-P[\mathbf{y}]\|^2 \leqslant \|\mathbf{y}_i-s_i\boldsymbol{\eta}_i-\mathbf{y}\|^2$$
$$=\|\mathbf{y}_i-\mathbf{y}\|^2+s_i^2\|\boldsymbol{\eta}_i\|^2+2s_i\boldsymbol{\eta}_i(\mathbf{y}-\mathbf{y}_i).$$

Since $P[\mathbf{y}]=\mathbf{y}$ and $P[\mathbf{y}_i-s_i\boldsymbol{\eta}_i]=\mathbf{y}_{i+1}$, we have

$$\|\mathbf{y}_{i+1}-\mathbf{y}\|^2 \leqslant \|\mathbf{y}_i-\mathbf{y}\|^2+s_i^2\|\boldsymbol{\eta}_i\|^2+2s_i\boldsymbol{\eta}_i(\mathbf{y}-\mathbf{y}_i).$$

■

We now present convergence results for three particular step sizes. Note that for the first two step sizes the subgradient algorithm produces a subsequence that converges to the optimum.

Proposition C.4

Let (i) g^* denote the optimal objective value for P;
(ii) $\{\lambda_i\}$ be any infinite sequence such that $\lambda_i > 0$, $\lim_{i\to\infty} \lambda_i = 0$, $\Sigma_{i=0}^{\infty} \lambda_i = \infty$; and
(iii) $s_i = \lambda_i / \|\eta_i\|$.

Then there is a subsequence $\{\mathbf{y}_{i_k}\}$ such that $g(\mathbf{y}_{i_k}) \to g^*$.

Proof. Select an infinite sequence $\{\alpha_k\}$ such that $\alpha_k > g^*$ and $\lim_{k\to\infty} \alpha_k = g^*$. For each α_k, let $\psi(\alpha_k) = \{y : g(\mathbf{y}) \leqslant \alpha_k, \mathbf{y} \varepsilon \Gamma\}$. Now $\psi(\alpha_k) \neq \phi$ since $g^* < \alpha_k$. Pick any point $\bar{\mathbf{y}} \varepsilon \psi(\alpha_k)$. Since g is continuous, there exists a $\delta > 0$ such that $\|\mathbf{y} - \bar{\mathbf{y}}\| < \delta$ for all $\mathbf{y} \varepsilon \psi(\alpha_k)$. We now show that there is some i for which $\mathbf{y}_i \varepsilon \psi(\alpha_k)$.

Assume the contrary: that $\mathbf{y}_i \not\varepsilon \psi(\alpha_k)$ for all i. Thus, for all i and $\mathbf{y} \varepsilon \psi(\alpha_k)$, $g(\mathbf{y}_i) > \alpha_k \geqslant g(\mathbf{y})$. For all i, $\boldsymbol{\eta}_i \neq 0$ since $\{\mathbf{y}_i\}$ is an infinite sequence, and $\hat{\mathbf{y}} = \bar{\mathbf{y}} + (\delta / \|\boldsymbol{\eta}_i\|)\boldsymbol{\eta}_i$ satisfies $\|\hat{\mathbf{y}} - \bar{\mathbf{y}}\| = \delta$, so $\hat{\mathbf{y}} \varepsilon \psi(\alpha_k)$. Thus for all i

$$\boldsymbol{\eta}_i(\hat{\mathbf{y}} - \mathbf{y}_i) \leqslant g(\hat{\mathbf{y}}) - g(\mathbf{y}_i) < 0. \tag{C.1}$$

By Proposition C.3

$$\|\mathbf{y}_{i+1} - \bar{\mathbf{y}}\|^2 \leqslant \|\mathbf{y}_i - \bar{\mathbf{y}}\|^2 + \lambda_i^2 + 2\left(\frac{\lambda_i}{\|\boldsymbol{\eta}_i\|}\right)\boldsymbol{\eta}_i(\bar{\mathbf{y}} - \mathbf{y}_i). \tag{C.2}$$

Now

$$\begin{aligned}\boldsymbol{\eta}_i(\bar{\mathbf{y}} - \mathbf{y}_i) &= \boldsymbol{\eta}_i\left[\hat{\mathbf{y}} - \mathbf{y}_i - \left(\frac{\delta}{\|\boldsymbol{\eta}_i\|}\right)\boldsymbol{\eta}_i\right] \\ &= \boldsymbol{\eta}_i(\hat{\mathbf{y}} - \mathbf{y}_i) - \delta\|\boldsymbol{\eta}_i\|. \end{aligned} \tag{C.3}$$

From (C.1)–(C.3) we obtain

$$\|\mathbf{y}_{i+1} - \bar{\mathbf{y}}\|^2 \leqslant \|\mathbf{y}_i - \bar{\mathbf{y}}\|^2 + \lambda_i^2 - 2\delta\lambda_i.$$

Since $\lambda_i \to 0$, we can choose an integer N such that, for all $n \geqslant N$, $\lambda_n \leqslant \delta$. Hence for all $n \geqslant N$

$$\|\mathbf{y}_{n+1} - \bar{\mathbf{y}}\|^2 \leqslant \|\mathbf{y}_n - \bar{\mathbf{y}}\|^2 - \delta\lambda_n. \tag{C.4}$$

Since $\Sigma_{i=0}^{\infty} \lambda_i = \infty$, we can choose an M such that $\Sigma_{n=N}^{N+M} \lambda_n > \|\mathbf{y}_N - \bar{\mathbf{y}}\|^2 / \delta$. Adding together the inequalities obtained from (C.4) by letting n take on all values from N to $N+M$, we obtain

$$\|\mathbf{y}_{N+M+1}-\bar{\mathbf{y}}\|^2 \leqslant \|\mathbf{y}_N-\bar{\mathbf{y}}\|^2-\delta \sum_{n=N}^{N+M} \lambda_\eta < 0,$$

a contradiction. Therefore for each α_k there is an i_k such that $\mathbf{y}_{i_k} \varepsilon \psi(\alpha_k)$. Since $\alpha_k \to g^*$, $g(\mathbf{y}_{i_k}) \to g^*$. ■

Proposition C.5

Let (i) g^* denote the optimal objective value for P;
(ii) $\{\lambda_i\}$ be any infinite sequence such that $\lambda_i > 0$, $\lim_{i\to\infty} \lambda_i = 0$, $\Sigma_{i=0}^{\infty} \lambda_i = \infty$; and
(iii) $s_i = \lambda_i$.

If there is a constant C such that $\|\boldsymbol{\eta}_i\| \leqslant C$ for all i, then there is a subsequence $\{\mathbf{y}_{i_k}\}$ such that $g(\mathbf{y}_{i_k}) \to g^*$.

Proof. The proof is nearly identical to that of Proposition C.4, except that in place of (C.2) we obtain, by Proposition C.3,

$$\|\mathbf{y}_{i+1}-\bar{\mathbf{y}}\|^2 \leqslant \|\mathbf{y}_i-\bar{\mathbf{y}}\| + \lambda_i^2\|\boldsymbol{\eta}_i\|^2 + 2\lambda_i\boldsymbol{\eta}_i(\bar{\mathbf{y}}-\mathbf{y}_i). \tag{C.5}$$

Also, we must choose an integer N such that, for all $n \geqslant N$, $\lambda_n \leqslant \delta/C^2$. ■

Proposition C.6

Let (i) $\bar{g}$ be a constant smaller than the optimal objective value g^*; and
(ii) $s_i = [g(\mathbf{y}_i) - \bar{g}]/\|\boldsymbol{\eta}_i\|^2$.

If there is a constant C such that $\|\boldsymbol{\eta}_i\| \leqslant C$ for all i, then there is some n such that $g(\mathbf{y}_n) \leqslant 2g^* - \bar{g} + \delta$ for any given $\delta \geqslant 0$.

Proof. Suppose that, for all i, $g(\mathbf{y}_i) > 2g^* - \bar{g} + \delta$, where $\delta > 0$ is given. Let $\mathbf{y}^* \varepsilon \Gamma$ be an optimal point. By Proposition C.3

$$\|\mathbf{y}_{i+1}-\mathbf{y}^*\| \leqslant \|\mathbf{y}_i-\mathbf{y}^*\|^2 + \frac{\left[g(\mathbf{y}_i)-\bar{g}\right]^2}{\|\boldsymbol{\eta}_i\|^2} + 2\left[\frac{g(\mathbf{y}_i)-\bar{g}}{\|\boldsymbol{\eta}_i\|^2}\right]\boldsymbol{\eta}_i(\mathbf{y}^*-\mathbf{y}_i).$$

Since $\boldsymbol{\eta}_i \varepsilon \partial g(\mathbf{y}_i)$, $\boldsymbol{\eta}_i(\mathbf{y}^* - \mathbf{y}_i) \leqslant g^* - g(\mathbf{y}_i)$. Thus

$$\begin{aligned}
\|\mathbf{y}_{i+1} - \mathbf{y}^*\|^2 &\leqslant \|\mathbf{y}_i - \mathbf{y}^*\|^2 + [\, g(\mathbf{y}_i) - \bar{g}\,] \frac{2g^* - g(\mathbf{y}_i) - \bar{g}}{\|\boldsymbol{\eta}_i\|^2} \\
&< \|\mathbf{y}_i - \mathbf{y}^*\|^2 - \frac{[\, g(\mathbf{y}_i) - \bar{g}\,]\delta}{C^2} \\
&\leqslant \|\mathbf{y}_i - \mathbf{y}^*\|^2 - \frac{(g^* - \bar{g})\delta}{C^2}. \qquad \text{(C.6)}
\end{aligned}$$

We can choose an integer N large enough that

$$\frac{C^2\|\mathbf{y}_1 - \mathbf{y}^*\|^2}{(g^* - \bar{g})\delta} < N.$$

Adding together the inequalities obtained from (C.6) by letting i take on all values from 1 to N, we obtain

$$\|\mathbf{y}_{N+1} - \mathbf{y}^*\|^2 < \|\mathbf{y}_1 - \mathbf{y}^*\|^2 - \frac{N(g^* - \bar{g})\delta}{C^2} < 0,$$

a contradiction. ■

Note that, if g^* is known, then, by taking $\bar{g}$ arbitrarily close to g^*, the procedure must necessarily obtain an iterate whose objective value is arbitrarily close to g^*.

NOTES AND REFERENCES

The results in this appendix can be found in Ref. [102]. When the optimal objective value g^* is known, additional convergence results are available (see Poljak [153, 154]).

APPENDIX **D**

Projection Operation for Subgradient Algorithm

In this appendix we present a simple algorithm to solve the following quadratic program;

$$\begin{aligned} \min \quad & \tfrac{1}{2}\mathbf{z}\overline{\mathbf{D}}\mathbf{z} - \mathbf{a}\mathbf{z} && \text{(D.1)} \\ \text{s.t.} \quad & \textstyle\sum_j z_j = c && \text{(D.2)} \\ & \mathbf{0} \leqslant \mathbf{z} \leqslant \mathbf{u}, && \text{(D.3)} \end{aligned}$$

where $\overline{\mathbf{D}}$ is a positive diagonal matrix and $\mathbf{u}$ is a nonnegative vector.

We rewrite (D.1)–(D.3) in a slightly different form as follows:

$$\begin{aligned} \min \quad & \tfrac{1}{2}\mathbf{z}\overline{\mathbf{D}}\mathbf{z} - \mathbf{a}\mathbf{z} & & \text{(D.4)} \\ \text{s.t.} \quad & \textstyle\sum_j z_j - c = 0 & (\lambda) & \quad \text{(D.5)} \\ & z_j - u_j \leqslant 0 & (w_j) & \quad \text{(D.6)} \\ & -z_j \leqslant 0 & (v_j), & \quad \text{(D.7)} \end{aligned}$$

where λ, w_j, and v_j are the Kuhn-Tucker multipliers associated with the three types of constraints. The Kuhn-Tucker conditions for (D.4)–(D.7) may be stated as follows:

$$d_j z_j - a_j + w_j - v_j + \lambda = 0 \qquad \text{for all} \quad j, \tag{D.8}$$

$$w_j(z_j - u_j) = 0 \qquad \text{for all} \quad j, \tag{D.9}$$

$$v_j z_j = 0 \qquad \text{for all} \quad j, \tag{D.10}$$

$$w_j, v_j \geqslant 0 \qquad \text{for all} \quad j, \tag{D.11}$$

plus (D.5), (D.6), and (D.7),

where d_j is the jth diagonal element of $\overline{\mathbf{D}}$. Consider the following solution as a function of λ:

$$\left.\begin{aligned} z_j(\lambda) &= \max\left\{\min\left[\frac{a_j - \lambda}{d_j}, u_j\right], 0\right\}, \\ w_j(\lambda) &= \max\{a_j - \lambda - d_j u_j, 0\}, \\ v_j(\lambda) &= \max\{\lambda - a_j, 0\}. \end{aligned}\right\} \tag{D.12}$$

For any selection of λ, this solution clearly satisfies (D.6), (D.7), and (D.11). We show that this solution will also satisfy (D.8), (D.9), and (D.10).

Proposition D.1

The solution given by (D.12) satisfies (D.9).

Proof.

Case 1 $a_j - \lambda - d_j u_j < 0 \Rightarrow w_j = 0 \Rightarrow w_j(z_j - u_j) = 0.$

Case 2 $a_j - \lambda - d_j u_j \geqslant 0 \Rightarrow (a_j - \lambda)/d_j \geqslant u_j \Rightarrow z_j = u_j \Rightarrow w_j(z_j - u_j) = 0.$ ■

Proposition D.2

The solution given by (D.12) satisfies (D.10).

Proof.

Case 1 $a_j - \lambda < 0 \Rightarrow (a_j - \lambda)/d_j < 0 \Rightarrow z_j = 0 \Rightarrow v_j z_j = 0.$

Case 2 $a_j - \lambda \geqslant 0 \Rightarrow v_j = 0 \Rightarrow v_j z_j = 0.$ ■

Proposition D.3

The solution given by (D.12) satisfies (D.8).

Proof.

Case 1 $(a_j-\lambda)/d_j \geqslant u_j \Rightarrow z_j = u_j, a_j-\lambda-d_ju_j \geqslant 0$, and $a_j-\lambda \geqslant 0$.
$a_j-\lambda-d_ju_j \geqslant 0 \Rightarrow w_j = a_j-\lambda-d_ju_j$.
$a_j-\lambda \geqslant 0 \Rightarrow v_j = 0$.
Thus

$$d_jz_j - a_j + w_j - v_j + \lambda = d_ju_j - a_j + a_j - \lambda - d_ju_j - 0 + \lambda = 0.$$

Case 2 $0<(a_j-\lambda)/d_j<u_j \Rightarrow z_j=(a_j-\lambda)/d_j, a_j-\lambda-d_ju_j<0$, and $a_j-\lambda > 0$.
$a_j-\lambda-d_ju_j<0 \Rightarrow w_j=0$.
$a_j-\lambda>0 \Rightarrow v_j=0$.
Thus

$$d_jz_j - a_j + w_j - v_j + \lambda = d_j\left(\frac{a_j-\lambda}{d_j}\right) - a_j + 0 - 0 + \lambda = 0.$$

Case 3 $(a_j-\lambda)/d_j \leqslant 0 \Rightarrow z_j = 0$ and $a_j-\lambda \leqslant 0$.
$a_j-\lambda \leqslant 0 \Rightarrow a_j-\lambda-d_ju_j \leqslant 0$ and $v_j=\lambda-a_j$.
$a_j-\lambda-d_ju_j \leqslant 0 \Rightarrow w_j=0$.
Thus

$$d_jz_j - a_j + w_j - v_j + \lambda = 0 - a_j + 0 - (\lambda - a_j) + \lambda = 0.$$

■

Hence to solve (D.1)–(D.3) one need only find the appropriate λ such that (D.5) is satisfied. Let

$$g(\lambda) = \sum_j z_j(\lambda) = \sum_j \max\left\{ \min\left(\frac{a_j-\lambda}{d_j}, u_j\right), 0\right\}.$$

Then we must find λ^* such that $g(\lambda^*)=c$. Note that, since $d_j>0$ and $u_j \geqslant 0, z_j(\lambda)$ may be expressed as follows:

$$z_j(\lambda) = \begin{cases} u_j, & \lambda \leqslant a_j - d_ju_j; \\ \dfrac{a_j-\lambda}{d_j}, & a_j - d_ju_j < \lambda \leqslant a_j; \\ 0, & \lambda > a_j. \end{cases}$$

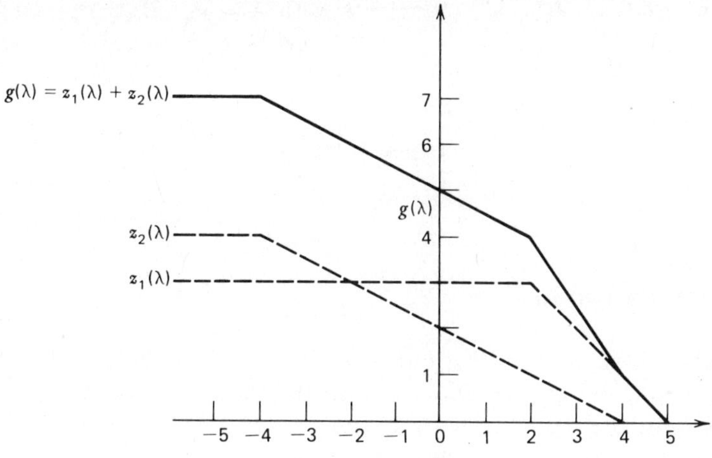

FIGURE D.1 Illustration of $g(\lambda)$ $(a_1=5, d_1=1, u_1=3, a_2=4, d_2=2, u_2=4)$.

Clearly each $z_j(\lambda)$ is piecewise linear and monotonically nonincreasing. Since the sum of such functions preserves this property, $g(\lambda)$ is piecewise linear and monotonically nonincreasing. A typical g is illustrated in Figure D.1.

Suppose z has n components. Then the breakpoints for the piecewise linear function $g(\lambda)$ occur at the $2n$ points $a_j - d_j u_j$ and a_j for $j=1,\ldots,n$. Let $y_1,\ldots,y_{2n}$ denote these breakpoints, where $y_1 \leqslant y_2 \leqslant \cdots \leqslant y_{2n}$. Then, for $\lambda \leqslant y_1, g(\lambda)=\Sigma_j u_j$ and, for $\lambda \geqslant y_{2n}, g(\lambda)=0$. Hence (D.1)–(D.3) has a feasible solution if and only if $0 \leqslant c \leqslant \Sigma_j u_j$.

We now present an algorithm for obtaining the value λ^* such that $g(\lambda^*)=c$. The procedure consists of a binary search to bracket λ^* between two breakpoints, followed by a linear interpolation.

ALGORITHM FOR λ^*

0 *Initialization.* If $c > \Sigma_j u_j$ or $c < 0$, terminate with no feasible solution; otherwise set $l \leftarrow 1, r \leftarrow 2n$, $L \leftarrow \Sigma_j u_j$, and $R \leftarrow 0$, and let $y_1 \leqslant y_2 \leqslant \cdots \leqslant y_{2n}$ be the breakpoints.

1 *Test for Bracketing.* If $r-l=1$, go to step 4; otherwise set $m \leftarrow [(l+r)/2]_I$, where $[k]_I$ is the greatest integer $\leqslant k$.

2 *Compute New Value.* Set

$$C \leftarrow \sum_j \max\left\{ \min\left(\frac{a_j - y_m}{d_j}, u_j \right), 0 \right\}.$$

3 *Update.* If $C=c$, terminate with $\lambda^* \leftarrow y_m$. If $C>c$, set $l \leftarrow m$, $L \leftarrow C$, and go to step 1. If $C<c$, set $r \leftarrow m, R \leftarrow C$, and go to step 1.

4 *Interpolate.* Terminate with

$$\lambda^* \leftarrow y_l + \frac{(y_r - y_l)(c-L)}{(R-L)}.$$

NOTES AND REFERENCES

The basic idea of this algorithm may be traced to the paper by Held, Wolfe, and Crowder [97]. These results can also be found in [101]. Related results have been presented by Bitran and Hax [24] and by McCallum [142].

APPENDIX E

A Product Form Representation of the Inverse of a Multicommodity Cycle Matrix

In this appendix we present the formulae required for maintaining the inverse of the cycle matrix associated with a multicommodity network flow problem. Recall that every basis for MP (see Chapter 6) can be placed in the form

$$
\mathbf{B}^* = \overbrace{}^{\text{key columns}}\overbrace{}^{\text{nonkey columns}}
\left[\begin{array}{c:c:c:c:c:c:c}
\mathbf{B}^1 & & & \mathbf{R}^1 & & & \\ \hdashline
 & \ddots & & & \ddots & & \\ \hdashline
 & & \mathbf{B}^K & & & \mathbf{R}^K & \\ \hdashline
\mathbf{P}^1 & \cdots & \mathbf{P}^K & \mathbf{T}^1 & \cdots & \mathbf{T}^K & \\ \hdashline
\mathbf{S}^1 & \cdots & \mathbf{S}^K & \mathbf{U}^1 & \cdots & \mathbf{U}^K & \mathbf{I}
\end{array}\right]
\begin{array}{l}
\} \bar{I} \\ \vdots \\ \} \bar{I}, \\ \} q \\ \} p
\end{array}
$$

$$
\underbrace{}_{\bar{I}} \quad \underbrace{}_{\bar{I}} \quad \underbrace{}_{n_1} \quad \underbrace{}_{n_K} \quad \underbrace{}_{n_0}
$$

where the first $K\bar{I}$ columns, which will be called key columns, are chosen so that $\mathrm{Det}(\mathbf{B}^k)\neq 0$ for $k=1,\ldots,K$. Consider the following nonsingular matrix:

$$\mathbf{L}=\left[\begin{array}{c:c:c:c:c:c:c}
\mathbf{I} & & & (-\mathbf{B}^1)^{-1}\mathbf{R}^1 & & & \\ \hdashline
 & \ddots & & & \ddots & & \\ \hdashline
 & & \mathbf{I} & & & (-\mathbf{B}^K)^{-1}\mathbf{R}^K & \\ \hdashline
 & & & \mathbf{I} & & & \\ \hdashline
 & & & & \ddots & & \\ \hdashline
 & & & & & \mathbf{I} & \\ \hdashline
 & & & & & & \mathbf{I}
\end{array}\right].$$

Postmultiplying $\mathbf{B}^*$ by $\mathbf{L}$ yields

$$\mathbf{B}^*\mathbf{L}=\left[\begin{array}{c:c:c:c:c:c:c}
\mathbf{B}^1 & & & & & & \\ \hdashline
 & \ddots & & & & & \\ \hdashline
 & & \mathbf{B}^K & & & & \\ \hdashline
\mathbf{P}^1 & \cdots & \mathbf{P}^K & \mathbf{T}^1-\mathbf{P}^1(\mathbf{B}^1)^{-1}\mathbf{R}^1 & \cdots & \mathbf{T}^K-\mathbf{P}^K(\mathbf{B}^K)^{-1}\mathbf{R}^K & \\ \hdashline
\mathbf{S}^1 & \cdots & \mathbf{S}^K & \mathbf{U}^1-\mathbf{S}^1(\mathbf{B}^1)^{-1}\mathbf{R}^1 & \cdots & \mathbf{U}^K-\mathbf{S}^K(\mathbf{B}^K)^{-1}\mathbf{R}^K & \mathbf{I}
\end{array}\right].$$

$$\text{Let } \mathbf{Y}=\left[\begin{array}{c:c:c:c}
\mathbf{T}^1-\mathbf{P}^1(\mathbf{B}^1)^{-1}\mathbf{R}^1 & \cdots & \mathbf{T}^K-\mathbf{P}^K(\mathbf{B}^K)^{-1}\mathbf{R}^K & \\ \hdashline
\mathbf{U}^1-\mathbf{S}^1(\mathbf{B}^1)^{-1}\mathbf{R}^1 & \cdots & \mathbf{U}^K-\mathbf{S}^K(\mathbf{B}^K)^{-1}\mathbf{R}^K & \mathbf{I}
\end{array}\right],$$

and we call $\mathbf{Y}$ the *working basis*.

Let $\mathbf{B}_i^*$, $\mathbf{L}_i$, and $\mathbf{Y}_i$ denote the basis, transformation matrix, and working basis at iteration i. Then we wish to determine an expression for $\mathbf{Y}_{i+1}^{-1}$ in terms of $\mathbf{Y}_i^{-1}$. Let

$$\mathcal{B}_i^*=\mathbf{B}_i^*\mathbf{L}_i. \tag{E.1}$$

Then

$$(\mathbf{B}_i^*)^{-1}=\mathbf{L}_i(\mathcal{B}_i^*)^{-1}. \tag{E.2}$$

Likewise,

$$\mathcal{B}_{i+1}^*=\mathbf{B}_{i+1}^*\mathbf{L}_{i+1}, \text{ and}$$

$$(\mathcal{B}_{i+1}^*)^{-1}=\mathbf{L}_{i+1}^{-1}(\mathbf{B}_{i+1}^*)^{-1}. \tag{E.3}$$

To maintain the proper partitioning, it may be necessary, before the pivoting operation is performed, to interchange a nonkey column with a key column that is to leave the basis. In this case we require two working basis changes to accomplish the update. Even though $\mathbf{B}_i^*$ and $\mathbf{B}_{i+1}^*$ differ only in that two columns have been interchanged, the working bases $\mathbf{Y}_i$ and $\mathbf{Y}_{i+1}$ will, in general, be completely different. Therefore,

$$(\mathbf{B}_{i+1}^*)^{-1} = \mathbf{E}(\mathbf{B}_i^*)^{-1}, \tag{E.4}$$

where $\mathbf{E}$ is either an *elementary column matrix* (i.e., a matrix that differs from the identity matrix in only one column) or a permutation matrix. Substituting (E.4) into (E.3) yields

$$(\mathcal{B}_{i+1}^*)^{-1} = \mathbf{L}_{i+1}^{-1}\mathbf{E}(\mathbf{B}_i^*)^{-1}. \tag{E.5}$$

Substituting (E.2) into (E.5) yields

$$(\mathcal{B}_{i+1}^*)^{-1} = \mathbf{L}_{i+1}^{-1}\mathbf{E}\mathbf{L}_i(\mathcal{B}_i^*)^{-1}. \tag{E.6}$$

Consider the following partitionings:

$$(\mathcal{B}_{i+1}^*)^{-1} = \underset{\quad\; K\bar{I} \qquad\quad \bar{J}}{\left[\begin{array}{c:c} \bar{\mathbf{B}}_{i+1} & \\ \hdashline \bar{\mathbf{A}}_{i+1} & \mathbf{Y}_{i+1}^{-1} \end{array}\right]} \begin{array}{l} \}K\bar{I} \\ \}\bar{J} \end{array}, \quad (\mathcal{B}_i^*)^{-1} = \underset{\quad K\bar{I} \quad \bar{J}}{\left[\begin{array}{c:c} \bar{\mathbf{B}}_i & \\ \hdashline \bar{\mathbf{A}}_i & \mathbf{Y}_i^{-1} \end{array}\right]} \begin{array}{l} \}K\bar{I} \\ \}\bar{J} \end{array},$$

$$\mathbf{L}_{i+1} = \underset{\quad K\bar{I} \quad\;\; \bar{J}}{\left[\begin{array}{c:c} \mathbf{I} & \mathbf{V}_{i+1} \\ \hdashline & \mathbf{I} \end{array}\right]} \begin{array}{l} \}K\bar{I} \\ \}\bar{J} \end{array}, \quad \mathbf{L}_i = \underset{\quad K\bar{I} \quad \bar{J}}{\left[\begin{array}{c:c} \mathbf{I} & \mathbf{V}_i \\ \hdashline & \mathbf{I} \end{array}\right]} \begin{array}{l} \}K\bar{I} \\ \}\bar{J} \end{array},$$

and

$$\mathbf{E} = \underset{\quad K\bar{I} \quad \bar{J}}{\left[\begin{array}{c:c} \mathbf{E}_1 & \mathbf{E}_2 \\ \hdashline \mathbf{E}_3 & \mathbf{E}_4 \end{array}\right]} \begin{array}{l} \}K\bar{I} \\ \}\bar{J} \end{array}. \tag{E.7}$$

Then we note that

$$\mathbf{L}_{i+1}^{-1} = \left[\begin{array}{c:c} \mathbf{I} & -\mathbf{V}_{i+1} \\ \hdashline & \mathbf{I} \end{array}\right].$$

From (E.6) and straightforward matrix multiplication using the partitioned matrices given above, we obtain

$$\mathbf{Y}_{i+1}^{-1}=(\mathbf{E}_4+\mathbf{E}_3\mathbf{V}_i)\mathbf{Y}_i^{-1}. \tag{E.8}$$

Recall that the working basis itself has special structure and may be placed in the following form:

$$\mathbf{Y}=\underbrace{\left[\begin{array}{c|c}\mathbf{Q} & \\ \hline \mathbf{D} & \mathbf{I}\end{array}\right]}_{q \quad \bar{J}-q}\begin{array}{l}\}q\\ \}\bar{J}-q\end{array}$$

with

$$\mathbf{Y}^{-1}=\left[\begin{array}{c|c}\mathbf{Q}^{-1} & \\ \hline -\mathbf{D}\mathbf{Q}^{-1} & \mathbf{I}\end{array}\right]\begin{array}{l}\}q\\ \}\bar{J}-q\end{array}. \tag{E.9}$$

We shall say that a row of (E.8) is *active* if the corresponding row is in the partition of **Y** corresponding to **Q** and is *inactive* otherwise.

In determining the updating formulae, we must examine major cases with subcases.

Case 1 The leaving column is nonkey. For this case **E** from (E.7) takes the following form:

$$\left[\begin{array}{c|c}\mathbf{E}_1 & \mathbf{E}_2\\ \hline \mathbf{E}_3 & \mathbf{E}_4\end{array}\right]=\left[\begin{array}{c|c}\mathbf{I} & \mathbf{E}_2\\ \hline & \mathbf{E}_4\end{array}\right].$$

Since $\mathbf{E}_3$ is a zero matrix, (E.8) simplifies to

$$\mathbf{Y}_{i+1}^{-1}=\mathbf{E}_4\mathbf{Y}_i^{-1}, \tag{E.10}$$

where $\mathbf{E}_4$ is an elementary column matrix. Consider the following four subcases.

Subcase 1a Both the entering and leaving columns are structural columns (i.e., an x_j^k as opposed to a slack column). We partition $\mathbf{E}_4$ to be compatible with $\mathbf{Y}_i^{-1}$ as shown in (E.9), that is,

$$\mathbf{E}_4 = \underbrace{}_{q}\underbrace{}_{\bar{J}-q}\left[\begin{array}{c:c} \mathbf{E}_5 & \\ \hdashline \mathbf{E}_6 & \mathbf{I} \end{array}\right]\begin{array}{l} \}q \\ \}\bar{J}-q \end{array}. \tag{E.11}$$

Then, using (E.9) and (E.11) in (E.10), we have

$$\mathbf{Y}_{i+1}^{-1} = \left[\begin{array}{c:c} \mathbf{Q}_{i+1}^{-1} & \\ \hdashline -\mathbf{D}_{i+1}\mathbf{Q}_{i+1}^{-1} & \mathbf{I} \end{array}\right] = \left[\begin{array}{c:c} \mathbf{E}_5 & \\ \hdashline \mathbf{E}_6 & \mathbf{I} \end{array}\right]\left[\begin{array}{c:c} \mathbf{Q}_i^{-1} & \\ \hdashline -\mathbf{D}_i\mathbf{Q}_i^{-1} & \mathbf{I} \end{array}\right].$$

Hence

$$\mathbf{Q}_{i+1}^{-1} = \mathbf{E}_5\mathbf{Q}_i^{-1}, \tag{E.12}$$

where $\mathbf{E}_5$ is an elementary column matrix.

Subcase 1b The entering column is a structural column, and the leaving column is a slack column. For this case we first make the row corresponding to the leaving variable active and then use (E.10). Suppose the leaving variable is s_n. Then, after row and column interchanges in the last $\bar{J}-q$ rows and columns, Y_i takes the form

$$\mathbf{Y}_i = \left[\begin{array}{c:c:c} \mathbf{Q}_i & & \\ \hdashline \mathbf{b} & 1 & \\ \hdashline \mathbf{D}_i^* & & \mathbf{I} \end{array}\right]\begin{array}{l} \}q \\ \}1 \\ \}\bar{J}-q-1 \end{array}$$
$$\quad\;\; q \quad 1 \quad \bar{J}-q-1$$

$$= \left[\begin{array}{c:c} \mathbf{Q}_i^* & \\ \hdashline \mathfrak{D}_i^* & \mathbf{I} \end{array}\right]\begin{array}{l} \}q+1 \\ \}\bar{J}-q-1 \end{array} \tag{E.13}$$
$$\quad q+1 \quad \bar{J}-q-1$$

where the $(q+1)$st row corresponds to arc n.

Then, from (E.13),

$$\mathbf{Y}_i^{-1}=\left[\begin{array}{c|c}(\mathbf{Q}_i^*)^{-1} & \\ \hline -\mathfrak{D}_i^*(\mathbf{Q}_i^*)^{-1} & \mathbf{I}\end{array}\right]. \tag{E.14}$$

Note that

$$(\mathbf{Q}_i^*)^{-1}=\left[\begin{array}{c|c}\mathbf{Q}_i^{-1} & \\ \hline -\mathbf{b}\mathbf{Q}_i^{-1} & \mathbf{I}\end{array}\right]. \tag{E.15}$$

To make a currently inactive row active, we perform two operations. First, we append a row and column to $\mathbf{Q}_i^{-1}$, which has a one in the diagonal element and zeros elsewhere. Second, we premultiply by an *elementary row matrix* (i.e., a matrix that differs from an identity matrix in only one row). Then

$$(\mathbf{Q}_i^*)^{-1}=\Bigg[\underbrace{\begin{array}{c}\mathbf{I}\\ \hline -\mathbf{b}\end{array}}_{q}\Bigg|\underbrace{\begin{array}{c} \\ \hline 1\end{array}}_{1}\Bigg]\left[\begin{array}{c|c}\mathbf{Q}_i^{-1} & \\ \hline & 1\end{array}\right]\begin{array}{l}\}q\\ \}1\end{array}, \tag{E.16}$$

which conforms with (E.15).

Partitioning $\mathbf{E}_4$ to be compatible with $\mathbf{Y}_i$ as shown in (E.13), we obtain

$$\mathbf{Q}_{i+1}^{-1}=\mathbf{E}_5\left[\begin{array}{c|c}\mathbf{I} & \\ \hline -\mathbf{b} & 1\end{array}\right]\left[\begin{array}{c|c}\mathbf{Q}_i^{-1} & \\ \hline & 1\end{array}\right]. \tag{E.17}$$

Subcase 1ç The entering column is a slack column, and the leaving column is a structural column. For this case we first execute the pivot operation and then make inactive the row corresponding to the entering slack variable. If $\mathbf{E}_4$ is partitioned to be compatible with $\mathbf{Y}_i$, then, from (E.10),

$$\mathbf{Y}_{i+1}^{-1}=\left[\begin{array}{c|c}\mathbf{E}_5 & \\ \hline \mathbf{E}_6 & \mathbf{I}\end{array}\right]\left[\begin{array}{c|c}\mathbf{Q}_i^{-1} & \\ \hline -\mathbf{D}_i\mathbf{Q}_i^{-1} & \mathbf{I}\end{array}\right]\begin{array}{l}\}q\\ \}\bar{J}-q\end{array}$$

$$=\left[\begin{array}{c|c}(\mathbf{Q}_{i+1}^*)^{-1} & \\ \hline -\mathbf{D}_{i+1}^*(\mathbf{Q}_{i+1}^*)^{-1} & \mathbf{I}\end{array}\right]\begin{array}{l}\}q\\ \}\bar{J}-q\end{array}. \tag{E.18}$$

From (E.18)

$$(\mathbf{Q}_{i+1}^{*})^{-1}=\mathbf{E}_5\mathbf{Q}_i^{-1}. \tag{E.19}$$

Let s_m denote the entering variable, and suppose the active row corresponding to arc m appears in the rth column of $(\mathbf{Q}_{i+1}^{*})^{-1}$. Furthermore, suppose the leaving column appears in the lth column of $\mathbf{Q}_i$ and hence is associated with the lth row of $(\mathbf{Q}_{i+1}^{*})^{-1}$. Let $(\mathcal{Q}_{i+1}^{*})^{-1}$ denote the matrix formed from $(\mathbf{Q}_{i+1}^{*})^{-1}$ by interchanging rows q and l and columns q and r. Then

$$(\mathcal{Q}^{*}{}_{i+1})^{-1}=\left[\begin{array}{c:c} \underbrace{\begin{array}{c}\mathbf{Q}_{i+1}^{-1} \\ \hdashline -\mathbf{b}\mathbf{Q}_{i+1}^{-1}\end{array}}_{q-1} & \underbrace{\begin{array}{c} \\ \hdashline 1\end{array}}_{1} \end{array}\right]\begin{array}{l}\}q-1\\ \}1\end{array}. \tag{E.20}$$

Define the matrix $\mathbf{D}_q$ as follows:

$$
\begin{array}{rl}
 & \qquad\qquad\qquad\quad p\text{th column} \\
p\text{th row}\rightarrow & \underbrace{\left[\begin{array}{c|c|c|c|c|c|c}
1 & & & & & & \\ \hline
 & \ddots & & & & & \\ \hline
 & & 1 & & & & \\ \hline
 & & & & 1 & & \\ \hline
 & & & & & \ddots & \\ \hline
 & & & & & & 1 \\
 & & & & & &
\end{array}\right]}_{q}\left.\vphantom{\begin{array}{c}1\\1\\1\\1\\1\\1\\1\end{array}}\right\}q-1.
\end{array}
$$

Here $\mathbf{D}_p$ is known as a *row cancel matrix*, while $\mathbf{T}_p=\mathbf{D}_p'$ (the transpose of $\mathbf{D}_p$) is called a *column cancel matrix*. Then from (E.20)

$$\mathbf{Q}_{i+1}^{-1}=\mathbf{D}_q(\mathcal{Q}_{i+1}^{*})^{-1}\mathbf{T}_q.$$

But $(\mathcal{Q}_{i+1}^{*})^{-1}$ was obtained from $(\mathbf{Q}_{i+1}^{*})^{-1}$ by interchanging rows q and l and columns q and r. Therefore

$$\mathbf{Q}_{i+1}^{-1}=\mathbf{D}_l(\mathbf{Q}_{i+1}^{*})^{-1}\mathbf{T}_r. \tag{E.21}$$

Substituting (E.19) into (E.21) yields

$$\mathbf{Q}_{i+1}^{-1} = \mathbf{D}_l \mathbf{E}_5 \mathbf{Q}_i^{-1} \mathbf{T}_r. \tag{E.22}$$

Subcase 1d Both the entering and leaving columns are slack columns. First, for this case, we make the row corresponding to the leaving variable active; second, we perform a pivot; and, finally, we make inactive the row corresponding to the entering slack variable. Suppose the leaving variable is s_n. Then, after row and column interchanges in the last $\bar{J}-q$ rows and columns, $\mathbf{Y}_i$ takes the form

$$\underbrace{\left[\begin{array}{c|c|c} \mathbf{Q}_i & & \\ \hline \mathbf{b} & 1 & \\ \hline \mathbf{D}_i^* & & \mathbf{I} \end{array}\right]}_{q \quad 1 \quad \bar{J}-q-1} \begin{array}{l} \}q \\ \}1 \\ \}\bar{J}-q-1 \end{array} = \left[\begin{array}{c|c} \mathbf{Q}_i^* & \\ \hline \mathcal{D}_i^* & \mathbf{I} \end{array}\right] \begin{array}{l} \}q+1 \\ \}\bar{J}-q-1 \end{array}, \quad \begin{array}{cc} q+1 & \bar{J}-q-1 \end{array} \tag{E.23}$$

where the $(q+1)$st row corresponds to arc n. Then, as in Subcase 1b, we obtain

$$(\mathbf{Q}_{i+1}^*)^{-1} = \mathbf{E}_5 (\mathbf{Q}_i^*)^{-1}, \tag{E.24}$$

where

$$\mathbf{Y}_{i+1}^{-1} = \left[\begin{array}{c|c} \mathbf{Q}_{i+1}^* & \\ \hline \mathbf{D}_{i+1} & \mathbf{I} \end{array}\right]^{-1} = \mathbf{E}_4 \mathbf{Y}_i^{-1}.$$

Let s_m denote the entering variable, and suppose that the active row corresponding to arc m appears in the rth row of $\mathbf{Q}_i$. Then let $(\mathcal{Q}_{i+1}^*)^{-1}$ denote the matrix formed from $(\mathbf{Q}_{i+1}^*)^{-1}$ by interchanging columns $q+1$ and r. Then

$$(\mathcal{Q}_{i+1}^*)^{-1} = \left[\begin{array}{c|c} \mathbf{Q}_{i+1} & \\ \hline \mathbf{d} & 1 \end{array}\right]^{-1} \begin{array}{l} \}q \\ \}1 \end{array},$$

where the $(q+1)$st row corresponds to arc m. Then, as shown in Subcase 1c,

$$\mathbf{Q}_{i+1}^{-1} = \mathbf{D}_{q+1} (\mathcal{Q}_{i+1}^*)^{-1} \mathbf{T}_{q+1}.$$

But $(\mathcal{Q}_{i+1}^*)^{-1}$ was obtained from $(\mathbf{Q}_{i+1}^*)^{-1}$ by interchanging columns $q+1$ and r. Therefore

$$\mathbf{Q}_{i+1}^{-1}=\mathbf{D}_{q+1}(\mathbf{Q}_{i+1}^{*})^{-1}\mathbf{T}_r. \qquad \text{(E.25)}$$

Hence from (E.16), (E.24), and (E.25) we obtain

$$\mathbf{Q}_{i+1}^{-1}=\mathbf{D}_{p+1}\mathbf{E}_5\left[\begin{array}{c|c}\mathbf{I} & \\ \hline -\mathbf{b} & 1\end{array}\right]\left[\begin{array}{c|c}\mathbf{Q}_i^{-1} & \\ \hline & 1\end{array}\right]\mathbf{T}_r. \qquad \text{(E.26)}$$

Case 2 The leaving column is a key column. Then the corresponding arc is in one of the spanning trees, say $\mathcal{T}^{k^*}$. If this arc is removed from the basis, the structure of $\mathcal{T}^{k^*}$ may be destroyed. Therefore we attempt to switch the leaving column with some nonkey column, that is, from the partitioned basis $\mathbf{B}^*$ we wish to interchange a column from $\mathbf{R}^{k^*}$ with a certain column from $\mathbf{B}^{k^*}$ (i.e., the leaving column). Suppose the leaving column is in the lth position of $\mathbf{B}^{k^*}$. Consider some column from $\mathbf{R}^{k^*}$, say r. Then r can replace the lth column of $\mathbf{B}^{k^*}$, and the resulting matrix will be nonsingular if and only if $y_l \neq 0$, where $\mathbf{y}=(\mathbf{B}^{k^*})^{-1}\mathbf{R}^{k^*}$. Let $\boldsymbol{\beta}_l$ denote the lth row of $(\mathbf{B}^{k^*})^{-1}$. Then $y_l=\boldsymbol{\beta}_l\mathbf{R}^{k^*}(l)$.

Define $\boldsymbol{\gamma}=\boldsymbol{\beta}_l\mathbf{R}^{k^*}$. We now distinguish two subcases.

Subcase 2a $\boldsymbol{\gamma}\neq\mathbf{0}$. Suppose $\gamma_j\neq 0$. Then the jth column of $\mathbf{R}^{k^*}$ can be interchanged with the lth column of $\mathbf{B}^{k^*}$, and the partitioning of $\mathbf{B}^*$ can be preserved. After this interchange Subcase 1a can be followed to effect the basis change. Consider the following permutation matrix:

$$\mathbf{E}=\left[\begin{array}{ccc|c|ccc|c|ccc}
1 & & & & & & & & & & \\
 & \ddots & & & & & & & & & \\
 & & 1 & & & & & & & & \\
 & & & & & & & 1 & & & \\
 & & & & 1 & & & & & & \\
 & & & & & \ddots & & & & & \\
 & & & & & & 1 & & & & \\
 & & & 1 & & & & & & & \\
 & & & & & & & & 1 & & \\
 & & & & & & & & & \ddots & \\
 & & & & & & & & & & 1
\end{array}\right]\begin{array}{l} \\ \\ \\ \leftarrow \text{row}(k^*-1)\bar{I}+l \\ \\ \\ \\ \leftarrow \text{row}\,\bar{I}K+n^*+j \\ \\ \\ \\ \end{array},$$

$\uparrow$ column $(k^*-1)\bar{I}+l$ $\qquad$ $\uparrow$ column $\bar{I}K+n^*+j$

where

$$n^* = \begin{cases} 0 & \text{if } k^* = 1, \\ \displaystyle\sum_{k=1}^{k=k^*-1} v_k & \text{otherwise.} \end{cases}$$

Then $\mathbf{B}_{i+1}^*$ (the matrix formed from $\mathbf{B}_i^*$ by interchanging column $(k^*-1)\bar{I}+l$ with column $\bar{I}K+n^*+j$) is given by $\mathbf{B}_i^*\mathbf{E}$. Then $(\mathbf{B}_{i+1}^*)^{-1} = \mathbf{E}^{-1}(\mathbf{B}_i^*)^{-1}$. But $\mathbf{E}=\mathbf{E}^{-1}$. Therefore

$$(\mathbf{B}_{i+1}^*)^{-1} = \mathbf{E}(\mathbf{B}_i^*)^{-1}, \tag{E.27}$$

as given in (E.4). Let $\mathbf{E}$ of (E.27) be partitioned as shown in (E.7). Then

$$\mathbf{E}_3 = \begin{array}{|c|c|c|} \hline & & \\ \hline & 1 & \\ \hline & & \\ \hline \end{array} \leftarrow \text{row } n^*+j$$

$$\uparrow$$

$$\text{column}(k^*-1)\bar{I}+l$$

and

$$\mathbf{E}_4 = \begin{array}{|c|c|c|} \hline \mathbf{I} & & \\ \hline & 0 & \\ \hline & & \mathbf{I} \\ \hline \end{array} \leftarrow \text{row } n^*+j$$

$$\uparrow$$

$$\text{column}\, n^*+j$$

Since $\mathbf{V}_i$ is defined as

$$\mathbf{V}_i = \left[\begin{array}{c:c:c:c} (-\mathbf{B}^1)^{-1}\mathbf{R}^1 & & & \\ \hdashline & \ddots & & \\ \hdashline & & (-\mathbf{B}^K)^{-1}\mathbf{R}^K & \\ \end{array}\right] \begin{array}{l} \}\bar{I} \\ \vdots \\ \}\bar{I} \end{array},$$

$$\underbrace{\qquad}_{v_1} \quad \cdots \quad \underbrace{\qquad}_{v_K} \quad \underbrace{\quad}_{v_0}$$

$\mathbf{E}_4+\mathbf{E}_3V_i$ from (E.8) is

$$\mathbf{E}_5=\left[\begin{array}{c|c|c|c|c} \mathbf{I} & & & & \\ \hline & \mathbf{I} & & & \\ \hline & & -\gamma & & \\ \hline & & & \mathbf{I} & \\ \hline & & & & \mathbf{I} \end{array}\right] \leftarrow \text{row } j \text{ of } k^*\text{th block.}$$

Then

$$\mathbf{Y}_{i+1}^{-1}=\mathbf{E}_5\mathbf{Y}_i^{-1}. \qquad \text{(E.28)}$$

Partitioning $\mathbf{E}_5$ to be compatible with (E.9), we get

$$\mathbf{E}_5=\left[\begin{array}{c|c} \mathbf{E}_6 & \\ \hline & \mathbf{I} \end{array}\right]\begin{array}{l}\}q \\ \}\bar{J}-q\end{array}, \qquad \text{(E.29)}$$

where $\mathbf{E}_6$ is an elementary row matrix. Therefore using (E.9) and (E.29) in (E.28) yields

$$\left[\begin{array}{c|c} \mathbf{Q}_{i+1}^{-1} & \\ \hline -\mathbf{D}_i\mathbf{Q}_{i+1}^{-1} & \mathbf{I} \end{array}\right]=\left[\begin{array}{c|c} \mathbf{E}_6 & \\ \hline & \mathbf{I} \end{array}\right]\left[\begin{array}{c|c} \mathbf{Q}_i^{-1} & \\ \hline -\mathbf{D}_i\mathbf{Q}_i^{-1} & \mathbf{I} \end{array}\right] \qquad \text{(E.30)}$$

which implies

$$\mathbf{Q}_{i+1}^{-1}=\mathbf{E}_6\mathbf{Q}_i^{-1}. \qquad \text{(E.31)}$$

Subcase 2b $\gamma=\mathbf{0}$. Then the entering and leaving columns must both be associated with commodity k^*, and a direct pivot is possible. Otherwise the partitioning of $\mathbf{B}^*$ would be impossible. Hence there is no change in the cycle matrix.

APPENDIX **F**

NETFLO

NETFLO is a FORTRAN code that solves minimal cost network flow problems with lower bounds. Mathematically, these problems take the form

$$\begin{aligned} \min \quad & \mathbf{c}\bar{\mathbf{x}} \\ \text{s.t.} \quad & \mathbf{A}\bar{\mathbf{x}} = \bar{\mathbf{r}} \\ & \bar{\mathbf{v}} \leqslant \bar{\mathbf{x}} \leqslant \bar{\mathbf{u}}, \end{aligned}$$

where $\mathbf{A}$ is an $\bar{I} \times \bar{J}$ node-arc incidence matrix, $\mathbf{c}$ is a $1 \times \bar{J}$ vector of unit costs, $\bar{\mathbf{r}}$ is an $\bar{I} \times 1$ vector of requirements, $\bar{\mathbf{v}}$ is a $\bar{J} \times 1$ vector of lower bounds, and $\bar{\mathbf{u}}$ is a $\bar{J} \times 1$ vector of upper bounds. The vector $\bar{\mathbf{r}}$ is called the requirements vector. If, for node i, $\bar{\mathbf{r}}_i > 0$, then node i is a supply point and has supply equal to $\mathbf{r}_i$. If, for node i, $\bar{\mathbf{r}}_i < 0$, then node i is a demand point and the demand is $|\bar{\mathbf{r}}_i|$. Transshipment points have $\bar{\mathbf{r}}_i = 0$.

F.1 INITIAL SOLUTION

We first transform the problem so that the lower bounds are zero. Let $\bar{\mathbf{x}} = \mathbf{x} + \bar{\mathbf{v}}$. Substituting for $\bar{\mathbf{x}}$, we obtain

$$\min \quad \mathbf{c}\mathbf{x} + \alpha \tag{F.1}$$

$$\text{s.t.} \quad \mathbf{A}\mathbf{x} = \mathbf{r} \tag{F.2}$$

$$\mathbf{0} \leqslant \mathbf{x} \leqslant \mathbf{u}, \tag{F.3}$$

where $\alpha = \mathbf{c}\bar{\mathbf{v}}$, $\mathbf{r} = \bar{\mathbf{r}} - \mathbf{A}\mathbf{v}$, and $\mathbf{u} = \bar{\mathbf{u}} - \bar{\mathbf{v}}$. The network associated with (F.1)–(F.3) will be denoted by $[\mathcal{N}, \mathcal{A}]$.

In this appendix we introduce the notation $x_{p,q}$, $c_{p,q}$, and $u_{p,q}$ to denote the flow, unit cost, and capacity, respectively, for arc (p,q). Also, for any node $q \varepsilon \mathcal{N}$, let $B_q = \{(p,q):(p,q) \varepsilon \mathcal{A}\}$; B_q is frequently called the "before set" and is simply the set of all arcs whose "to" node is q.

Having transformed the problem, we apply a heuristic procedure to obtain the initial basic feasible solution. The main idea of this procedure is to quickly find low-cost paths through the network that will transport a large amount of commodity to the demand nodes.

We first enlarge the network. For each node $q \varepsilon \mathcal{N}$ with an induced supply, we add an artificial arc (q,a) with $c_{q,a} = +\infty$. These arcs are all made part of the spanning tree, and each is assigned an initial flow equal to the induced supply at the node that is connected to the root, that is, $x_{q,a} = r_q$. The flow $x_{q,a}$ will be decreased if a set of arcs is found that allows for the distribution of r_q to one or more demand nodes.

We then form a list L of the nodes with induced demand, ordered by magnitude of demand, with the node having the largest demand appearing first on the list. For each node q in L, we define a quantity Q_q, which we call the unsatisfied demand. For the demand nodes we initially put in L, we let $Q_q = -r_q$. We attempt to build backward chains (directed paths) beginning at each demand node and terminating at some supply node. Each chain consists initially of a single node and may be extended by the addition of new nodes and connecting arcs. The node most recently added to the chain will be referred to as the lowest node on the chain. Chains are extended only at the lowest node. Eventually each chain is connected to the spanning tree either by an artificial arc from the root to the lowest node in the chain or by an arc $(p,q) \varepsilon \mathcal{A}$, where p is an induced supply node and q is the lowest node in the chain.

The procedure consists of two phases. In phase 1, part of a spanning tree is formed so as to satisfy the induced demand via chains from sources to demands and via artificial arcs. In phase 2, arcs are added with a flow of zero, so as to complete the spanning tree. After phases 1 and 2 have been completed, the spanning tree contains all nodes in the network and all basic and nonbasic variables have feasible values.

ALG F.1 FIND AN INITIAL FEASIBLE BASIS

(Phase 1)

1 *Select First Node in Demand List*

a. [List empty?] If L is empty, go to Step 5; otherwise let q be the first node in L.

b. [Demand at q satisfied?] If $Q_q = 0$, go to Step 4.

2 *Find Supply Nodes to Satisfy Demand*

a. [Find arcs that may be set to upper bound] Let $D_q^1 = \{(p,q):(p,q)\varepsilon B_q, r_p > 0, u_{p,q} \leqslant \min[Q_q, x_{p,a}]\}$.

b. [Find arcs that may become basic] Let $D_q^2 = \{(p,q):(p,q)\varepsilon B_q, r_q > 0, Q_q < u_{p,q}, Q_q \leqslant x_{p,a}\}$.

c. [Select candidate with least cost] If $D_q^1 \cup D_q^2 = \phi$, go to Step 3; otherwise find $c_{p^*,q} = \min\{c_{p,q}:(p,q)\varepsilon D_q^1 \cup D_q^2\}$.

d. [Either set (p^*,q) to the upper bound or make it basic] If $(p^*,q)\varepsilon D_q^1, x_{p^*,q} \leftarrow u_{p^*,q}, Q_q \leftarrow Q_q - u_{p^*,q}, x_{p^*,a} \leftarrow x_{p^*,a} - u_{p^*,q}$, and go to Step 1; otherwise $x_{p^*,q} \leftarrow S_q, x_{p^*,a} \leftarrow x_{p^*,a} - Q_q, Q_q \leftarrow 0$, connect the chain with q as lowest node to the tree via (p^*,q), remove q from L, and go to Step 1.

3 *Find Transshipment Nodes to Transfer Demand*

a. [Determine candidates] Let $D_q^3 = \{(p,q):(p,q)\varepsilon B_q, r_p = 0$, and p not part of a chain$\}$.

b. [Select candidate with least cost] If $D_q^3 = \phi$, go to Step 4; otherwise find $c_{p^*,q} = \min\{c_{p,q}:(p,q)\varepsilon D_q^3\}$.

c. [Either set (p^*,q) to the upper bound or make it basic] If $u_{p^*,q} < Q_q, x_{p^*,q} \leftarrow u_{p^*,q}$, and $Q_p \leftarrow Q_q - u_{p^*,q}$, place p^* in the first position of L with $Q_{p^*} \leftarrow u_{p^*,q}$, begin a new chain with p^*, and go to Step 1; otherwise $x_{p^*,q} \leftarrow Q_q$, remove q from L, place p^* in the first position of L with $Q_{p^*} \leftarrow Q_q, Q_q \leftarrow 0$, extend the chain with q as lowest node to p^* via (p^*,q), and go to Step 1.

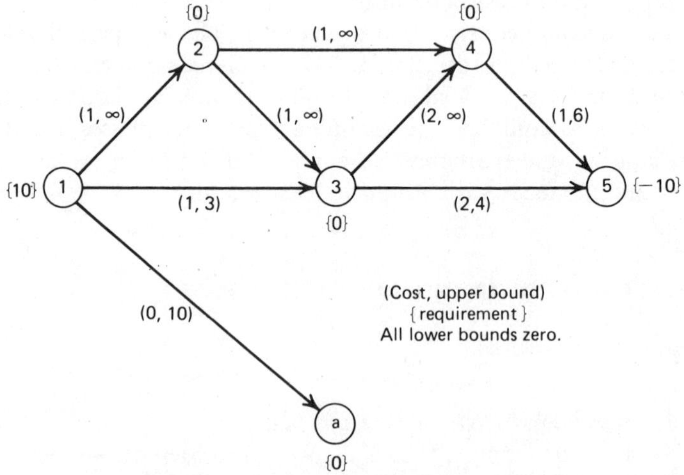

FIGURE F.1 Sample network for start procedure.

4 *Connect q to Tree with an Artificial Arc.* Remove q from L. Create an artificial arc, (a,q), with $c_{a,q}=u_{a,q}=\infty, x_{a,q}\leftarrow Q_q$, connect the chain with q as lowest node to the tree via (a,q), and go to Step 1.

(Phase 2)

5 *Initialize Node Counter.* $q\leftarrow 1$.

6 *Test for Termination.* If all nodes of $\mathfrak{N}$ are connected to the tree, terminate.

7 *Save Starting Node for Current Search.* $q'\leftarrow q$.

8 *Find Low-Cost Connectable Arc to Add to Tree.*

a. [Determine candidates] Let $D_q^4=\{(p,q):(p,q)\varepsilon B_q$ and only one of p and q is not connected to the tree$\}$.

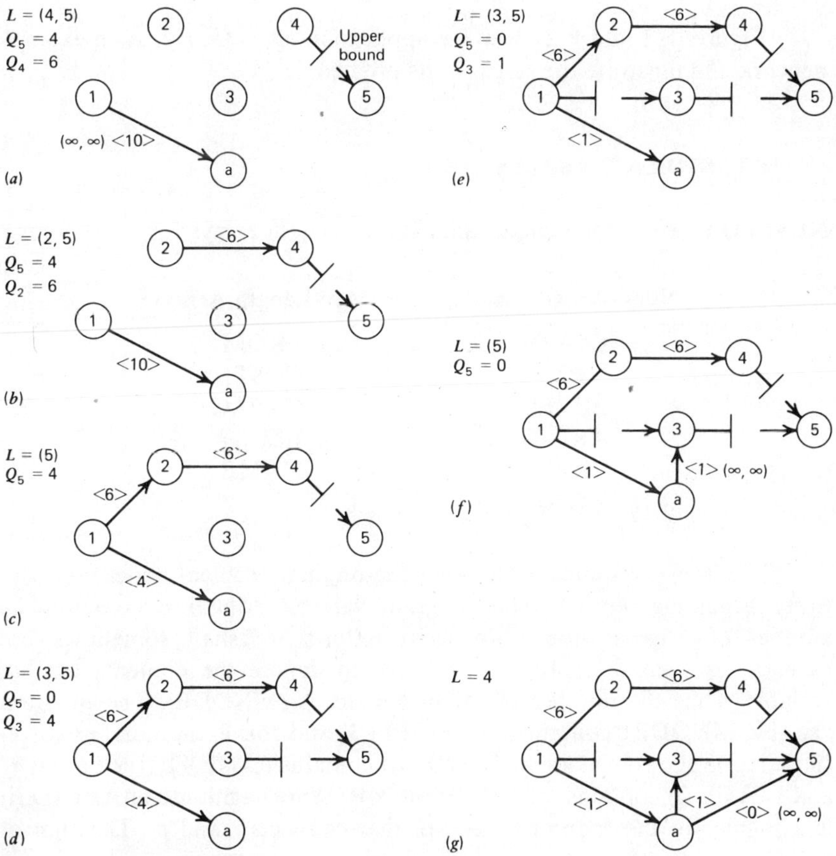

FIGURE F.2 Stages in the progress of the start procedure.

b. [Select candidate with least cost] If $D_q^4 = \phi$, go to Step 9; otherwise find $c_{p^*,q} = \min\{c_{p,q} : (p,q) \varepsilon D_q^r\}$.
c. [Make (p^*,q) basic] $x_{p^*,q} \leftarrow 0$ and connect (p^*,q) to the tree.
d. [Increment node counter] $q \leftarrow q+1$. If $q > |\bar{I}|, q \leftarrow 1$. Go to Step 6.

9 *Increment Node Counter and Test*

a. [Increment counter] $q \leftarrow q+1$. If $q > |\bar{I}|, q \leftarrow 1$.
b. [Have all nodes been examined?] If $q \neq q'$, go to Step 8.

10 *Connect Isolated Nodes*

a. [Determine isolated nodes] Let $I = \{q : q \varepsilon \mathfrak{N}$ and q not connected to the tree$\}$.
b. [Connect to tree with an artificial arc] For each $q \varepsilon I$, create an artificial arc, (a,q), with $c_{a,q} = u_{a,q} = \infty$ and $x_{a,q} \leftarrow 0$ and connect (a,q) to the tree.

Figures F.1 and F.2 show the application of ALG. F.1 to an example network and illustrate the stages in its progress.

F.2 STORAGE ARRAYS

NETFLO uses six node-length and five arc-length arrays for data storage.

Node-Length Arrays	Arc-Length Arrays
DOWN	FROM
NEXT	COST
LEVEL	CAPAC
ARCID	FLOOR
FLOW	NAME
DUAL	

The user is required to number the original problem nodes consecutively beginning with 1. The program variable MNODE contains the number of original nodes. In what follows we shall sometimes use parentheses around a program variable to denote the contents of that variable. Thus the number of original nodes is (MNODE). The program variable MNODP1 contains (MNODE)+1, and this is the number corresponding to the root node a. DOWN (i) contains p_i, NEXT (i) contains t_i, and LEVEL (i) contains d_i (see Appendix B). The magnitude of ARCID (i) is a pointer to the spanning tree arc that connects i and p_i. The sign of

ARCID (i) is minus if the connecting arc is (i,p_i) and is plus if the connecting arc is (p_i,i). FLOW (i) contains the value of $x_{p,q}$, where (p,q) is the spanning tree arc connecting i and p_i. Before node i becomes part of the initial spanning tree, DUAL (i) contains a pointer to the first arc in the node-length arrays that is a member of B_i. After node i has become part of the spanning tree, DUAL (i) contains the value of the dual variable.

The arc-length arrays are indexed by integers beginning with 0. The last arc in the expanded network has the index that is contained in the program variable MARC. The arc with index 0 serves as a general reference for all arcs of the form (a,p) with $c_{a,p}=+\infty$ and $u_{a,p}=+\infty$, that is, if the arc corresponding to ARCID (i) is of the form (a,p), with a the root node, then ARCID (i) will be 0. Following this, the arcs are indexed so that, in order, the corresponding arcs are the arcs of $B_1, B_2, \ldots, B_{(\text{MNODE})}$. The arcs within the set corresponding to B_i appear in the same order as input by the user in data. If any $B_i=\phi$, a dummy arc is present and serves as a placeholder within this structure. The index of the last arc of $B_{(\text{MNODE})}$ is given by the program variable MREG. Following the arc with index (MREG) is the set of all arcs of the form $(0,a)$ with $c_{p,a}=0$ and $u_{p,a}=r_p$, that is, the slack arcs. If there are no slacks, a dummy placeholder arc is present. The last one of these arcs has the index given by the program variable MSLK. Following next is a single arc whose index is (MARC), which serves as a general reference for all excess arcs of the form (p,a) with $c_{p,a}=+\infty$, and $u_{p,a}=+\infty$, that is, if the arc corresponding to ARCID (i) is as stated, ARCID (i) will contain $-$(MARC). For $1 \leqslant i \leqslant$ (MSLK), the magnitude of FROM (i) contains p, where i is the index of (p,q). If q is odd, the sign of FROM (i) is minus; if q is even, it is plus. The slack arcs continue this pattern, that is, if the sign of FROM(MREG) $=-$, and i is the index of a slack arc (p,a), then FROM (i) $=+p$, and vice-versa. If any B_i is null and the index of the dummy placeholder is j, then FROM (j) $=+i$ if i is even and $=-i$ if i is odd. Similarly, if there are no slack arcs, FROM(MSLK) $=+$(MNODP1) if (MNODP1) is even and $=-$(MNODP1) if (MNODP1) is odd. Thus a search of the array FROM involving changes of sign can be used to locate the arcs of any B_i for $1 \leqslant i \leqslant$ MNODE and all slack arcs. Dummy arcs can be recognized since they will be of the form (i,i).

Let i be the index of arc (p,q). COST (i) contains $c_{p,q}$. The magnitude of CAPAC (i) contains $u_{p,q}$. If (p,q) is in upper bound status, the sign of CAPAC (i) will be minus, otherwise plus. FLOOR (i) contains $v_{p,q}$. NAME (i) contains the character name assigned to arc (p,q) by the user through input data and is not, strictly speaking, necessary, being used only for user convenience on output.

F.3 OPTIONAL PROGRAM FEATURES

This code is designed for use in environments ranging from purely experimental to commercial. Some features desirable in a given environment may be unnecessary or may even interfere in another environment. To allow for maximum flexibility in the selection of features, much of the code has been designed so that, with a text editor, only desired features may be selected or enabled. This avoids the penalty of run-time testing. To be suitable for this purpose, a text editor must possess the capability to replace all occurrences of a specified character string by a single blank character. The features available and the string to be replaced by a single character for selected options are given below, together with a brief description.

1 Timing

C(TIMING)

This feature enables the production of timing statistics for evaluation of program performance.

2 Optimization History

C(DUMP OPTIMIZATION HISTORY)

This feature, together with the dump control feature, may be used to determine the optimization path characteristics of a particular problem.

3 Flow History

C(DUMP FLOWS)

This feature, together with the dump control feature, may be used to obtain a complete description of the arcs in upper bound status and the basic arcs and their flows.

4 Cost History

C(DUMP COST)

This feature, together with the dump control feature, may be used to determine the objective function improvement characteristics of a particular problem. An optimal feasible solution will automatically produce this cost figure.

5 Array Dumps

C(DUMP ARRAYS)

This feature, together with the dump control feature, may be used in debugging. Complete details of storage array contents are produced under dump control specifications.

6 Echo Arc Data

C(ECHO ARC DATA)

Since large problems require a large number of data to be input for arc specification, these data are normally input without echoing it back on output. If such an echo is desired, this feature may be enabled.

7 Dual Variables

C(DUAL VARIABLES)

In some problems the dual variables have a physical meaning. If, at optimality, the value of these variables is desired, this feature may be enabled.

8 Arc Flows

C(ARC FLOWS)

In most problems this output is desirable. When enabled, the flows on the individual arcs are output at optimality.

9 Node-to-Node Flows

C(NODE-TO-NODE FLOW)

In some problems a convex cost function may be replaced by a piecewise linear approximation. In such cases multiple arcs appear between nodes. Usually it is the total flow between nodes that is the desired output at optimality. These total flows are calculated and output when this feature is enabled.

10 Degenerate Pivots

C(DEGENERATE PIVOTS)

In some problems the analyst may desire to know how many basis or bound changes result in zero change in flow. Also of interest may be the number of such changes from the starting solution until a change in flow occurs. These statistics are computed and output at optimality when this feature is enabled.

11 Data Unit Control

C(DATA UNIT CONTROL)

In some environments data files may be generated by other programs on tape or disk units, and it may be desirable to input these files on more than one input unit. The units to be used for problem data input are specified by data on input unit 5 when this feature is enabled.

12 Dump Control

C(DUMP CONTROL)

When this feature is enabled, the user may specify on which iterations (basis or bound changes) the dumps for optional features 2–5 will occur and whether or not these dumps will occur at optimality.

F.4 USER DATA

NETFLO has the capability of solving successive problems in one run. The following is a description of the data required to specify each problem. Additional input data will be required if certain optional features are enabled. In what follows, I10 indicates an integer field of 10 characters right-justified and A10 indicates a character field of 10 characters.

F.4.1 Problem Data

Card Group	Composition	Description and Format
1	One card	Number of nodes I10
2	Card set + one blank card	Number of node requirement I10 I10
3	Card set	Number of arcs to node 1, Number of arcs to node 2· · · I10 I10 I10 I10 I10 I10 I10 I10
4	Card set + one blank card	Name, from, to, unit cost, upper bound, lower bound A10 I10 I10 I10 I10 I10

Node requirements need be input only for supply nodes (+) and demand nodes (−). The blank cards at the end of sets 2 and 4 are required to signal end-of-set. If the user does not know the information for card group 3, estimates should be supplied. Overestimates are preferable to underestimates.

F.4.2 Unit Control Data

If the data unit control feature has been enabled, one additional card is read before the first set of problem data. Its description and format are as follows:

Group 1 unit	Group 2 unit	Group 3 unit	Group 4 unit	Change flag
I10	I10	I10	I10	I10

As the problem data are read, each group is read from the unit specified as above. The change flag value may be 0 or 1. If 0, the units specified on this card will remain as problem data units for all following problems, and no more unit control data will be read. If 1, another unit control data card will be read after the following problem is finished and before the succeeding problem is begun. If the units specified are 0 or the fields are blank, unit 5 is used. Units 3 and 4 are available for problem data units in the program, and others may easily be introduced by simply modifying the code.

F.4.3 Dump Control Data

If the dump control feature has been enabled, one additional card is read after each set of problem data. Its description and format are as follows:

Start dump iteration	Stop dump iteration	Increment interval	Optimality dump flag	Change flag
I10	I10	I10	I10	I10

The iterations (basis or bound changes) are numbered consecutively after the initial spanning tree and flows are obtained. Each of dump features 2–5 when enabled will cause output when the iteration count reaches the start dump iteration and at iterations at increment intervals thereafter, until the stop dump iteration is passed. If optimality is reached and the optimality dump flag is 1, the output for the enabled dump features will be given. If the stop dump iteration is reached and the change flag is 1, another dump control card is read. This continues until a dump control card with change flag of 0 is read or optimality is reached. At optimality, if any dump control cards for this problem remain, they will be read through before proceeding to the next problem.

F.5 OUTPUT

At the beginning of each run, messages are printed identifying the optional features that have been enabled. For each problem a problem number is printed. All data are echoed back except for group 4 input data. The iteration number at optimality and the cost or an infeasibility indication, whichever is appropriate, is printed. When the program terminates, a message appears indicating a stop number. The stop number indicates the reason for termination as follows:

Stop	Meaning
1	End-of-file reading data unit control card.
2	End-of-file reading number of nodes.
3	Data specifies number of nodes as negative or zero.
4	Too many nodes; expand dimensions if possible.
5	End-of-file reading node requirements data.
6	Node requirements data have negative node number.
7	Requirements data for a particular node have been specified on a previous data card.
8	Infeasibility; net requirements for the network are negative.
9	End-of-file reading number of arcs to each node.
10	Too many arcs; expand dimensions if possible.
11	End-of-file reading arc specification data.
12	Node numbers on arc specification card too large or negative or zero.
13	Upper or lower bounds on arc flow improper.
14	Too many arcs; expand dimensions if possible.
15	Too many arcs; expand dimensions if possible.
16	Infeasibility; net requirements induced by lower bounds on arc flow are negative.
17	End-of-file reading dump control card.
18	End-of-file reading dump control card.
19	End-of-file reading dump control card.

Stop 2 or Stop 1 (if data unit control is enabled) would ordinarily be considered normal termination.

All of the above constitute the normal output with no optional features and will appear on unit 6. Output for optional features 1, 6, and 10 will also appear on unit 6. Output for optional features 2–5 and 7–9 will appear on unit 7.

F.6 TIMING AND COMPARISONS

The program NETGEN, a generator for large-scale network test problems, was obtained from Professor Klingman at the University of Texas. The first 35 test problems listed in Klingman, Napier, and Stutz [122] were generated and used for testing purposes. Timing information was obtained and is presented in Table F.1 along with the times for RNET. RNET was developed by Professors Grigoriadis and Hsu at Rutgers University and produced the shortest times that we have seen on these 35 test problems.

Table F.1 Computational Experience with the Test Problems on a CDC Cyber 73

			Time (sec)	
Problem Number	Number of Nodes	Number of Arcs	RNET	NETFLO
Transportation Problems				
1	100×100	1300	2.108	2.055
2	100×100	1500	1.558	2.126
3	100×100	2000	2.715	2.555
4	100×100	2200	2.354	2.450
5	100×100	2900	2.988	2.698
6	150×150	3150	4.573	5.445
7	150×150	4500	5.940	5.856
8	150×150	5155	6.487	6.085
9	150×150	6075	6.258	7.436
10	150×150	6300	6.457	7.149
Total time			41.438	43.855
Assignment Problems				
11	200×200	1500	2.011	4.220
12	200×200	2250	2.854	4.091
13	200×200	3000	3.690	4.297
14	200×200	3750	4.141	4.405
15	200×200	4500	4.035	5.161
Total time			16.731	22.174
Capacitated Network Problems				
16	400	1306	1.387	2.453
17	400	2443	1.773	2.521
18	400	1306	1.247	2.484
19	400	2443	1.837	2.268
20	400	1416	1.811	2.863
21	400	2836	2.240	2.320

Table F.1

Problem Number	Number of Nodes	Number of Arcs	Time (sec) RNET	Time (sec) NETFLO
22	400	1416	1.724	2.861
23	400	2836	2.136	2.149
24	400	1382	1.873	2.521
25	400	2676	2.394	4.675
26	400	1382	1.501	1.682
27	400	2676	1.693	2.704
Total time			21.616	31.501
Uncapacitated Network Problems				
28	1000	2900	5.315	9.785
29	1000	3400	5.757	9.479
30	1000	4400	4.912	10.886
31	1000	4800	5.854	11.781
32	1500	4342	6.200	18.745
33	1500	4385	7.750	18.769
34	1500	5107	7.423	22.373
35	1500	5730	9.825	21.977
Total time			53.036	123.795

```
      PROGRAM NETFLO(INPUT,OUTPUT,TAPE5=INPUT,TAPE6=OUTPUT,
     *TAPE3,TAPE4,TAPE7,TAPE8)
C     INTEGER XOR
      INTEGER   NODE(NNNN,6)
      INTEGER   LNODE,MNODE
      INTEGER   DOWN(1),NEXT(1),LEVEL(1),ARCID(1),FLOW(1),DUAL(1)
      EQUIVALENCE (NODE(1,1),DOWN(1))
     *           ,(NODE(1,2),NEXT(1))
     *           ,(NODE(1,3),LEVEL(1))
     *           ,(NODE(1,4),ARCID(1))
     *           ,(NODE(1,5),FLOW(1))
     *           ,(NODE(1,6),DUAL(1))
      INTEGER   CAT(1)
      EQUIVALENCE (DUAL(1),CAT(1))
C
C
C     NODE FORMAT
C
C
C     1 - PREDECESSOR OR DOWN POINTER
C     2 - THREAD OR NEXT POINTER
C     3 - LEVEL NUMBER
C     4 - ASSOCIATED ARC IDENTIFIER
C     (+=ORIENTATION OPPOSITE TO DOWN POINTER)
C     (-=ORIENTATION SAME AS DOWN POINTER)
C     5 - FLOW ON ARC
C     6 - DUAL VARIABLE VALUE
C
      INTEGER   ARC0(AAAA,5),ARC(AAAA,5)
      EQUIVALENCE (ARC0(2,1),ARC(1,1))
      INTEGER   LARC,MARC
      INTEGER   FROM(1),COST(1),CAPAC(1),FLOOR(1),NAME(1)
      EQUIVALENCE (ARC(1,1),FROM(1))
     *           ,(ARC(1,2),COST(1))
     *           ,(ARC(1,3),CAPAC(1))
     *           ,(ARC(1,4),FLOOR(1))
     *           ,(ARC(1,5),NAME(1))
C
C     ARC FORMAT
C
C     1 - FROM NODE
C     2 - COST
C     3 - CAPACITY(-,IF AT UB)
C     4 - LOWER BOUND
C     5 - NAME
C
C     ARC(0) = ARC0(1) IS GENERAL REFERENCE FOR ARTIFICIAL ARCS
C
      INTEGER   TO
      INTEGER   SLACK,ARTIF,DUMMY,XCESS
      INTEGER   PRICE0,TOO
      INTEGER   TRY,PRICE
      INTEGER   NEWARC,NEWPR
      INTEGER   NEWFRM,NEWTO
      INTEGER   ZERO,BIG
      INTEGER   THD
      INTEGER   DW(2),CH(2),DWN,CHG
      INTEGER   THETA,JTHETA,KTHETA,POSS,JPOSS
      INTEGER   FRM,LVJ
      INTEGER   FM,LST
      INTEGER   DWE
      INTEGER   FLW,AID
```

```
      INTEGER   Q1,Q2
      INTEGER   DIR,REF
      INTEGER   U1,U2,U3,U4
C(DATA UNIT CONTROL)     INTEGER   U(4),UCHG
C(DUMP FLOWS)      INTEGER   SYSFLW,ARCFLW,ARTFLW
C(NODE-TO-NODE FLOW)     INTEGER   SUMFLW
C(ARC FLOWS)      INTEGER   CST
      LOGICAL   INFEAS
      LOGICAL   OPTIM
      LOGICAL   DMP
      LOGICAL   PPR
      LOGICAL   PRES
C
      DATA ITDOPT/ 1 /
      DATA LNODE/ NNNN /
      DATA LARCP1/ AAAA /
      DATA ZERO/ 0 /
      DATA BIG/  1000000000  /
C(DATA UNIT CONTROL)     DATA U/5,5,5,5/
C     DATA U/3,3,3,4/
      DATA U1,U2,U3,U4/5,5,5,5/
C
C(IBM)     DATA SLACK/ 4HSLCK /
C(IBM)     DATA ARTIF/ 4HARTF /
C(IBM)     DATA DUMMY/ 4HDMMY /
C(IBM)     DATA XCESS/ 4HEXCS /
      DATA SLACK/ 10HSLACK ARC  /
      DATA ARTIF/ 10HARTIFICIAL /
      DATA DUMMY/ 10HDUMMY ARC  /
      DATA XCESS/ 10HEXCESS ARC /
C
 1000 FORMAT(8I10)
 1001 FORMAT(1X,8I10)
 1002 FORMAT(A10,5I10)
 1003 FORMAT(1X,A10,5I10)
 1004 FORMAT(F10.0)
 1005 FORMAT(1X,F10.5)
 2001 FORMAT(1X,I5,1H,,4I15,1X,A10)
 2002 FORMAT(1X,I5,1H,,6I15)
C(TIMING)8900 FORMAT(18H ELAPSED TIME    =,F10.2,8H SECONDS)
C(TIMING)8901 FORMAT(18H SETUP TIME      =,F10.2,8H SECONDS)
C(TIMING)8902 FORMAT(18H I/O TIME        =,F10.2,8H SECONDS)
C(TIMING)8903 FORMAT(18H SUMMARY TIME    =,F10.2,8H SECONDS)
C(TIMING)8904 FORMAT(18H PRICING TIME    =,F10.2,8H SECONDS)
C(TIMING)8905 FORMAT(18H RATIO TEST TIME =,F10.2,8H SECONDS)
C(TIMING)8906 FORMAT(18H UPDATE TIME     =,F10.2,8H SECONDS)
C(TIMING)8907 FORMAT(18H SOLUTION TIME   =,F10.2,8H SECONDS)
C(TIMING)8908 FORMAT(18H PROBLEM TIME    =,F10.2,8H SECONDS)
C(TIMING)8909 FORMAT(1H+,40X,1H(,F5.2,18H OF SOLUTION TIME))
 8910 FORMAT(8H0PROBLEM,I5)
C(TIMING)8911 FORMAT(1H+,40X,1H(,F8.5,23H SECONDS PER ITERATION))
C
C     NSTOP - EXIT CONDITION
C     MNODE - NUMBER OF NODES
C     NET -TOTAL BALANCE IN NETWORK
C     MSORC - NUMBER OF SOURCES
C     MSINK - NUMBER OF SINKS
C     MARC - NUMBER OF ARCS
C     MTREE - NUMBER OF BRANCHES ON TREE(EXCLUDING ROOT)
C     THD - POINTER MOVING ALONG THREAD
C     TRY - VARIABLE ENCOUNTERED DURING SETUP OR PRICING
C     PRICE - REDUCED  COST FOR TRY
```

```
C     NEWARC  - BEST VARIABLE TO ENTER
C     NEWPR - PRICE FOR NEWARC
C     DW - DOWN PTRS FOR RATIO TEST(FROM STEM,TO STEM)
C     CH - PATH CONDITIONS FOR RATIO TEST (FROM STEM,TO STEM)
C     DWN - POINTER MOVING ALONG DOWN PATH
C     CHG - PATH CONDITION
C     THETA - MINIMUM RATIO IN RATIO TEST
C     JTHETA - UB (-) OR LB(+1) CONDITION FOR MIN THETA
C     KTHETA - MIN THETA OCCURS ON FROM STEM(1) OR TO STEM (2)
C     POSS - CANDIDATE FOR MIN THETA
C     JPOSS - UB (-1) OR LB(+1) CONDITION FOR CANDIDATE FOR MIN THETA
C     DWE - ROOT OF CYCLE
C
C(TIMING)     WRITE(6,8700)
C(TIMING)8700 FORMAT(15H TIMING ENABLED)
C(DUMP OPTIMIZATION HISTORY)     WRITE(6,8701)
C(DUMP OPTIMIZATION HISTORY)8701 FORMAT(34H DUMP OPTIMIZATION HISTORY ENABLED)
C(DUMP FLOWS)     WRITE(6,8702)
C(DUMP FLOWS)8702 FORMAT(19H DUMP FLOWS ENABLED)
C(DUMP COST)     WRITE(6,8703)
C(DUMP COST)8703 FORMAT(18H DUMP COST ENABLED)
C(DUMP ARRAYS)     WRITE(6,8704)
C(DUMP ARRAYS)8704 FORMAT(20H DUMP ARRAYS ENABLED)
C(ECHO ARC DATA)     WRITE(6,8706)
C(ECHO ARC DATA)8706 FORMAT(22H ECHO ARC DATA ENABLED)
C(DEGENERATE PIVOTS)     WRITE(6,8707)
C(DEGENERATE PIVOTS)8707 FORMAT(26H DEGENERATE PIVOTS ENABLED)
C(DUMP CONTROL)     WRITE(6,8708)
C(DUMP CONTROL)8708 FORMAT(21H DUMP CONTROL ENABLED)
C(DATA UNIT CONTROL)     WRITE(6,8709)
C(DATA UNIT CONTROL)8709 FORMAT(26H DATA UNIT CONTROL ENABLED)
C(DUAL VARIABLES)     WRITE(6,8710)
C(DUAL VARIABLES)8710 FORMAT(23H DUAL VARIABLES ENABLED)
C(NODE-TO-NODE FLOW)     WRITE(6,8711)
C(NODE-TO-NODE FLOW)8711 FORMAT(26H NODE-TO-NODE FLOW ENABLED)
C(ARC FLOWS)     WRITE(6,8712)
C(ARC FLOWS)8712 FORMAT(18H ARC FLOWS ENABLED)
C     ALLOW FOR ARTIFICIAL ARC
      LARC = LARCP1-1
C     UNUSED NODE NUMBER
      LNODP1 = LNODE+1
C     REWIND ALTERNATE DATA UNITS
C(DATA UNIT CONTROL)     REWIND 3
C(DATA UNIT CONTROL)     REWIND 4
      NPROB = 0
      KARD = 0
C(DATA UNIT CONTROL)     UCHG = 1
    5 CONTINUE
      PPR = .TRUE.
      PRES = .FALSE.
      NPROB = NPROB+1
C(DATA UNIT CONTROL)     IF(UCHG.EQ.0) GO TO 7
C     INPUT UNIT NUMBERS FOR DATA AND CHANGE SIGNAL
C(DATA UNIT CONTROL)     NSTOP = 1
C(DATA UNIT CONTROL)      READ(5,1000) U1,U2,U3,U4,UCHG
C
C     UNIT NUMBERS FOR DATA (4I10),CHANGE UNITS NEXT PROBLEM (I10)
C
C(DATA UNIT CONTROL)     IF(EOF(5)) 999,6
C(DATA UNIT CONTROL)   6 CONTINUE
C(DATA UNIT CONTROL)     KARD = KARD+1
C(DATA UNIT CONTROL)     WRITE(6,8910) NPROB
```

```
C(DATA UNIT CONTROL)     PPR = .FALSE.
C(DATA UNIT CONTROL)     WRITE(6,1001) U1,U2,U3,U4,UCHG
C     IF ZERO, USE DEFAULT VALUES
C(DATA UNIT CONTROL)     IF(U1.EQ.0) U1 = U(1)
C(DATA UNIT CONTROL)     IF(U2.EQ.0) U2 = U(2)
C(DATA UNIT CONTROL)     IF(U3.EQ.0) U3 = U(3)
C(DATA UNIT CONTROL)     IF(U4.EQ.0) U4 = U(4)
C(DATA UNIT CONTROL)   7 CONTINUE
C     INPUT NUMBER OF REGULAR NODES
C(TIMING)     T0 = -SECOND(X)
C(TIMING)     T1 = T0
C(TIMING)     T2 = T0
      NSTOP = 2
       READ(U1,1000) I
C
C     NUMBER OF NODES(I10)
C
      IF(EOF(U1)) 999,8
    8 CONTINUE
      KARD = KARD+1
      IF(PPR) WRITE(6,8910) NPROB
      WRITE(6,1001) I
C(TIMING)     T1 = T1+SECOND(X)
      NSTOP = 3
      IF(I.LE.0) GO TO 999
      MNODE = I
      MNODP1 = MNODE+1
      MNODP2 = MNODE+2
      NSTOP = 4
      IF(MNODP1.GT.LNODE) GO TO 999
C     INITIALIZE NODE ARRAY
      DO 10 J10=1,MNODP1
      DO 10 I10=1,5
   10 NODE(J10,I10) = 0
C     INITIALIZE ARTIFICIAL
      FROM(ZERO) = MNODP1
      COST(ZERO) = BIG
      CAPAC(ZERO) = 0
      FLOOR(ZERO) = 0
      NAME(ZERO) = ARTIF
C
C     DURING SETUP WE USE THE ABSENCE OF THE UPPER BOUND FLAG
C     ON AN ARC TO INDICATE THAT THE ARC IS ELIGIBLE TO BECOME
C     PART OF THE STARTING BASIS, OTHERWISE IT IS NOT.
C
      NET = 0
      MSORC = 0
      MARC = 0
C     INITIALIZE SUPPLY AND DEMAND LISTS WITH SELF-POINTER
      NEXT(MNODP1) = MNODP1
      DOWN(MNODP1) = MNODP1
C     INPUT NON-ZERO NODAL BALANCES
   15 CONTINUE
C(TIMING)     T1 = T1-SECOND(X)
      NSTOP = 5
       READ(U2,1000) I,J
C
C     NODE NUMBER(I10),BALANCE(I10)/BLANK= ESCAPE
C
      IF(EOF(U2)) 999,16
   16 CONTINUE
      KARD = KARD+1
```

```
C(TIMING)     T1 = T1+SECOND(X)
      NSTOP = 6
      IF(I) 999,60,20
   20 CONTINUE
      IF(I.GT.MNODE) GO TO 999
      NSTOP = 7
      IF(FLOW(I).NE.0) GO TO 999
      FLOW(I) = J
      NET = NET+J
      IF(J.LE.0) GO TO 15
      MSORC = MSORC+1
C     SAVE ORIGINAL SUPPLY IN LEVEL
      LEVEL(I) = J
      NEXT(I) = NEXT(MNODP1)
      NEXT(MNODP1) = I
      GO TO 15
   60 CONTINUE
C     TEST FOR FEASIBILITY
      NSTOP = 8
      IF(NET.LT.0) GO TO 999
C     INPUT NUMBER OF ARCS TO EACH NODE(IN NUMERIC ORDER)
C(TIMING)     T1 = T1-SECOND(X)
      NSTOP = 9
       READ(U3,1000) (CAT(I),I=1,MNODE)
C
C     NUMBER OF ARCS TO EACH NODE (8I10)
C
      IF(EOF(U3)) 999,61
   61 CONTINUE
      KARD = KARD+1
C(ECHO ARC DATA)     WRITE(6,1001) (CAT(I),I=1,MNODE)
C(TIMING)     T1 = T1+SECOND(X)
C
C     RESERVE LOCATIONS FOR INPUT ARCS BY FILLING WITH DUMMIES
C      CAT(N) WILL POINT TO THE NEXT OPEN LOCATION FOR STORING
C      ARCS WHOSE TERMINAL NODE IS N.
C
      NSTOP = 10
      I = 1
      J = 0
      DO 80 J80 = 1,MNODE
      I = -I
      K70 = MAX0(1,CAT(J80))
      IF(J+K70.GT.LARC) GO TO 999
      CAT(J80) = ISIGN(J+1,I)
      DO 70 I70=1,K70
      J = J+1
C     NOTE THAT DUMMY ARCS HAVE SAME INITIAL AND TERMINAL NODE
      FROM(J) = ISIGN(J80,I)
      COST(J) = 0
      CAPAC(J) = -BIG
      FLOOR(J) = 0
      NAME(J) = DUMMY
   70 CONTINUE
   80 CONTINUE
      MARC = J+1
      IF(MARC.GT.LARC) GO TO 999
      FROM(MARC) = ISIGN(MNODP1,-I)
C     INPUT REGULAR ARCS
      KOSTO = 0
   90 CONTINUE
C(TIMING)     T1 = T1-SECOND(X)
```

```
      NSTOP = 11
       READ(U4,1002)    N,I,J,K,L,M
C
C     NAME,FROM NODE,TO NODE,COST,CAPACITY,LOWER BOUND
C      (A10) , (I10) , (I10),(I10), (I10) ,    (I10)
C     /BLANK = ESCAPE
C
      IF(EOF(U4)) 999,91
   91 CONTINUE
      KARD = KARD+1
C(ECHO ARC DATA)     WRITE(6,1003)    N,I,J,K,L,M
C(TIMING)     T1 = T1+SECOND(X)
      NSTOP = 12
      IF(I) 999,150,100
  100 CONTINUE
      IF(I.GT.MNODE) GO TO 999
      IF(J.GT.MNODE) GO TO 999
      IF(J.LE.0) GO TO 999
      NSTOP = 13
      IF(L.GE.BIG) GO TO 999
      IF(L.EQ.0) L = BIG
      IF(L.LT.0) L = 0
      IF(M.GE.BIG) GO TO 999
      IF(M.LT.0) GO TO 999
      IF(M.GT.L) GO TO 999
      II = CAT(J)
      JJ = IABS(II)
C     TEST TO SEE IF CATEGORY IS FULL
      KK = ISIGN(LNODP1,II)
      IF(XOR(KK,FROM(JJ)).GT.0) GO TO 140
C     MOVE REST OF ARCS DOWN TO ACCOMODATE
      NSTOP = 14
      IF(MARC.EQ.LARC) GO TO 999
      MARC = MARC+1
      K120 = MARC-JJ
      M120 = MARC
      DO 120 J120=1,K120
      L120 = M120-1
      DO 110 I110=1,5
  110 ARC(M120,I110) = ARC(L120,I110)
      M120 = L120
  120 CONTINUE
      DO 130 J130=J,MNODE
  130 CAT(J130) = CAT(J130)+ISIGN(1,CAT(J130))
C     INSERT NEW ARC
  140 CONTINUE
      FROM(JJ) = ISIGN(I,II)
      COST(JJ) = K
      KOSTO = KOSTO+K*M
      CAPAC(JJ) = L-M
      FLOOR(JJ) = M
      FLOW(I) = FLOW(I)-M
      FLOW(J) = FLOW(J)+M
      NAME(JJ) = N
      CAT(J) = ISIGN(JJ+1,II)
C     MARK FROM NODE WITH ARC RUNNING FROM IT FLAG
      ARCID(I) = -1
      GO TO 90
  150 CONTINUE
C
C     ELIMINATE NON-ESSENTIAL DUMMY ARCS
C     CAT(N) WILL POINT TO LOCATION OF FIRST ARC TERMINAL AT NODE N
```

```
C
      I = LNODP1
      K = 0
      L = 0
      MARC = MARC-1
      DO 190 J190=1,MARC
      J = FROM(J190)
      IF(XOR(I,J).GT.0) GO TO 160
      I = -I
      L = L+1
      CAT(L) = K+1
      GO TO 170
  160 IF(IABS(J).EQ.L) GO TO 190
  170 K = K+1
      IF(K.EQ.J190) GO TO 190
      DO 180 I180=1,5
  180 ARC(K,I180) = ARC(J190,I180)
  190 CONTINUE
      MARC = K
      MREG = K
      NSTOP = 15
      IF(MARC+MAX0(1,MSORC)+1.GT.LARC) GO TO 999
C
C     ADD REGULAR SLACKS
C
      I = -FROM(MARC)
      THD = NEXT(MNODP1)
      NEXT(MNODP1) = MNODP1
      IF(THD.NE.MNODP1) GO TO 192
C     NO REGULAR SLACKS, ADD DUMMY
      MARC = MARC+1
      FROM(MARC) = ISIGN(MNODP1,I)
      COST(MARC) = 0
      CAPAC(MARC) = -BIG
      FLOOR(MARC) = 0
      NAME(MARC) = DUMMY
      GO TO 194
C     FOLLOW LIST
  192 CONTINUE
      MARC = MARC+1
      NAME(MARC) = SLACK
      FROM(MARC) = ISIGN(THD,I)
      COST(MARC) = 0
      CAPAC(MARC) = LEVEL(THD)
      LEVEL(THD) = 0
      FLOOR(MARC) = 0
      NXT = NEXT(THD)
      NEXT(THD) = 0
      THD = NXT
      IF(THD.NE.MNODP1) GO TO 192
  194 CONTINUE
      MSLK = MARC
C     ADD EXCESS ARC AT END OF REGULAR SLACKS
      MARC = MARC+1
      FROM(MARC) = ISIGN(MNODP2,-I)
      COST(MARC) = BIG
      CAPAC(MARC) = 0
      FLOOR(MARC) = 0
      NAME(MARC) = XCESS
C
C     LOCATE SOURCES AND SINKS FOR PROBLEM
C     ADJUSTED FOR LOWER BOUNDS
```

```
C
      NET = 0
      MTREE = 0
      THD = MNODP1
      DO 200 I200 = 1,MNODE
      J = FLOW(I200)
      NET = NET+J
      IF(J) 30,200,50
C     SINK
   30 CONTINUE
      FLOW(I200) = -J
C     LINK DEMANDS IN DECREASING SIZE ORDER
      DWN = MNODP1
   35 CONTINUE
      NXT = DOWN(DWN)
      IF(FLOW(NXT)+J.LE.0) GO TO 40
      DWN = NXT
      GO TO 35
   40 CONTINUE
      DOWN(DWN) = I200
      DOWN(I200) = NXT
      LEVEL(I200) = -1
      GO TO 200
C     SOURCE
   50 CONTINUE
      MTREE = MTREE+1
      ARCID(I200) = -MARC
      FLOW(I200) = J
      NEXT(THD) = I200
      DOWN(I200) = MNODP1
      NEXT(I200) = MNODP1
      LEVEL(I200) = 1
      DUAL(I200) = BIG
      THD = I200
  200 CONTINUE
C     CHECK FOR FEASIBILITY
      NSTOP = 16
      IF(NET.LT.0) GO TO 999
C
C     ADVANCED START
C
C(DUMP FLOWS)      MART = 0
C(DUMP FLOWS)      ARCFLW = 0
C(DUMP FLOWS)      SYSFLW = 0
C     SELECT HIGHEST RANK DEMAND ON LIST
      T9=SECOND(X)
 1010 CONTINUE
      TO = DOWN(MNODP1)
C     IS LIST EXHAUSTED
      IF(TO.EQ.MNODP1) GO TO 210
 1020 CONTINUE
C     SET TO LINK TO ARTIFICIAL
      NEWARC = 0
      NEWPR = BIG
C     ANY DEMAND LEFT
      IF(FLOW(TO).EQ.0) GO TO 1110
C
C     LOOK FOR SOURCES FIRST
C
      TRY = CAT(TO)
      FRM = FROM(TRY)
      LST = ISIGN(LNODP1,FRM)
```

```
 1030 CONTINUE
C     IS IT UNAVAILABLE
      IF(CAPAC(TRY).LE.0) GO TO 1050
      FM = IABS(FRM)
C     IS IT FROM A NON-SOURCE
      IF(LEVEL(FM).NE.1) GO TO 1050
      IF(ARCID(FM).EQ.0) GO TO 1050
      PRICE = COST(TRY)
C     IS COST WORSE
      IF(PRICE.GE.NEWPR) GO TO 1050
C     DOES CAPACITY EXCEED DEMAND
      IF(CAPAC(TRY).GT.FLOW(TO)) GO TO 1040
C     IS THERE NOT ENOUGH SUPPLY FOR CAPACITY
      IF(FLOW(FM).LT.CAPAC(TRY)) GO TO 1050
      NEWARC = -TRY
      NEWPR = PRICE
      IF(NEWPR.EQ.0) GO TO 1055
      GO TO 1050
 1040 CONTINUE
C     IS THERE NOT ENOUGH SUPPLY FOR DEMAND
      IF(FLOW(FM).LT.FLOW(TO)) GO TO 1050
      NEWARC = TRY
      NEWPR = PRICE
      IF(NEWPR.EQ.0) GO TO 1055
 1050 CONTINUE
      TRY = TRY+1
      FRM = FROM(TRY)
      IF(XOR(FRM,LST).GT.0) GO TO 1030
      IF(NEWARC.EQ.0) GO TO 1070
 1055 CONTINUE
C     ARE WE SENDING DEMAND
      IF(NEWARC.GT.0) GO TO 1060
C     SEND CAPACITY
      NEWARC = -NEWARC
      FM = IABS(FROM(NEWARC))
C     GET CAPACITY
      FLW = CAPAC(NEWARC)
C(DUMP FLOWS)     ARCFLW = ARCFLW+FLW
C(DUMP FLOWS)     SYSFLW = SYSFLW+FLW
C     MARK UNAVAILABLE
      CAPAC(NEWARC) = -FLW
C     ADJUST FLOWS
      FLOW(FM) = FLOW(FM)-FLW
      FLOW(TO) = FLOW(TO)-FLW
      GO TO 1020
C     SEND DEMAND
 1060 CONTINUE
C     MARK UNAVAILABLE
      CAPAC(NEWARC) = -CAPAC(NEWARC)
C     ADJUST FLOWS
      FM = IABS(FROM(NEWARC))
      FLOW(FM) = FLOW(FM)-FLOW(TO)
C(DUMP FLOWS)     ARCFLW = ARCFLW+FLOW(TO)
      K = BIG
      GO TO 1115
C
C     LOOK FOR TRANSSHIPMENT POINTS
C
 1070 CONTINUE
      TRY = CAT(TO)
      FRM = FROM(TRY)
 1080 CONTINUE
```

```
C     IS IT UNAVAILABLE
      IF(CAPAC(TRY).LE.0) GO TO 1090
      FM = IABS(FRM)
C     IS IT ALREADY LINKED
      IF(LEVEL(FM).NE.0) GO TO 1090
      PRICE = COST(TRY)
C     IS COST WORSE
      IF(PRICE.GE.NEWPR) GO TO 1090
      NEWARC = TRY
      NEWPR = PRICE
      IF(NEWPR.EQ.0) GO TO 1095
 1090 CONTINUE
      TRY = TRY+1
      FRM = FROM(TRY)
      IF(XOR(FRM,LST).GT.0) GO TO 1080
      IF(NEWARC.EQ.0) GO TO 1110
 1095 CONTINUE
      FM = IABS(FROM(NEWARC))
C     DOES CAPACITY EXCEED DEMAND
      IF(CAPAC(NEWARC).GT.FLOW(TO)) GO TO 1100
C     GET CAPACITY
      FLW = CAPAC(NEWARC)
C     MARK UNAVAILABLE
      CAPAC(NEWARC) = -FLW
C     ADJUST FLOWS
      FLOW(FM) = FLW
      FLOW(TO) = FLOW(TO)-FLW
C     LINK IN TO DEMAND LIST
      DOWN(FM) = TO
      DOWN(MNODP1) = FM
C     START NEW CHAIN
      LEVEL(FM) = -1
      GO TO 1010
 1100 CONTINUE
C     MARK UNAVAILABLE
      CAPAC(NEWARC) = -CAPAC(NEWARC)
C     PASS ALONG FLOW DEMAND
      FLOW(FM) = FLOW(TO)
C     LINK IN NEW DEMAND NODE
      DOWN(FM) = DOWN(TO)
      DOWN(TO) = FM
      DOWN(MNODP1) = FM
      NEXT(FM) = TO
      ARCID(TO) = NEWARC
      LEVEL(FM) = LEVEL(TO)-1
      DUAL(TO) = NEWPR
      MTREE = MTREE+1
      GO TO 1010
C     ADD CHAIN TO TREE
 1110 CONTINUE
C(DUMP FLOWS)     MART = MART+1
      K = 0
 1115 CONTINUE
C(DUMP FLOWS)     SYSFLW = SYSFLW+FLOW(TO)
C     REMOVE FROM DEMAND LIST
      DOWN(MNODP1) = DOWN(TO)
C     LINK IN AS LEFTMOST BRANCH
      FM = IABS(FROM(NEWARC))
      ARCID(TO) = NEWARC
      DUAL(TO) = NEWPR
      DOWN(TO) = FM
      I = NEXT(FM)
```

```
      NEXT(FM) = TO
      J = LEVEL(FM)-LEVEL(TO)+1
      THD = FM
 1120 CONTINUE
C     MOVE ALONG CHAIN
      THD = NEXT(THD)
C     ADJUST LEVEL AND DUAL VARIABLES
      L = LEVEL(THD)
      LEVEL(THD) = L+J
      K = K-DUAL(THD)
      DUAL(THD) = K
      IF(L.NE.-1) GO TO 1120
      NEXT(THD) = I
      MTREE = MTREE+1
      GO TO 1010
  210 CONTINUE
C(DUMP FLOWS)     ARTFLW = SYSFLW-ARCFLW
C(TIMING)     T1 = T1-SECOND(X)
C(DUMP FLOWS)      WRITE(6,8015) SYSFLW,ARCFLW,ARTFLW,MART
C(DUMP FLOWS)8015 FORMAT(6X,10HTOTAL FLOW,I15/
C(DUMP FLOWS)    *5X,11HSOURCE FLOW,I15/
C(DUMP FLOWS)    *1X,15HARTIFICIAL FLOW,I15,5X,I10,12H ARTIFICIALS)
C(TIMING)     T1 = T1+SECOND(X)
C     SET UP TO EXPAND TREE
      TO = 1
      TRY = 1
      FRM = FROM(TRY)
  220 CONTINUE
C     DO WE NEED TO EXPAND TREE TO REACH ALL NODES
      IF(MTREE.EQ.MNODE) GO TO 285
      TOO = TO
      NEWPR = BIG
C     SEARCH FOR LEAST COST CONNECTABLE ARC IN CURRENT TO-GROUP
  230 CONTINUE
      LVJ = LEVEL(TO)
      LST = ISIGN(LNODP1,FRM)
  235 CONTINUE
      IF(CAPAC(TRY).LE.0) GO TO 260
      M = COST(TRY)
      IF(NEWPR.LT.M) GO TO 260
      FM = IABS(FRM)
      IF(LEVEL(FM).EQ.0) GO TO 240
      IF(LVJ.NE.0) GO TO 260
C     TO END IS CONNECTABLE
      I = FM
      J = TO
      K = M
      L = TRY
      GO TO 250
  240 CONTINUE
      IF(LVJ.EQ.0) GO TO 260
C     FROM END IS CONNECTABLE
      I = TO
      J = FM
      K = -M
      L = -TRY
  250 CONTINUE
      NEWPR = M
  260 CONTINUE
      TRY = TRY+1
      FRM = FROM(TRY)
C     IS TO-GROUP EXHAUSTED
```

```
      IF(XOR(FRM,LST).GT.0) GO TO 235
C     PREPARE FOR NEXT TO-GROUP
      TO = TO+1
      IF(TO.NE.MNODP1) GO TO 270
      TO = 1
      TRY = 1
      FRM = FROM(TRY)
  270 CONTINUE
      IF(NEWPR.NE.BIG) GO TO 280
      IF(TO.NE.TOO) GO TO 230
C     NOT ALL NODES CONNECTABLE - CHECK FOR ISOLATED POINTS
      DO 275 I275=1,MNODE
      IF(LEVEL(I275).NE.0) GO TO 275
C     TEST FOR ARCS RUNNING FROM IT
      IF(ARCID(I275).EQ.-1) GO TO 274
C     CHECK FOR DUMMY - NO ARCS RUNNING TO IT
      J275 = CAT(I275)
      IF(IABS(FROM(J275)).NE.I275) GO TO 274
      WRITE(6,8021) I275
 8021 FORMAT(6H NODE ,I10,12H IS ISOLATED)
C     ADD ARTIFICIAL
  274 CONTINUE
      ARCID(I275) = 0
      FLOW(I275) = 0
      NEXT(I275) = NEXT(MNODP1)
      NEXT(MNODP1) = I275
      DOWN(I275) = MNODP1
      LEVEL(I275) = 1
      DUAL(I275) = -BIG
  275 CONTINUE
      GO TO 285
  280 CONTINUE
      ARCID(J) = L
      DOWN(J) = I
      NEXT(J) = NEXT(I)
      NEXT(I) = J
      LEVEL(J) = LEVEL(I)+1
      DUAL(J) = DUAL(I)-K
      NEWARC = IABS(L)
      CAPAC(NEWARC) = -CAPAC(NEWARC)
      MTREE = MTREE+1
      GO TO 220
  285 CONTINUE
C     CLEAR UPPER BOUND FLAGS ON BASIC ARCS
      DO 290 I290=1,MNODE
      J290 = IABS(ARCID(I290))
      CAPAC(J290) = -CAPAC(J290)
  290 CONTINUE
C     CLEAR OUT UPPER BOUND FLAGS ON DUMMY ARCS
      DO 295 I295=1,MARC
      IF(CAPAC(I295)+BIG.EQ.0)
     *CAPAC(I295) = 0
  295 CONTINUE
C     SET UPPER BOUND FOR ARTIFICIAL AND EXCESS
      CAPAC(ZERO) = BIG
      CAPAC(MARC) = BIG
C(TIMING)      T2 = T2+SECOND(X)-T1
C(TIMING)      WRITE(6,8901) T2
C     DUMP CONTROL PARAMETERS
C(TIMING)      T1 = T1-SECOND(X)
C(DUMP CONTROL)      NSTOP = 17
C(DUMP CONTROL)      READ(5,1000) ITDMP1,ITDMP2,ITDMP3,ITDOPT,ITDCHG
```

```
C
C     DUMP ITER START,DUMP ITER END,DUMP ITER STEP,DUMP OPT,CHANGE
C        (I10)     ,       (I10)  ,       (I10)    ,  (I10) ,   (I10)
C
C(DUMP CONTROL)      IF(EOF(5)) 999,296
C(DUMP CONTROL) 296 CONTINUE
C(DUMP CONTROL)      KARD = KARD+1
C(DUMP CONTROL)      WRITE(6,1001) ITDMP1,ITDMP2,ITDMP3,ITDOPT,ITDCHG
C(TIMING)      T1 = T1+SECOND(X)
C(TIMING)      T6 = -SECOND(X)
C(TIMING)      T3 = 0.
C(TIMING)      T4 = 0.
C(TIMING)      T8 = 0.
C
C     INITIALIZE PRICING
C
      TO = 1
      TRY = 1
      FRM = FROM(TRY)
      ITER = 0
      OPTIM = .FALSE.
      DMP = .FALSE.
C(DUMP CONTROL)      ITDMPD = 0
C(DEGENERATE PIVOTS)      IDEG = 0
C(DEGENERATE PIVOTS)      KDEG = BIG
C
C     NEW ITERATION
C
C(TIMING)      T5 = -SECOND(X)
  300 CONTINUE
C(TIMING)      T5 = T5+SECOND(X)
      ITER = ITER+1
C(DUMP CONTROL)      DMP = .FALSE.
C(DUMP CONTROL)      IF(ITER.NE.ITDMP1) GO TO 301
C(DUMP CONTROL)      DMP = .TRUE.
C(DUMP CONTROL)      ITDMP1 = ITDMP1+MAX0(1,ITDMP3)
C(DUMP CONTROL) 301 CONTINUE
C(DUMP CONTROL)      IF(ITER.NE.ITDMP2) GO TO 303
C(DUMP CONTROL)      ITDMP1 = 0
C(DUMP CONTROL)      IF(ITDCHG.EQ.0) GO TO 303
C     READ NEXT DUMP CONTROL CARD
C(TIMING)      T1 = T1-SECOND(X)
C(DUMP CONTROL)      NSTOP = 18
C(DUMP CONTROL)      READ(5,1000) ITDMP1,ITDMP2,ITDMP3,ITDOPT,ITDCHG
C
C     DUMP ITER START,DUMP ITER END,DUMP ITER STEP,DUMP OPT,CHANGE
C        (I10)     ,       (I10)  ,       (I10)    ,  (I10) ,   (I10)
C
C(DUMP CONTROL)      IF(EOF(5)) 999,302
C(DUMP CONTROL) 302 CONTINUE
C(DUMP CONTROL)      KARD = KARD+1
C(DUMP CONTROL)      WRITE(6,1001) ITDMP1,ITDMP2,ITDMP3,ITDOPT,ITDCHG
C(TIMING)      T1 = T1+SECOND(X)
C(DUMP CONTROL) 303 CONTINUE
C(DUMP CONTROL)      IF(DMP) GO TO 795
  305 CONTINUE
C
C     PRICING
C
C     NOTE THAT WE ARE PRICING OUT BASIC ARCS
C(TIMING)      T3 = T3-SECOND(X)
C(DUMP OPTIMIZATION HISTORY)      IF(DMP) WRITE(7,8009) ITER
```

```
      T00 = T0
      NEWPR = 0
  310 CONTINUE
      PRICE0 = -DUAL(T0)
      LST = ISIGN(LNODP1,FRM)
  320 CONTINUE
      FM = IABS(FRM)
      PRICE = DUAL(FM)+PRICE0-COST(TRY)
      IF(CAPAC(TRY)) 325,330,326
  325 CONTINUE
      PRICE = -PRICE
  326 CONTINUE
      IF(PRICE.LE.NEWPR) GO TO 330
      NEWARC = TRY
      NEWPR = PRICE
      NEWTO = T0
  330 CONTINUE
      TRY = TRY+1
      FRM = FROM(TRY)
      IF(XOR(FRM,LST).GT.0) GO TO 320
      T0 = T0+1
      IF(T0.NE.MNODP2) GO TO 350
      T0 = 1
      TRY = 1
      FRM = FROM(TRY)
  350 CONTINUE
      IF(NEWPR.NE.0) GO TO 360
      IF(T0.NE.T00) GO TO 310
C
C     OPTIMAL INDICATION
C
C(TIMING)     SEC = SECOND(X)
C(TIMING)     T3 = T3+SEC
C(TIMING)     T6 = T6+SEC
C(TIMING)     T7 = T6+T2
C(DUMP OPTIMIZATION HISTORY)     IF(DMP) WRITE(7,8000)
      IF(ITDOPT.NE.0) DMP = .TRUE.
      OPTIM = .TRUE.
      T9=SECOND(X)-T9
      PRINT9999,T9
9999  FORMAT(*0TIME = *,F10.3,/)
      GO TO 795
  360 CONTINUE
      NEWFRM = IABS(FROM(NEWARC))
C
C     RATIO TEST
C
C(TIMING)     SEC = SECOND(X)
C(TIMING)     T3 = T3+SEC
C(TIMING)     T4 = T4-SEC
C(DUMP OPTIMIZATION HISTORY)     IF(DMP) WRITE(7,8003) NEWPR,NAME(NEWARC)
C(DUMP OPTIMIZATION HISTORY)8003 FORMAT(1H+,21X,5HPRICE,I15,1X,A10,7H ENTERS)
      THETA = IABS(CAPAC(NEWARC))
      JTHETA = 0
C     SET FOR CYCLE SEARCH
      CH(2) = ISIGN(LARCP1,CAPAC(NEWARC))
      CH(1) = -CH(2)
      DW(1) = NEWFRM
      DW(2) = NEWTO
      LDIFF = LEVEL(NEWFRM)-LEVEL(NEWTO)
      KTHETA = 1
      IF(LDIFF) 380,450,390
```

```
  380 KTHETA = 2
  390 CONTINUE
      DWN = DW(KTHETA)
      CHG = CH(KTHETA)
      K440 = IABS(LDIFF)
      DO 440 I440=1,K440
      IF(XOR(CHG,ARCID(DWN)).GT.0) GO TO 410
C     INCREASING FLOW
      I = IABS(ARCID(DWN))
      POSS = CAPAC(I)-FLOW(DWN)
      JPOSS = -DWN
      GO TO 420
C     DECREASING FLOW
  410 POSS = FLOW(DWN)
      JPOSS = DWN
C     FIND MIN
  420 CONTINUE
      IF(THETA.LE.POSS) GO TO 430
      THETA = POSS
      JTHETA = JPOSS
      IF(THETA.EQ.0) GO TO 530
  430 CONTINUE
      DWN = DOWN(DWN)
  440 CONTINUE
      DW(KTHETA) = DWN
C     AT COMMON LEVEL
  450 CONTINUE
C     SEARCH FOR CYCLE END
  460 CONTINUE
      IF(DW(1).EQ.DW(2)) GO TO 520
      DO 510 L510=1,2
      DWN = DW(L510)
      IF(XOR(CH(L510),ARCID(DWN)).GT.0) GO TO 480
C     INCREASING FLOW
      I = IABS(ARCID(DWN))
      POSS = CAPAC(I)-FLOW(DWN)
      JPOSS = -DWN
      GO TO 490
C     DECREASING FLOW
  480 POSS = FLOW(DWN)
      JPOSS = DWN
C     FIND MIN
  490 CONTINUE
      IF(THETA.LE.POSS) GO TO 500
      THETA = POSS
      JTHETA = JPOSS
      KTHETA = L510
      IF(THETA.EQ.0) GO TO 530
  500 CONTINUE
      DW(L510) = DOWN(DWN)
  510 CONTINUE
      GO TO 460
  520 DWE = DW(1)
C     RATIO TEST COMPLETE
  530 CONTINUE
C(TIMING)     SEC = SECOND(X)
C(TIMING)     T4 = T4+SEC
C(TIMING)     T5 = T5-SEC
C(DUMP OPTIMIZATION HISTORY)     IF(JTHETA.EQ.0) GO TO 535
C(DUMP OPTIMIZATION HISTORY)     ITHETA = IABS(JTHETA)
C(DUMP OPTIMIZATION HISTORY)     I = IABS(ARCID(ITHETA))
C(DUMP OPTIMIZATION HISTORY)     IF(DMP) WRITE(7,8004) THETA,NAME(I)
```

```
C(DUMP OPTIMIZATION HISTORY)8004 FORMAT(1H+,60X,5HTHETA,I15,1X,A10,7H LEAVES)
C(DUMP OPTIMIZATION HISTORY)      IF((I.GT.0).AND.(I.LE.MREG)) GO TO 535
C(DUMP OPTIMIZATION HISTORY)      IF(DMP) WRITE(7,8014) ITHETA
C(DUMP OPTIMIZATION HISTORY)8014 FORMAT(1H+,99X,16HCONNECTING NODE ,I10)
C(DUMP OPTIMIZATION HISTORY) 535 CONTINUE
C(DEGENERATE PIVOTS)      IDEG = IDEG+1
      IF(THETA.EQ.0) GO TO 560
C(DEGENERATE PIVOTS)      IDEG = IDEG-1
C(DEGENERATE PIVOTS)      KDEG = MIN0(KDEG,ITER)
C     UPDATE FLOWS ON INTACT PORTION OF CYCLE
      DW(1) = NEWFRM
      DW(2) = NEWTO
      IF(JTHETA.NE.0)
     *DW(KTHETA) = IABS(JTHETA)
      DO 550 I550=1,2
      DWN = DW(I550)
      CHG = ISIGN(THETA,CH(I550))
  540 CONTINUE
      IF(DWN.EQ.DWE) GO TO 550
      FLOW(DWN) = FLOW(DWN)-CHG*ISIGN(1,ARCID(DWN))
      DWN = DOWN(DWN)
      GO TO 540
  550 CONTINUE
  560 CONTINUE
      IF(JTHETA.NE.0) GO TO 570
C     CHANGE OF BOUNDS ONLY
      CAPAC(NEWARC) = -CAPAC(NEWARC)
C(DUMP OPTIMIZATION HISTORY)      IF(DMP) WRITE(7,8010) NAME(NEWARC)
C(DUMP OPTIMIZATION HISTORY)8010 FORMAT(1H+,60X,A10,16H SWITCHES BOUNDS)
      GO TO 300
  570 CONTINUE
      ITHETA = IABS(JTHETA)
      IF(JTHETA.GT.0) GO TO 590
      J = IABS(ARCID(ITHETA))
C     SET OLD ARC TO UPPER BOUND
      CAPAC(J) = -CAPAC(J)
C(DUMP OPTIMIZATION HISTORY)      IF(DMP) WRITE(7,8011)
C(DUMP OPTIMIZATION HISTORY)8011 FORMAT(1H+,99X,15H AT UPPER BOUND)
  590 CONTINUE
C     FLOW ON NEWARC
      FLW = THETA
      IF(CAPAC(NEWARC).GT.0) GO TO 600
      CAPAC(NEWARC) = -CAPAC(NEWARC)
      FLW = CAPAC(NEWARC)-FLW
      NEWPR = -NEWPR
  600 CONTINUE
      IF(KTHETA.EQ.2) GO TO 610
      Q1 = NEWFRM
      Q2 = NEWTO
      AID = -NEWARC
      NEWPR = -NEWPR
      GO TO 620
  610 CONTINUE
      Q1 = NEWTO
      Q2 = NEWFRM
      AID = NEWARC
  620 CONTINUE
C
C     UPDATE TREE
C
      I = Q1
      J = DOWN(I)
```

```
      LSTAR = LEVEL(Q2)+1
      IF(THETA.EQ.0) GO TO 730
C     FLOWS NEED TO BE UPDATED
      CHG = ISIGN(THETA,CH(KTHETA))
  680 CONTINUE
C     UPDATE DUAL VARIABLE ON STEM
      DUAL(I) = DUAL(I)+NEWPR
C     UPDATE FLOW ON STEM
      N = FLOW(I)
      FLOW(I) = FLW
C     UPDATE ARC ID ON STEM
      DIR = ISIGN(1,ARCID(I))
      REF = IABS(ARCID(I))
      ARCID(I) = AID
C     PREPARE FOR LEVEL UPDATES
      LSAVE = LEVEL(I)
      LDIFF = LSTAR-LSAVE
      LEVEL(I) = LSTAR
      THD = I
  690 CONTINUE
      NXT = NEXT(THD)
      IF(LEVEL(NXT).LE.LSAVE) GO TO 700
C     UPDATE LEVEL
      LEVEL(NXT) = LEVEL(NXT)+LDIFF
C     UPDATE DUAL VARIABLE
      DUAL(NXT) = DUAL(NXT)+NEWPR
      THD = NXT
      GO TO 690
  700 CONTINUE
      K = J
  710 CONTINUE
      L = NEXT(K)
      IF(L.EQ.I) GO TO 720
      K = L
      GO TO 710
  720 CONTINUE
C     TEST FOR LEAVING ARC
      IF(I.EQ.ITHETA) GO TO 790
C     PREPARE FOR NEXT UPDATE ON STEM
      FLW = N-CHG*DIR
      AID = -ISIGN(REF,DIR)
C     MOVE DOWN STEM
      NEXT(K) = NXT
      NEXT(THD) = J
      K = I
      I = J
      J = DOWN(J)
      DOWN(I) = K
      LSTAR = LSTAR+1
      GO TO 680
  730 CONTINUE
C     ONLY FLOW ON NEW ARC CHANGES
  740 CONTINUE
C     UPDATE DUAL VARIABLE ON STEM
      DUAL(I) = DUAL(I)+NEWPR
C     UPDATE FLOW ON STEM
      N = FLOW(I)
      FLOW(I) = FLW
C     UPDATE ARC ID ON STEM
      DIR = ISIGN(1,ARCID(I))
      REF = IABS(ARCID(I))
      ARCID(I) = AID
```

```
C     PREPARE FOR LEVEL UPDATES
      LSAVE = LEVEL(I)
      LDIFF = LSTAR-LSAVE
      LEVEL(I) = LSTAR
      THD = I
  750 CONTINUE
      NXT = NEXT(THD)
      IF(LEVEL(NXT).LE.LSAVE) GO TO 760
C     UPDATE LEVEL
      LEVEL(NXT) = LEVEL(NXT)+LDIFF
C     UPDATE DUAL VARIABLE
      DUAL(NXT) = DUAL(NXT)+NEWPR
      THD = NXT
      GO TO 750
  760 CONTINUE
      K = J
  770 CONTINUE
      L = NEXT(K)
      IF(L.EQ.I) GO TO 780
      K = L
      GO TO 770
  780 CONTINUE
C     TEST FOR LEAVING ARC
      IF(I.EQ.ITHETA) GO TO 790
C     PREPARE FOR NEXT UPDATE ON STEM
      FLW = N
      AID = -ISIGN(REF,DIR)
C     MOVE DOWN STEM
      NEXT(K) = NXT
      NEXT(THD) = J
      K = I
      I = J
      J = DOWN(J)
      DOWN(I) = K
      LSTAR = LSTAR+1
      GO TO 740
  790 CONTINUE
      NEXT(K) = NXT
      NEXT(THD) = NEXT(Q2)
      NEXT(Q2) = Q1
      DOWN(Q1) = Q2
      GO TO 300
C
C     SUMMARY
C
  795 CONTINUE
C(DUMP CONTROL)     IF(ITDMPD.EQ.ITER) GO TO 870
C(DUMP CONTROL)     ITDMPD = ITER
C(TIMING)      T8 = T8-SECOND(X)
C(DUMP FLOWS)     IF(DMP) WRITE(7,8009) ITER
      INFEAS = .FALSE.
      KOST = KOSTO
      DO 830 I830=1,MNODE
      I = IABS(ARCID(I830))
C(DUMP FLOWS)     IF(DMP) WRITE(7,8007) NAME(I),FLOW(I830)
C(DUMP FLOWS)8007 FORMAT(1X,A10,6H BASIC,6H  FLOW,I15)
C(DUMP FLOWS)     IF((I.GT.0).AND.(I.LE.MREG)) GO TO 810
C(DUMP FLOWS)     IF(DMP) WRITE(7,8013) I830
C(DUMP FLOWS)8013 FORMAT(1H+,40X,16HCONNECTING NODE ,I10)
C(DUMP FLOWS) 810 CONTINUE
      IF((FLOW(I830).NE.0).AND.(COST(I).EQ.BIG)) INFEAS = .TRUE.
      KOST = KOST+COST(I)*FLOW(I830)
```

```
  830 CONTINUE
      DO 840 I840=1,MSLK
      IF(CAPAC(I840).GE.0) GO TO 840
      J840 = -CAPAC(I840)
      KOST = KOST+COST(I840)*J840
C(DUMP FLOWS)     IF(DMP) WRITE(7,8006) NAME(I840),J840
C(DUMP FLOWS)8006 FORMAT(1X,A10,6H AT UB,6H  FLOW,I15)
C(DUMP FLOWS)     IF(I840.LE.MREG) GO TO 840
C(DUMP FLOWS)     K840 = IABS(FROM(I840))
C(DUMP FLOWS)     IF(DMP) WRITE(7,8013) K840
  840 CONTINUE
C(DUMP COST)     IF(DMP) WRITE(7,8009) ITER
      IF(OPTIM) WRITE(6,8009) ITER
      IF(INFEAS) GO TO 850
      IF(OPTIM) WRITE(6,8005) KOST
C(DUMP COST)     IF(DMP) WRITE(7,8005) KOST
 8005 FORMAT(1H+,21X,4HCOST,I16)
      GO TO 860
  850 CONTINUE
      IF(OPTIM) WRITE(6,8008)
C(DUMP COST)     IF(DMP) WRITE(7,8008)
 8008 FORMAT(1H+,21X,10HINFEASIBLE)
  860 CONTINUE
C(DUMP ARRAYS)     IF(DMP) WRITE(7,8009) ITER
C(DUMP ARRAYS)     IF(DMP) WRITE(7,2002)(I,(NODE(I,J),J=1,6),I=1,MNODP1)
C(DUMP ARRAYS)     IF(DMP) WRITE(7,2001) (I,(ARC(I,J),J=1,5),I=ZERO,MARC)
C(TIMING)     T8 = T8+SECOND(X)
C(DUMP CONTROL) 870 CONTINUE
      IF(.NOT.OPTIM) GO TO 305
      IF(INFEAS) GO TO 892
C(TIMING)     T8 = T8-SECOND(X)
      WRITE(6,8009) ITER
 8009 FORMAT(10H ITERATION,I10)
      WRITE(6,8000)
 8000 FORMAT(1H+,20X,19H OPTIMAL INDICATION)
C(DEGENERATE PIVOTS)     WRITE(6,8012) IDEG
C(DEGENERATE PIVOTS)8012 FORMAT(1X,I10,18H DEGENERATE PIVOTS)
C(DEGENERATE PIVOTS)     IF(KDEG.NE.BIG) WRITE(6,8022) KDEG
C(DEGENERATE PIVOTS)8022 FORMAT(40H FIRST NON-DEGENERATE PIVOT AT ITERATION,I10)

C(DUAL VARIABLES)     WRITE(7,8016)
C(DUAL VARIABLES)8016 FORMAT(15H0DUAL VARIABLES)
C(DUAL VARIABLES)     J880 = 1+(MNODE-1)/8
C(DUAL VARIABLES)     L880 = 0
C(DUAL VARIABLES)     DO 880 I880=1,J880
C(DUAL VARIABLES)     K880 = L880+1
C(DUAL VARIABLES)     L880 = MIN0(MNODE,L880+8)
C(DUAL VARIABLES)     WRITE(7,8017) (I,I=K880,L880)
C(DUAL VARIABLES)     WRITE(7,8018) (DUAL(I),I=K880,L880)
C(DUAL VARIABLES)8017 FORMAT(1H0,5H NODE,8I15,A1)
C(DUAL VARIABLES)8018 FORMAT(1X ,5HVALUE,8I15,A1)
C(DUAL VARIABLES) 880 CONTINUE
C(NODE-TO-NODE FLOW)     PRES = .TRUE.
C(ARC FLOWS)     PRES = .TRUE.
      IF(.NOT.PRES) GO TO 885
C     MOVE BASIC FLOWS TO CAPACITY CELLS FOR BASIC ARCS
      DO 884 I884=1,MNODE
      J884 = IABS(ARCID(I884))
      CAPAC(J884) = -FLOW(I884)
  884 CONTINUE
  885 CONTINUE
C(NODE-TO-NODE FLOW)     WRITE(7,8019)
```

```
C(NODE-TO-NODE FLOW)8019 FORMAT(19HONODE-TO-NODE FLOWS/
C(NODE-TO-NODE FLOW)    *41H0 FROM   TO          FLOW          COST)
C     SUM FLOWS IN EACH TO-GROUP
C(NODE-TO-NODE FLOW)      TO = 1
C(NODE-TO-NODE FLOW)      TRY = 1
C(NODE-TO-NODE FLOW)      FRM = FROM(TRY)
C(NODE-TO-NODE FLOW) 886 CONTINUE
C(NODE-TO-NODE FLOW)      LST = ISIGN(LNODP1,FRM)
C     CLEAR OUT ALL FLOW AND DUAL VARIABLE CELLS
C     FOR SUMS OF FLOWS AND COSTS, RESPECTIVELY
C(NODE-TO-NODE FLOW)      DO 887 I887=1,MNODE
C(NODE-TO-NODE FLOW)      FLOW(I887) = 0
C(NODE-TO-NODE FLOW)      DUAL(I887) = 0
C(NODE-TO-NODE FLOW) 887 CONTINUE
C(NODE-TO-NODE FLOW)      SUMFLW = 0
C(NODE-TO-NODE FLOW) 888 CONTINUE
C(NODE-TO-NODE FLOW)      FLW = MAX0(0,-CAPAC(TRY))+FLOOR(TRY)
C(NODE-TO-NODE FLOW)      IF(FLW.EQ.0) GO TO 889
C(NODE-TO-NODE FLOW)      FM = IABS(FRM)
C(NODE-TO-NODE FLOW)      FLOW(FM) = FLOW(FM)+FLW
C(NODE-TO-NODE FLOW)      DUAL(FM) = DUAL(FM)+FLW*COST(TRY)
C(NODE-TO-NODE FLOW)      SUMFLW = SUMFLW+FLW
C(NODE-TO-NODE FLOW) 889 CONTINUE
C(NODE-TO-NODE FLOW)      TRY = TRY+1
C(NODE-TO-NODE FLOW)      FRM = FROM(TRY)
C(NODE-TO-NODE FLOW)      IF(XOR(FRM,LST).GT.0) GO TO 888
C     OUTPUT NON-ZERO VALUES
C(NODE-TO-NODE FLOW)      IF(SUMFLW.EQ.0) GO TO 891
C(NODE-TO-NODE FLOW)      DO 890 I890=1,MNODE
C(NODE-TO-NODE FLOW)      IF(FLOW(I890).EQ.0) GO TO 890
C(NODE-TO-NODE FLOW)      WRITE(7,8020) I890,TO,FLOW(I890),DUAL(I890)
C(NODE-TO-NODE FLOW)8020 FORMAT(1X,2I5,2I15)
C(NODE-TO-NODE FLOW) 890 CONTINUE
C(NODE-TO-NODE FLOW) 891 CONTINUE
C(NODE-TO-NODE FLOW)      TO = TO+1
C(NODE-TO-NODE FLOW)      IF(TO.NE.MNODP1) GO TO 886
C(ARC FLOWS)      WRITE(7,8023)
C(ARC FLOWS)8023 FORMAT(10HOARC FLOWS/
C(ARC FLOWS)     *52H0ARC          FROM   TO          FLOW          COST)
C(ARC FLOWS)      TO = 1
C(ARC FLOWS)      TRY = 1
C(ARC FLOWS)      FRM = FROM(TRY)
C(ARC FLOWS)8886 CONTINUE
C(ARC FLOWS)      LST = ISIGN(LNODP1,FRM)
C(ARC FLOWS)8888 CONTINUE
C(ARC FLOWS)      FLW = MAX0(0,-CAPAC(TRY))+FLOOR(TRY)
C(ARC FLOWS)      IF(FLW.EQ.0) GO TO 8889
C(ARC FLOWS)      FM = IABS(FRM)
C(ARC FLOWS)      CST = FLW*COST(TRY)
C(ARC FLOWS)      WRITE(7,8024) NAME(TRY),FM,TO,FLW,CST
C(ARC FLOWS)8024 FORMAT(1X,A10,1X,2I5,2I15)
C(ARC FLOWS)8889 CONTINUE
C(ARC FLOWS)      TRY = TRY+1
C(ARC FLOWS)      FRM = FROM(TRY)
C(ARC FLOWS)      IF(XOR(FRM,LST).GT.0) GO TO 8888
C(ARC FLOWS)      TO = TO+1
C(ARC FLOWS)      IF(TO.NE.MNODP1) GO TO 8886
C(TIMING)      T8 = T8+SECOND(X)
  892 CONTINUE
C
C     END OF PROBLEM
C
```

```
C(TIMING)     T8 = T8-SECOND(X)
C(TIMING)     WRITE(6,8904) T3
C(TIMING)     P3 = T3/T6
C(TIMING)     WRITE(6,8909) P3
C(TIMING)     WRITE(6,8905) T4
C(TIMING)     P4 = T4/T6
C(TIMING)     WRITE(6,8909) P4
C(TIMING)     WRITE(6,8906) T5
C(TIMING)     P5 = T5/T6
C(TIMING)     WRITE(6,8909) P5
C(TIMING)     WRITE(6,8907) T6
C(TIMING)     T6 = T6/FLOAT(ITER)
C(TIMING)     WRITE(6,8911) T6
C(TIMING)     WRITE(6,8908) T7
C(TIMING)     T7 = T7/FLOAT(ITER)
C(TIMING)     WRITE(6,8911) T7
C(TIMING)     T8 = T8+SECOND(X)
C(TIMING)     WRITE(6,8903) T8
C     READ LEFT-OVER DUMP CONTROL CARDS, IF ANY
C(DUMP CONTROL) 894 CONTINUE
C(DUMP CONTROL)      IF(ITDCHG.EQ.0) GO TO 896
C(TIMING)     T1 = T1-SECOND(X)
C(DUMP CONTROL)      NSTOP = 19
C(DUMP CONTROL)       READ(5,1000) ITDMP1,ITDMP2,ITDMP3,ITDOPT,ITDCHG
C
C     DUMP ITER START,DUMP ITER END,DUMP ITER STEP,DUMP OPT,CHANGE
C       (I10)     ,       (I10)  ,       (I10)    ,  (I10) ,   (I10)
C
C(DUMP CONTROL)      IF(EOF(5)) 999,895
C(DUMP CONTROL) 895 CONTINUE
C(DUMP CONTROL)      KARD = KARD+1
C(DUMP CONTROL)      WRITE(6,1001) ITDMP1,ITDMP2,ITDMP3,ITDOPT,ITDCHG
C(TIMING)     T1 = T1+SECOND(X)
C(DUMP CONTROL)      GO TO 894
C(DUMP CONTROL) 896 CONTINUE
C(TIMING)     WRITE(6,8902) T1
C(TIMING)     T0 = T0+SECOND(X)
C(TIMING)     WRITE(6,8900) T0
      GO TO 5
C
C     EXIT
C
  999 CONTINUE
      WRITE(6,1999) NSTOP,KARD
 1999 FORMAT(5H STOP,I5/1X,I10,16H DATA CARDS READ)
      STOP
      END
C     INTEGER FUNCTION XOR(I,J)
C     XOR = I*J
C     IF(XOR.EQ.0) XOR = I+J
C     RETURN
C     END
```

References

1 Aashtiani, H. A., and T. L. Magnanti, "Implementing Primal-Dual Network Flow Algorithms," Technical Report OR 055-76, Operations Research Center, M.I.T., Boston, Mass., 1976.

2 Abdulaal, M., and L. J. LeBlanc, "Multimodal Network Equilibrium," Technical Report 77013, Department of Industrial Engineering and Operations Research, Southern Methodist University, Dallas, Tex., 1977.

3 Ali, A., R. Helgason, J. Kennington, and H. Lall, "Primal-Simplex Network Codes: State-of-the-Art Implementation Technology," *Networks*, **8**, 315–339 (1978).

4 Ali, A., R. Helgason, J. Kennington, and H. Lall, "Computational Comparison among Three Multicommodity Network Flow Algorithms," to appear in *Operations Research*.

5 Armstrong, R. D., D. D. Klingman, and D. H. Whitman, "Implementation and Analysis of a Variant of the Dual Method for the Capacitated Transshipment Problem," paper presented at Joint National Meeting of ORSA/TIMS in Los Angeles, 1978.

6 Assad, A. A., "Multicommodity Network Flows: A Survey," *Networks*, **8**, **1**, 37–92 (1978).

7 Balachandran, V., and G. L. Thompson, "An Operator Theory of Parametric Programming for the Generalized Transportation Problem: I. Basic Theory," *Naval Research Logistics Quarterly*, **22**, 1, 79–100 (1975).

8 Balachandran, V., and G. L. Thompson, "An Operator Theory of Parametric Programming for the Generalized Transportation Problem: II. Rim, Cost, and Bound Operators," *Naval Research Logistics Quarterly*, **22**, 1, 101–126 (1975).

9 Balachandran, V., and G. L. Thompson, "An Operator Theory of Parametric Programming for the Generalized Transportation Problem: III. Weight Operators," *Naval Research Logistics Quarterly*, **22**, 2, 297–316 (1975).

10 Balachandran, V., and G. L. Thompson, "An Operator Theory of Parametric Programming for the Generalized Transportation Problem: IV. Global Operators," *Naval Research Logistics Quarterly*, **22**, 2, 317–340 (1975).

11 Barr, R. S., F. Glover, and D. Klingman, "An Improved Version of the Out-of-Kilter Method and a Comparative Study of Computer Codes," *Mathematical Programming*, **7**, 1, 60–87 (1974).

12 Barr, R. S., F. Glover, and D. Klingman, "The Alternating Basis Algorithm for Assignment Problems," *Mathematical Programming*, **13**, 1, 1–13 (1977).

13 Barr, R. S., "Algorithmic and Computer Code Developments for Combinational and Network Problems," unpublished dissertation, The University of Texas, Austin, Tex., 1978.

14 Barr, R. S., F. Glover, and D. Klingman, "A New Optimization Method for Large-Scale Fixed Charge Transportation Problems," Technical Report, Cox School of Business, Southern Methodist University, Dallas, Tex., 1978.

15 Barr, R. S., F. Glover, and D. Klingman, "The Generalized Alternating Path Algorithm for Transportation Problems," *European Journal of Operational Research*, **2**, 137–144 (1978).

16 Barr, R. S., J. Elam, F. Glover, and D. Klingman, "A Network Augmenting Path Basis Algorithm for Transshipment Problems," to appear in A. V. Fiacco and K. O. Kortanek, Eds., *Extremal Methods and Systems Analysis*, Springer-Verlag, New York, 1979.

17 Barr, R. S., F. Glover, and D. Klingman, "Enhancements of Spanning Tree Labelling Procedures for Network Optimization," *INFOR*, **17**, 1, 16–34 (1979).

18 Barr, R. S., and J. Turner, "Optimal Microdata File Merging through Large-Scale Network Technology," to appear in a special issue of *Mathematical Programming Studies*.

19 Bartle, R. G., *The Elements of Real Analysis*, John Wiley and Sons, New York, 1964.

20 Bazaraa, M. S., and J. J. Jarvis, *Linear Programming and Network Flows*, John Wiley and Sons, New York, 1977.

21 Beale, E. M. L., "An Algorithm for Solving the Transportation Problem When the Shipping Cost over Each Route is Convex," *Naval Research Logistics Quarterly*, **6**, 1, 43–56 (1959).

22 Bennett, J. M., "An Approach to Some Structured Linear Programming Problems," *Operations Research*, **14**, 636–645 (1966).

23 Berge, C., *Topological Spaces*, Macmillan Co., New York, 1963.

24 Bitran, G. R., and A. C. Hax, "On the Solution of Convex Knapsack Problems with Bounded Variables," to appear in the *Proceedings of the IX International Symposium on Mathematical Programming*, Budapest, 1976.

25 Bradley, G. H., G. G. Brown, and G. W. Graves, "Design and Implementation of Large-Scale Primal Transshipment Algorithms," *Management Science*, **24**, 1, 1–34 (1977).

26 Cantoni, A., "Optimal Curve Fitting with Piecewise Linear Functions," *IEEE Transactions on Computers*, **C-20**, 1, 59–67 (1971).

27 Cantor, D. G., and M. Gerla, "Optimal Routing in a Packet Switched Computer Network," *IEEE Transactions on Computers*, **C-23**, 10, 1062–1069 (1974).

28 Charnes, A., "Optimality and Degeneracy in Linear Programming," *Econometrica*, **20**, 2, 160–170 (1952).

29 Charnes, A., and W. W. Cooper, "Nonlinear Network Flows and Convex Programming over Incidence Matrices," *Naval Research Logistics Quarterly*, **5**, 3, 231–240 (1958).

30 Charnes, A., and W. W. Cooper, *Management Models and Industrial Applications of Linear Programming*, Vols. 1 and 2, John Wiley and Sons, New York, 1961.

31 Charnes, A., and W. W. Cooper, "Multicopy Traffic Network Models," in R. Herman, Ed., *Theory of Traffic Flow*, Elsevier Publishing Co., Amsterdam, 1961, pp. 85–96.

32 Chen, H., and C. G. DeWald, "A Generalized Chain Labeling Algorithm for Solving Multicommodity Flow Problems," *Computers and Operations Research*, **1**, 437–465 (1974).

33 Chen, S., and R. Saigal, "A Primal Algorithm for Solving a Capacitated Network Flow Problem with Additional Linear Constraints," *Networks*, **7**, 1, 59–79 (1977).

34 Clasen, R. J., "The Numerical Solution of Network Problems Using the Out-of-Kilter Algorithm," Memo, RM-5456-PR, The Rand Corp., Santa Monica, Calif., 1968.

35 Collins, M., L. Cooper, R. Helgason, J. Kennington, and L. J. LeBlanc, "Solving the Pipe Network Analysis Problem Using Optimization Techniques," *Management Science*, **24**, 7, 747–760 (1978).

36 Cooper, L., and D. Steinberg, *Methods and Applications of Linear Programming*, W. B. Saunders Co., Philadelphia, 1974.

37 Cooper, L., and J. Kennington, "Steady-State Analysis of Nonlinear Resistive Electrical Networks Using Optimization Techniques," Technical Report IEOR 77012, Department of Industrial Engineering and Operations Research, Southern Methodist University, Dallas, Tex., 1977.

38 Cooper, L., and L. J. LeBlanc, "Stochastic Transportation Problems and Other Network Related Convex Problems," *Naval Research Logistics Quarterly*, **24**, 2, 327–336 (1977).

39 Cox, M. G., "An Algorithm for Approximating Convex Functions by Means of First Degree Splines," *The Computer Journal*, **14**, 272–275 (1971).

40 Cremeans, J. E., R. A. Smith, and G. R. Tyndall, "Optimal Multicommodity Network Flows with Resource Allocation," *Naval Research Logistics Quarterly*, **17**, 269–280 (1970).

41 Cunningham, W. H., "A Network Simplex Method," *Mathematical Programming*, **11**, 2, 105–116 (1976).

42 Dafermos, S. C., "An Extended Traffic Assignment Model with Applications to Two-Way Traffic," *Transportation Science*, **5**, 4, 366–389 (1971).

43 Dafermos, S. C., "The Traffic Assignment Problem for Multiclass-User Transportation Networks," *Transportation Science*, **6**, 1, 73–87 (1971).

44 Dantzig, G. B., "Application of the Simplex Method to a Transportation Problem," in T. C. Koopmans, Ed., *Activity Analysis of Production and Allocation*, John Wiley and Sons, New York, 1951.

45 Dantzig, G. B., A. Orden, and P. Wolfe, "The Generalized Simplex Method for Minimizing a Linear Form under Linear Inequality Restraints," *Pacific Journal of Mathematics*, **5**, 2, 183–195 (1955).

46 Dantzig, G. B., and P. Wolfe, "Decomposition Principle for Linear Programs," *Operations Research*, **8**, 101–111 (1960).

47 Dantzig, G. B., *Linear Programming and Extensions*, Princeton University Press, Princeton, N. J., 1963.

48 Davis, P. J., *Interpolation and Approximation*, Blaisdell, New York, 1963.

49 Dembo, R. S., and J. G. Klincewicz, "A Second-Order Algorithm for Network Flow Problems with Convex Separable Costs," Technical Report 21, School of Business, Yale University, New Haven, Conn., 1978.

50 Dennis, J. B., *Mathematical Programming and Electrical Networks*, John Wiley and Sons, New York, 1959.

51 Dial, R., F. Glover, D. Karney, and D. Klingman, "A Computational Analysis of Alternative Algorithms and Labeling Techniques for Finding Shortest Path Trees," Research Report CCS 291, Center for Cybernetic Studies, The University of Texas, Austin, Tex., 1977.

52 Dwyer, P. S., "A General Treatment of Upper Unbounded and Bounded Hitchcock Problems," *Naval Research Logistics Quarterly*, **21**, 3, 445–464 (1974).

53 Elam, J., F. Glover, and D. Klingman, "A Strongly Convergent Primal-Simplex Algorithm for Generalized Networks," *Mathematics of Operations Research*, **4**, 1, 39–59 (1979).

54 Evans, J. R., "Theoretical and Computational Aspects of Integer Multicommodity Network Flow Problems," unpublished dissertation, Department of Industrial and Systems Engineering, Georgia Institute of Technology, Atlanta, Ga., 1975.

55 Evans, J. R., "A Combinatorial Equivalence Between a Class of Multicommodity Flow Problems and the Capacitated Transportation Problem," *Mathematical Programming*, **10**, 3, 401–404 (1976).

56 Evans, J. R., "Maximal Flow in Probabilistic Graphs—The Discrete Case," *Networks*, **6**, 2, 161–184 (1976).

57 Evans, J. R., J. J. Jarvis, and R. A. Duke, "Graphic Matroids and the Multicommodity Transportation Problem," *Mathematical Programming*, **13**, 323–328 (1977).

58 Evans, J. R., "A Single-Commodity Transformation for Certain Multicommodity Networks," *Operations Research*, **26**, 4, 673–681 (1978).

59 Evans, J. R., "The Simplex Method for Integral Multicommodity Networks," *Naval Research Logistics Quarterly*, **25**, 1, 31–38 (1978).

60 Evans, J. R., and J. J. Jarvis, "Network Topology and Integral Multicommodity Flow Problems," *Networks*, **8**, 2, 107–120 (1978).

61 Ferland, J. A., "Minimum Cost Multicommodity Circulation Problem with Convex Arc Costs," *Transportation Science*, **8**, 4, 355–360 (1974).

62 Florian, M., "An Improved Linear Approximation Algorithm for the Network Equilibrium (Packet Switching) Problem," Technical Report, Centre de Recherche sur les Transports, Universite de Montreal, Quebec, Canada, 1977.

63 Florian, M., S. Nguyen, and F. Soumis, "Two Methods for Accelerating an Equilibrium Traffic Assignment Algorithm," Technical Report 251, Department d'Informatique, University of Montreal, Quebec, Canada, 1977.

64 Ford, L. R., and D. R. Fulkerson, "A Suggested Computation for Maximal Multicommodity Network Flow," *Management Science*, **5**, 97–101 (1958).

65 Ford, L. R., and D. R. Fulkerson, *Flows in Networks*, Princeton University Press, Princeton, N. J., 1962.

66 Ford, L. R., and D. R. Fulkerson, "Maximal Flow through a Network," *Canadian Journal of Mathematics*, **8**, 3, 399–404 (1956).

67 Frank, H., and I. T. Frisch, *Communication, Transmission, and Transportation Networks*, Addison-Wesley Publishing Co., Reading, Mass., 1971.

68 Frank, M., and P. Wolfe, "An Algorithm for Quadratic Programming," *Naval Research Logistics Quarterly*, **3**, 2, 95–110 (1956).

69 Fratta, L., M. Gerla, and L. Kleinrock, "The Flow Deviation Method: An Approach to Store-and-Forward Communication Network Design," *Networks*, **3**, 97–133 (1973).

70 Fulkerson, D. R., "An Out-of-Kilter Method for Minimal-Cost Flow Problems," *Journal of the Society of Industrial and Applied Mathematics*, **9**, 1, 18–27 (1961).

71 Gass, S. I., *Linear Programming: Methods and Applications*, McGraw-Hill Book Co., New York, 1969.

72 Gavish, B., P. Schweitzer, and E. Shlifer, "The Zero Pivot Phenomenon in Transportation and Assignment Problems and Its Computational Implications," *Mathematical Programming*, **12**, 2, 226–240 (1977).

73 Geoffrion, A. M., "Primal Resource-Directive Approaches for Optimizing Nonlinear Decomposable Systems," *Operations Research*, **18**, 3, 375–403 (1970).

74 Geoffrion, A. M., and G. W. Graves, "Multicommodity Distribution System Design by Benders Decomposition," *Management Science*, **20**, 5, 822–844 (1974).

75 Glover, F., and D. Klingman, "On the Equivalence of Some Generalized Network Problems to Pure Network Problems," *Mathematical Programming*, **4**, 3, 269–278 (1973).

76 Glover, F., D. Karney, and D. Klingman, "Implementation and Computational Comparisons of Primal, Dual, and Primal-Dual Computer Codes for Minimum Cost Network Flow Problems," *Networks*, **4**, 3, 191–212 (1974).

77 Glover, F., D. Karney, D. Klingman, and A. Napier, "A Computational Study on Start Procedures, Basis Change Criteria, and Solution Algorithms for Transportation Problems," *Management Science*, **20**, 5, 793–813 (1974).

78 Glover, F., and D. Klingman, "New Advances in the Solution of Large-Scale Network and Network-Related Problems," Technical Report CS 177, Center for Cybernetic Studies, The University of Texas, Austin, Tex., 1974.

79 Glover, F., D. Klingman, and J. Stutz, "Augmented Threaded Index Method for Network Optimization," *INFOR*, **12**, 3, 293–298 (1974).

80 Glover, F., D. Klingman, and J. Stutz, "Extensions of the Augmented Predecessor Index Method to Generalized Network Problems," *Transportation Science*, **7**, 4, 377–384 (1974).

81 Glover, F., J. Hultz, and D. Klingman, "Improved Computer-Based Planning Techniques," Research Report CCS 283, Center for Cybernetic Studies, The University of Texas, Austin, Tex., 1977.

82 Glover, F., and D. Klingman, "Network Applications in Industry and Government," *AIIE Transactions*, **9**, 4, 363–376 (1977).

83 Glover, F., and D. Klingman, "The Simplex Son Algorithm for LP/Embedded Network Problems," Technical Report CCS 317, Center for Cybernetic Studies, The University of Texas, Austin, Tex., 1977.

84 Glover, F., J. Hultz, D. Klingman, and J. Stutz, "Generalized Networks: A Fundamental Computer-Based Planning Tool," *Management Science*, **24**, 12, 1209–1220 (1978).

85 Glover, F., D. Karney, and D. Klingman, and R. Russell, "Solving Singly Constrained Transshipment Problems," *Transportation Science*, **12**, 4, 277–297 (1978).

86 Glover, F., and D. Klingman, "Some Recent Practical Misconceptions about the State-of-the-Art of Network Algorithms," *Operations Research*, **2**, 370–379 (1978).

87 Golden, B. L., "A Minimum-Cost Multicommodity Network Flow Problem Concerning Imports and Exports," *Networks*, **5**, 4, 331–356 (1975).

88 Golden, B. L. and T. L. Magnanti, "Deterministic Network Optimization: A Bibliography," *Networks*, **7**, 2, 149–183 (1977).

89 Graves, G. W., and R. D. McBride, "The Factorization Approach to Large-Scale Linear Programming," *Mathematical Programming*, **10**, 1, 91–110 (1976).

90 Grigoriadis, M. D., and W. W. White, "A Partitioning Algorithm for the Multicommodity Network Flow Problem," *Mathematical Programming*, **3**, 157–177 (1972).

91 Hadley, G., *Linear Programming*, Addison-Wesley Publishing Co., Reading, Mass., 1962.

92 Hall, M. A., "Hydraulic Network Analysis Using (Generalized) Geometric Programming," *Networks*, **6**, 105–130 (1976).

93 Hartman, J. K. and L. S. Lasdon, "A Generalized Upper Bounding Method for Doubly Coupled Linear Programs," *Naval Research Logistics Quarterly*, **17**, 4, 411–429 (1970).

94 Hartman, J. K., and L. S. Lasdon, "A Generalized Upper Bounding Algorithm for Multicommodity Network Flow Problems," *Networks*, **1**, 333–354 (1972).

95 Hatch, R. S., "Bench Marks Comparing Transportation Codes Based on Primal-Simplex and Primal-Dual Algorithms," *Operations Research*, **23**, 6, 1167–1171 (1975).

96 Held, M., and R. Karp, "The Traveling Salesman Problem and Minimum Spanning Trees," *Operations Research*, **18**, 1138–1162 (1970).

97 Held, M., P. Wolfe, and H. Crowder, "Validation of Subgradient Optimization," *Mathematical Programming*, **6**, 62–88 (1974).

98 Helgason, R. V., and J. L. Kennington, "A Product Form Representation of the Inverse of a Multicommodity Cycle Matrix," *Networks*, **7**, 297–322 (1977).

99 Helgason, R. V., and J. L. Kennington, "An Efficient Procedure for Implementing a Dual-Simplex Network Flow Algorithm," *AIIE Transactions*, **9**, 1, 63–68 (1977).

100 Helgason, R. V., and J. L. Kennington, "An Efficient Specialization of the Convex Simplex Method for Nonlinear Network Flow Problems," Technical Report IEOR 77017, Department of Industrial Engineering and Operations Research, Southern Methodist University, Dallas, Tex., 1978.

101 Helgason, R. V., J. L. Kennington, and H. Lall, "A Polynomically Bounded Algorithm for a Singly Constrained Quadratic Program," to appear in *Mathematical Programming*.

102 Helgason, R. V., "A Lagrangean Relaxation Approach to the Generalized Fixed Charge Multicommodity Minimal Cost Network Flow Problem," unpublished dissertation, Department of Operations Research, Southern Methodist University, Dallas, Tex., 1980.

103 Hitchcock, F. L., "The Distribution of a Product from Several Sources to Numerous Localities," *Journal of Mathematics and Physics*, **20**, 224–230 (1941).

104 Holloway, C. A., "An Extension of the Frank and Wolfe Method of Feasible Directions," *Mathematical Programming*, **6**, 14–27 (1974).

105 Hu, T. C., "Multicommodity Network Flows," *Operations Research*, **11**, 344–360 (1963).

106 Hu, T. C., "Minimum-Cost Flows in Convex-Cost Networks," *Naval Research Logistics Quarterly*, **13**, 1, 1–9 (1966).

107 Jarvis, J. J., "On the Equivalence Between the Node-Arc and Arc-Chain Formulation for the Multicommodity Maximal Flow Problem," *Naval Research Logistics Quarterly*, **15**, 525–529 (1969).

108 Jarvis, J. J., and P. D. Keith, "Multicommodity Flows with Upper and Lower Bounds," Working Paper, School of Industrial and Systems Engineering, Georgia Institute of Technology, Atlanta, Ga., 1974.

109 Jarvis, J. J., and O. M. Martinez, "A Sensitivity Analysis of Multicommodity Network Flows," *Transportation Science*, **11**, 4, 299–306 (1977).

110 Jewell, W. S., "Optimal Flow Through Networks with Gains," *Operations Research*, **10**, 4, 476–499 (1962).

111 Jewell, W. S., "Models for Traffic Assignment," *Transportation Research*, **1**, 31–46 (1967).

112 Johnson, E. L., "Programming in Networks and Graphs," Technical Report ORC 65-1, Operations Research Center, University of California at Berkeley, 1965.

113 Johnson, E. L., "Networks and Basic Solutions," *Operations Research*, **14**, 619–623 (1966).

114 Jorgensen, N. O., "Some Aspects of the Urban Traffic Assignment Problem," Institute of Transportation and Traffic Engineering Graduate Report, University of California at Berkeley, 1963.

115 Kantorovich, L. V., "Mathematical Methods in the Organization and Planning of Production," Publication House of the Leningrad State University, 1939. 68 pp. Translated in *Management Science*, **6**, 366–422 (1960).

116 Kaul, R. N., "An Extension of Generalized Upper Bounded Techniques for Linear Programming," ORC Report No. 65-27, Department of Operations Research, University of California at Berkeley, 1965.

117 Kennington, J. L., and E. Unger, "A New Branch-and-Bound Algorithm for the Fixed-Charge Transportation Problem," *Management Science*, **22**, 10, 1116–1126 (1976).

118 Kennington, J. L., "Solving Multicommodity Transportation Problems Using a Primal Partitioning Simplex Technique," *Naval Research Logistics Quarterly*, **24**, 2, 309–325 (1977).

119 Kennington, J. L., and M. Shalaby, "An Effective Subgradient Procedure for Minimal Cost Multicommodity Flow Problems," *Management Science*, **23**, 9, 994–1004 (1977).

120 Kennington, J. L., "A Survey of Linear Cost Multicommodity Network Flows," *Operations Research*, **26**, 2, 209–236 (1978).

121 Klein, M., "A Primal Method for Minimal Cost Flows with Applications to the Assignment and Transportation Problems," *Management Science*, **14**, 3, 205–220 (1967).

122 Klingman, D., A. Napier, and J. Stutz, "NETGEN: A Program for Generating Large Scale Capacitated Assignment, Transportation, and Minimum Cost Flow Network Problems," *Management Science*, **20**, 5, 814–821 (1974).

123 Klingman, D., "Finding Equivalent Network Formulations for Constrained Network Problems," *Management Science*, **23**, 7, 737–744 (1977).

124 Klingman, D., and R. Russell, "A Streamlined Simplex Approach to the Singly Constrained Transportation Problem," *Naval Research Logistics Quarterly*, **25**, 4, 681–696 (1978).

125 Koopmans, T. C., "Optimum Utilization of the Transportation System," in *Proceedings of the International Statistical Conference*, Washington, D. C., 1947 (reprinted in *Econometrica*, **17**, supplement, 1949).

126 Koopmans, T. C., and S. Reiter, "A Model of Transportation," in T. C. Koopmans, Ed., *Activity Analysis of Production and Allocation*, John Wiley and Sons, New York, 1951.

127 Kotiah, T. C. T., and D. I. Steinberg, "On the Possibility of Cycling with the Simplex Method," *Operations Research*, **26**, 2, 374–376 (1978).

128 Kuhn, H. W., and A. W. Tucker, "Nonlinear Programming," in *Proceedings of the Second Berkeley Symposium on Mathematical Statistics and Probability*, Berkeley, Calif., 1950, pp. 481–492.

129 Kuhn, H. W., "The Hungarian Method for the Assignment Problem," *Naval Research Logistics Quarterly*, **2**, 83–97 (1955).

130 Langley, R. W., "Continuous and Integer Generalized Flow Problems," unpublished dissertation, Department of Industrial and Systems Engineering, Georgia Institute of Technology, Atlanta, Ga., 1973.

131 Langley, R. W., J. L. Kennington, and C. M. Shetty, "Efficient Computational Devices for the Capacitated Transportation Problem," *Naval Research Logistics Quarterly*, **21**, 4, 637–647 (1974).

132 Lasdon, L. S., "Optimization Theory for Large Systems," Macmillan Co., London, England, 1972.

133 LeBlanc, L. J., and L. Cooper, "The Transportation-Production Problem," *Transportation Science*, **8**, 4, 344–354 (1974).

134 LeBlanc, L. J., E. Morlok, and W. Pierskalla, "An Efficient Approach to Solving the Road Network Equilibrium Traffic Assignment Problem," *Transportation Research*, **9**, 309–318 (1975).

135 LeBlanc, L. J., "The Conjugate Gradient Technique for Certain Quadratic Network Problems," *Naval Research Logistics Quarterly*, **23**, 4, 597–602 (1976).

136 LeBlanc, L. J., M. Abdulaal, and R. Helgason, "On the Improved Rate of Convergence of the Frank-Wolfe Algorithm for Large-Scale Convex Network Problems," paper presented at the National Meeting of ORSA/TIMS in Atlanta, 1977.

137 Leventhal, T., G. Nemhauser, and L. Trotter, Jr., "A Column Generation Algorithm for Optimal Traffic Assignment," *Transportation Science*, **7**, 2, 168–176 (1973).

138 Loomba, N. P., *Linear Programming: A Managerial Perspective*, Macmillan Co., New York, 1976.

139 Luenberger, D. G., *Introduction to Linear and Nonlinear Programming*, Addison-Wesley Publishing Co., Reading, Mass., 1973.

140 Maier, S. F., "A Compact Inverse Scheme Applied to a Multicommodity Network with Resource Constraints," in R. Cottle and J. Krarup, Eds., *Optimization Methods for Resource Allocation*, The English University Press, London, England, 1974, pp. 179–203.

141 Maurras, J. F., "Optimization of the Flow through Networks with Gains," *Mathematical Programming*, **3**, 135–144 (1972).

142 McCallum, C. J., "An Algorithm for Certain Quadratic Integer Programs," Technical Report, Bell Laboratories, Holmdel, N. J. (undated).

143 Menon, V. V., "The Minimal Cost Flow Problem with Convex Costs," *Naval Research Logistics Quarterly*, **12**, 2, 163–172 (1965).

144 Minieka, E., *Optimization Algorithms for Networks and Graphs*, Marcel Dekker, New York, 1978.

145 Mulvey, J. M., "Pivot Strategies for Primal-Simplex Network Codes," *Journal of the Association for Computing Machinery*, **25**, 2, 266–270 (1978).

146 Nguyen, S., "A Mathematical Programming Approach to Equilibrium Methods of Traffic Assignment with Fixed Demands," Technical Report, Centre de Recherche sur les Transports, Universite de Montreal, Montreal, Quebec, Canada, 1973.

147 Nguyen, S., "An Algorithm for the Traffic Assignment Problem," *Transportation Science*, **8**, 203–216 (1974).

148 Orchard-Hays, W., *Advanced Linear-Programming Computing Techniques*, McGraw-Hill Book Co., New York, 1968.

149 Orden, A., "The Transshipment Problem," *Management Science*, **2**, 2, 276–285 (1956).

150 Pavlidis, T., "Segmentation of Plane Curves," *IEEE Transactions on Computers*, **C-23**, 8, 860–870 (1974).

151 Peterson, E. R., "A Primal-Dual Traffic Assignment Algorithm," *Management Science*, **22**, 1, 87–95 (1975).

152 Phillips, G. M., "Algorithms for Piecewise Straight Line Approximations," *The Computer Journal*, **11**, 211–212 (1968).

153 Poljak, B. T., "A General Method of Solving Extremum Problems," *Soviet Mathematics*, **8**, 3, 593–597 (1967).

154 Poljak, B. T., "Minimization of Unsmooth Functionals," *USSR Computational Mathematics and Mathematical Physics*, **9**, 3, 14–29 (1969).

155 Rardin, R. L., and V. E. Unger, "Solving Fixed Charge Network Problems with Group Theory-Based Penalties," *Naval Research Logistics Quarterly*, **23**, 1, 67–84 (1976).

156 Robacker, J. T., "Notes on Linear Programming: Part XXXVII, Concerning Multicommodity Networks," Memo RM-1799, The Rand Corporation, Santa Monica, Calif., 1956.

157 Rockafellar, R. T., *Convex Analysis*, Princeton University Press, Princeton, N. J., 1972.

158 Rosenthal, R. E., "Scheduling Reservoir Releases for Maximum Hydropower Benefit by Nonlinear Programming on a Network," Technical Report, Management Science Program, University of Tennessee, Knoxville, Tenn., 1977.

159 Ross, G. T., and R. M. Soland, "A Branch and Bound Algorithm for the Generalized Assignment Problem," *Mathematical Programming*, **8**, 1, 91–103 (1975).

160 Rothschild, B., and A. Whinston, "Feasibility of Two Commodity Network Flows," *Operations Research*, **14**, 1121–1129 (1966).

161 Rothschild, B., and A. Whinston, "On Two Commodity Network Flows," *Operations Research*, **14**, 377–387 (1966).

162 Rothschild, B., and A. Whinston, "Multicommodity Network Flows," Working Paper, Krannert School, Purdue University, Lafayette, Ind., 1969.

163 Rutenberg, D. P., "Maneuvering Liquid Assets in a Multi-National Company: Formulation and Deterministic Solution Procedures," *Management Science*, **16**, 10, B671–B684 (1970).

164 Saigal, R., "Multicommodity Flows in Directed Networks," ORC 67-38, Operations Research Center, University of California, Berkeley, Calif., 1967.

165 Sakarovitch, M., "Two Commodity Network Flows and Linear Programming," *Mathematical Programming*, **4**, 1–20 (1973).

166 Shapiro, J. F., "A Note on the Primal-Dual and Out-of-Kilter Algorithms for Network Optimization Problems," *Networks*, **7**, 1, 81–88 (1977).

167 Sharp, J. F., J. C. Snyder, and J. H. Greene, "A Decomposition Algorithm for Solving the Multifacility Production-Transportation Problem with Nonlinear Production Costs," *Econometrica*, **19**, 490–506 (1970).

168 Simmons, D. M., *Linear Programming for Operations Research*, Holden-Day, San Francisco, 1972.

169 Simonnard, M. A., *Linear Programming*, Prentice-Hall, Englewood Cliffs, N. J., 1966.

170 Smythe, W. R., and L. A. Johnson, *Introduction to Linear Programming with Applications*, Prentice-Hall, Englewood Cliffs, N. J., 1966.

171 Srinivasan, V., and G. L. Thompson, "Accelerated Algorithms for Labeling and Relabeling of Trees, with Applications to Distribution Problems," *Journal of the Association for Computing Machinery*, **19**, 4, 712–726 (1972).

172 Srinivasan, V., and G. L. Thompson, "Benefit-Cost Analysis of Coding Techniques for the Primal Transportation Algorithm," *Journal of the Association for Computing Machinery*, **20**, 194–213 (1973).

173 Srinivasan, V., and G. L. Thompson, "Algorithms for Minimizing Total Cost, Bottleneck Time and Bottleneck Shipment in Transportation Problems," *Naval Research Logistics Quarterly*, **23**, 4, 567–596 (1976).

174 Srinivasan, V., and G. L. Thompson, "Cost Operator Algorithms for the Transportation Problem," *Mathematical Programming*, **12**, 3, 372–391 (1977).

175 Swoveland, C., "Decomposition Algorithms for the Multicommodity Distribution Problem," Working Paper 184, Western Management Science Institute, University of California, Los Angeles, Calif., 1971.

176 Swoveland, C., "A Two-Stage Decomposition Algorithm for a Generalized Multicommodity Flow Problem," *INFOR*, **11**, 232–244 (1973).

177 Tomlin, J. A., "Mathematical Programming Models for Traffic Network Problems," unpublished dissertation, Department of Mathematics, University of Adelaide, Australia, 1967.

178 Tomlin, J. A., "A Mathematical Programming Model for the Combined Distribution-Assignment of Traffic," *Transportation Science*, **5**, 2, 122–140 (1971).

179 Truemper, K., "An Efficient Scaling Procedure for Gain Networks," *Networks*, 6, 2, 151–160 (1976).

180 Truemper, K., "Unimodular Matrices of Flow Problems with Additional Constraints," *Networks*, 7, 4, 343–358 (1977).

181 Truemper, K., "Optimal Flows in Nonlinear Gain Networks," *Networks*, **8**, 1, 17–36 (1978).

182 Weigel, H. S. and J. E. Cremeans, "The Multicommodity Network Flow Model Revised to Include Vehicle per Time Period and Mode Constraints," *Naval Research Logistics Quarterly*, **19**, 77–89 (1972).

183 Weintraub, A., "Primal Algorithm to Solve Network Flow Problems with Convex Costs," *Management Science*, **21**, 1, 87–97 (1974).

184 Williams, A. C., "A Stochastic Transportation Problem," *Operations Research*, **11**, 5, 759–770 (1963).

185 Wolfe, P., "Convergence Theory in Nonlinear Programming," in V. Abadie, Ed., *Integer and Nonlinear Programming*, North Holland, Amsterdam, 1970.

186 Wollmer, R. D., "Multicommodity Networks with Resource Constraints: The Generalized Multicommodity Flow Problem," *Networks*, **1**, 245–263 (1972).

187 Zangwill, W. I., "The Convex Simplex Method," *Management Science*, **14**, 3, 221–238, (1967).

188 Zionts, S., *Linear and Integer Programming*, Prentice-Hall, Englewood Cliffs, N. J., 1974.

189 Zoutendijk, G., *Methods of Feasible Directions*, Elsevier Publishing Co., Amsterdam, 1960.

Index